Good Chemistry

Methodological, Ethical, and Social Dimensions

Good Chemistry

Methodological, Ethical, and Social Dimensions

By

Jan Mehlich

Feng Chia University, Taiwan
Email: janmehlich@gmx.de

ROYAL SOCIETY
OF **CHEMISTRY**

Print ISBN: 978-1-78801-743-5
EPUB ISBN: 978-1-83916-039-4

A catalogue record for this book is available from the British Library

The Royal Society of Chemistry is a charity, registered in England and Wales, Number 207890, and a company incorporated in England by Royal Charter (Registered No. RC000524), registered office: Burlington House, Piccadilly, London W1J 0BA, UK, Telephone: +44 (0) 20 7437 8656.

Visit our website at www.rsc.org/books

Printed in the United Kingdom by CPI Group (UK) Ltd, Croydon, CR0 4YY, UK

Preface

Jan Mehlich*

International School of Technology and Management, Feng Chia University,
Wenhua Rd. No.100, 407 Taichung, Taiwan
*E-mail: janmehlich@gmx.de

There were some striking moments during my time as a chemistry student that, I assume, influenced my further career significantly. Here is one: we had a weekly "Kolloquium", every Thursday afternoon, in which invited researchers and university scholars presented their research. Once, an expert on explosives introduced his progress on making extremely powerful nitrogen-rich compounds that could be used for mining, space rockets, and missiles. He measured the efficiency of his explosives by a *killing rate* within a certain radius. This was reason enough for some of our professors not to join this presentation. That was the first time that I seriously asked myself: *"Is there any kind of research that would be wrong and beyond what is permissible as academic freedom? Aren't we scientists and researchers always on the good side with our noble curiosity-driven mission of knowledge generation and human enlightenment?"*

A second eye-opener was a conversation with my grandmother. I was brave enough to go for the destined-to-fail challenge of explaining my research topic to her: microcontact printing, a soft-lithographic method with an elastomeric rubber stamp to facilitate the formation of self-assembled monolayers in confined areas under the exploitation of a chemical reaction in which the 'ink' molecules form covalent

Good Chemistry: Methodological, Ethical, and Social Dimensions
By Jan Mehlich
© Jan Mehlich 2021
Published by the Royal Society of Chemistry, www.rsc.org

bonds with complementary molecules on a substrate surface. My principle investigator (PI), Prof. Bart Jan Ravoo, was hired by the Organic Chemistry Institute of the WWU Münster in a newly created chair on *nanoscience*, a research field that received a lot of funding in the early 2000s. My grandmother, a science layman, didn't understand a word of what I was talking about, but she surprised me with her question of whether "nano" also has something to do with "atoms" and whether that is also "dangerous and radioactive" like the rest of "that chemistry stuff". Clearly, she didn't share my enthusiasm about the new insights and arising possibilities from our scientific research. Was that my mistake? Did I talk about it in the wrong way? Would it even be important to make laymen like my grandmother understand what all the millions of Euros of tax money spent for academic research are good for, or could we just ignore their concerns and irrational fears?

To give a third example of a kind of reflection that is related to my chemical research activities, but in no way ever taught in chemical education, I'd like to address the very pragmatic considerations that play a role when writing a thesis and that outweigh the abovementioned noble and curiosity-driven virtues of science. I wanted the results of three and half years of experimenting and collecting data to make sense! I wanted graphs and schemes to look good. And I wanted them to have enough explanatory power to convince my supervisors that I did good work and deserved a good grade. But, for some mysterious reason, most of my data sets resulted in graphs with ugly outliers, unexplainable kinks in what should be a simple curve, and deviations beyond what would be easy to overlook or smoothen by just choosing the right scale in the diagram. How bendable is the presentation and interpretation of scientific data? Is science still what we wish science to be – a quest for a universal truth that does not depend on the opinion or predisposition of the inquirer – when all scientific knowledge and its construction is so much an issue of how the scientist refines and explains her observations and records? When can an outlier be omitted? Am I cheating when I declare a series of measurements to have failed, repeat it the next day, obtain data that fits my expectations better, and communicate only that latter but not the former data set?

The higher education landscape is changing. More and more students in science majors across all disciplines have a chance to enjoy courses on scientific integrity, good research practice, or even the social and environmental impact of their envisioned professional occupations. I use the word 'enjoy', here, because I am sure I would have enjoyed that very different type of learning challenge. As my department didn't offer it, I reached out for other options. The philosophy department had a study group on science theory to which science students were invited. After two sessions I realised that this wasn't helpful at all: the philosophy students read W.V.O. Quine and discussed in which way the truth of a sentence like "Brutus killed Caesar" depends on the meaning of the word "kill". What does language and linguistic convention have to do with science? For me, the naïve naturalist, language is just a tool that constitutes limitations that we aim at overcoming by understanding the true nature of things, somehow in the same way in which Neo in the Matrix movie trilogy stops seeing the computer-generated world (of which language is a part) but sees the matrix (the code) instead. Meanwhile, after much more self-studying, I changed my attitude, but at that time I thought that the philosophy of science didn't have much to offer that would solve my real-world problems of performing scientific research.

I found another appealing offer from Münster University, a Masters course (as an 'additional qualification') in *Applied Ethics*. This was very different from chemistry and scientific research. Here, I learned reading and writing anew. Far from moral philosophy, practical ethics aims to solve real-world ethical conflicts in medical, environmental, political, social, technological, or professional contexts. Finally, I found fruitful and rationally backed up ways to think about and communicate arguments in favour or against certain decisions and practices. There are good reasons for limiting academic freedom! There are discourse strategies that support whistleblowing of observed wrongdoing without risking one's job or grade! There are established procedures that empower scientists and researchers to contribute their specific competences to making scientific and technological progress sustainable (and, of course, reflections that clarify what sustainable even means)! And many more.

With this combination (nanoscience and applied ethics) I got a job in the field of *technology assessment* (TA). TA is a discipline that tries to accompany technological development with reflections on ethical and social implications of progress, so that it is possible to guide it into "the right direction". The particular project I was involved in was the application of nanoparticles for medical diagnostics, a part of *nanomedicine*. This work was highly interdisciplinary, involving university scholars (nanoparticle research, biomedical science), hospital doctors, chemical companies, regulators, and patient organisations. With my chemical background, I was able to understand what the scientific researchers were talking about and why they say the things they say. Yet, I also realized that the scientific facts fail to answer questions of how we want to live or what a technological advancement might be good for. This experience reinforced my impression that many scientists and researchers lack certain discourse skills when it comes to figuring out, in an interdisciplinary manner, what our options are concerning guiding progress into a desirable and sustainable direction.

We wrote a report to the European Commission that was in charge of making new regulations about new drugs and medical treatment methods. That means, we had a *client* who expected us to write certain things in the report (for example, that nanomedicine is *good* after all, and that everything is taken care of and there is nothing to worry about). Science and technology ethics as a tool for generating public acceptance? This was not my idea! I decided to quit that job in order to pursue an academic career. After a postdoctoral research project on the social and ethical implications of Taiwan's nanotechnology initiative, I started teaching courses on science and technology ethics at universities in Taiwan. Besides many cultural differences and specificities that play important roles in discussing science and technology, I noticed that the practical and pragmatic questions that young scientists (by this I mean: students in science majors) ask are more or less the same around the world and across cultures: How can I do science well, and how can I use my competence and expertise to work as a responsible and positively impacting member of the local and global society? The answers are not always trivial. Certainly, they are also not random, a matter of opinion or taste, or necessarily tacit or incommensurable with the empiric character of scientific thinking. This is what this book aims to show.

I explain all this because I want the reader to know me at least to such an extent that it becomes clear why I may be justified in writing this book. I am a chemist by training. Additionally, I learned how to solve normative-ethical problems. Moreover, I worked in the wider context of socio-techno-scientific systems assessment of which chemists are a crucial and vital part. I hope that my experiences can serve as an insightful source for your learning and your own quest to find answers.

This book is only possible because I had the support from many colleagues and friends! It is the result of the tireless efforts of the working party *Ethics in Chemistry* of the *European Chemical Society* (EuChemS) that, since its constitution, regards the education of chemistry students in questions of ethical and social dimensions of chemistry as one of its core missions. In particular, I'd like to thank Hartmut Frank and Luigi Campanella for their support and input! Frank Moser highlighted many of the aspects presented in this book during a lecture tour through Switzerland in autumn 2017 (made possible by the SCNat, represented by Christophe Copéret and Leo Merz). Wide parts of this book are the result of conceptualizing and compiling an online course 'Good Chemistry' for the EuChemS. This project was gratefully supported by the EuChemS executive board, most notably Pilar Goya, David Cole–Hamilton, and Nineta Hrastelj. The team that created the content of the course consists of Iwona Maciejowska and Rachel Mamlok-Naaman of the EuChemS *Division of Chemical Education*, and Walter Zeller, Bill Byers and Paola Ambrogi of the *European Chemistry Thematic Network* (ECTN). Without all their input and discussions, this book would be different and, believe me, much more boring and drier. Helen Armes, Connor Sheppard and Lewis Pearce of RSC Publishing are thanked for their patience and great support during the creation and editing of the manuscript! This is my first book project, and if publishing is really always as smooth as with the RSC in this case, I will be motivated to do it again! Last but not least, I thank all the students in Taiwan and Europe that contributed with their critical questions and their positive and negative feedback to the constant revision and, hopefully, increasing quality of my idea of what *good chemistry* is.

Jan Mehlich
Taichung, Taiwan
February 2020

Contents

Good Chemistry: Methodological, Ethical, and Social Dimensions
By Jan Mehlich
© Jan Mehlich 2021
Published by the Royal Society of Chemistry, www.rsc.org

13 Risk, Uncertainty, and Precaution   315

14 Science Governance, Technology Assessment   337

15 Science Communication   363

16 Summary   382

1 Introduction

Overview

Summary: At first glance, chemistry and ethics have nothing much to do with each other. One is a modern natural science, empirical, analytic, firmly grounded in ever-refined and improved theories, with high creative potential and applicability for improving everyone's life quality. The other either reminds us of common-sense folk morality or of dry, dusty, intellectual armchair philosophy. Where do these two disciplines – or, attitudes for making sense of our life world – meet?

Professional chemists, like all practitioners, ask themselves many normative questions, often without noticing it: what does it mean to do my job well? What is good professional practice? What is the right thing to do in cases of conflicts or dilemmas? Does my work have an impact on society and the environment, and is there anything within my responsibility to do about it? Indeed, especially for young chemists, finding answers or solutions for these questions may be much harder than solving chemical challenges like reaction mechanisms, spectrometric analyses, or practical experimental laboratory work.

Before going deeper into these matters, this introductory chapter attempts to explain and clarify some important basic considerations upon which the rationale of this book is built. It will explain the structural division into the three parts methodology, good scientific practice, and the social/environmental impact. It will introduce the role of ethical and normative reflection and make sure that every reader understands that ethical integrity does not require a degree in moral philosophy, but a mindful attitude, clear reason, goal-oriented discourse skills, and the motivation to act professionally as a responsible member of society and, more than that, as an influential and impacting decision-maker in academia, industry, or public service.

Key Themes: What is professionalism? When is chemistry *good*? Right methodology; good scientific practice; social/environmental impact; ethics, morality, normativity; interplay of facts and norms; the ethical prism; discourse skills; rationale and structure of this book.

Good Chemistry: Methodological, Ethical, and Social Dimensions
By Jan Mehlich
© Jan Mehlich 2021
Published by the Royal Society of Chemistry, www.rsc.org

Learning Objectives:

After reading this chapter you will:

- Understand the logic structure of the book, judge the significance of its themes and topics, and be able to select those chapters that matter for your own personal purposes.
- Be able to explain the difference between ethics and morality and use the implications of these definitions in real-life discourses.
- Be aware of the touching points between chemistry, society, and ethics.
- Be prepared for many subsequent discussions using terminology and concepts that are, usually, unfamiliar to chemists and chemistry students.

1.1   Introductory Cases

Case 1.1: Knowledge or Opinion?

In a courtyard, an analytic chemist presents the results of a report that investigates the ground water contamination in a wildlife sanctuary that is most likely originating from a nearby lacquer and paint manufacturer. These findings shall serve as evidence for the company violating environmental regulations. To his surprise, the chemist's evidence-based arguments are countered by the defence as '*one of many opinions*'. The company's lawyers claim that the ultimate proof that the contamination is coming from the factory is lacking ('*correlation is not causation!*'), and that science, '*merely based on theory*', would not be able to contribute much to this case that is about the usefulness of legal frameworks, about desirable economic social benefits, and about the quality of life of the consumer.

Case 1.2: Truth or Dare

In the research group of Prof. X, two Master students A and B, a PhD candidate C, and a postdoc D share a lab. Prof. X asks postdoc D to supervise student A while the PhD student C is asked to take care of student B. Student A often feels ridiculed and even bullied by D who regards students as immature foolish kids, looking down upon them. Thus, A's trust in D is low. At the same time, A observes that PhD student C encourages B to manipulate experimental data ("It is just for your thesis! Like this it looks more beautiful! Don't worry, X won't notice it!"). Yet, A knows that C would use at least parts of B's experiments for a research paper. A plans to talk to Prof. X about this after a group meeting. In this meeting, however, Prof. X expresses his admiration for his "best students" B and C. Since A's own work, partly due to the poor supervision and support from D, is lagging, she doesn't dare start the conversation, worrying about being judged as a trouble-maker and risking her grade.

Case 1.3: Not My Business?

A chemist who is working for a globally operating trade organisation is processing a tricky case: several tons of thiodiglycol were ordered by a company in a country recently riven by civil war. The company claims to operate in the textile industry. Yet, its origins and structures appear hidden and dubious. The trader's chemical background knowledge is good enough to tell that thiodiglycol is, at the same time, a precursor for mustard gas and nerve gas. The seller of the chemical, excited by the excellent business prospect, claims that it is not his or the trader's responsibility to question the communicated application of the chemical (textile manufacturing). Yet, given the low trust in the credibility of the country's regime, the trader has severe pricks of conscience in agreeing to this transaction.

Case 1.4: The Moral Finger

In a class on research ethics, a student raises concerns about the empiric rigidity of ethical claims. *"Science is truth, but ethics is just opinion, kind of random!"* Thus, he claims, a course like this is use-less, because it can't teach knowledge like a chemistry class, but only doctrine and ideology, like the morality finger that warns us not to do wrong. As he doesn't enjoy the class at all, he stands up and leaves.

1.2   What Is Good Chemistry?

Before a book is published, it goes through a review process. The reviewers of this book's initial conceptual proposal welcomed it and expressed that such a book was timely and necessary! The only point of controversy was (and maybe still is) the book's title: *Good Chemistry*. Why not *Ethics in Chemistry*? Isn't this a book about ethics?

No, it is not! Not primarily, at least. It is as much about ethics as a book about materials in construction engineering is about chemistry. Here, we want to apply ethical reflections – which is significantly different from *ethics*, as we will see soon – for a very practical purpose: as chemists we wish to do a good job, to do something meaningful and impacting, and to have a somehow positive or beneficial impact on the world and the people in it. This motivation is presupposed, here. Some may have other core motivations to pursue a career as a chemist, for example good income, a Nobel prize, or the intention to synthesise a powerful pathogen that will eradicate mankind and free the planet from this plague. These attitudes are not this book's business! This book addresses all those who, at least from time to time, wonder what it means to be a good chemist, and all those who got in touch with conflicts and dilemmas in the context of chemical activity that can only be solved in non-chemical ways, that means with strategies that are not taught in typical chemistry textbooks.

If *good* is an adjective that may legitimately be attributed to chemistry, then there must be something in chemistry as such that justifies this kind of judgment. The book title hints at this justification when we read the word *good* as a noun: chemistry as a public good, a common good, an endeavour that a society undertakes and affords because it is believed that it somehow pays off. The good$_{noun}$ *chemistry* is something we would not want to miss because it increases our life quality, facilitates innovation, creates business opportunities and jobs, or greatly advances humanism by enlightening our knowledge of the world we find ourselves living in. Yet, all these missions can, potentially, fail: chemical progress may have adverse effects on society and environment, chemical industry's labour force may encounter safety issues, or a scientistic-technocratic worldview may undermine the values that constitute our cultural and societal integrity and cohesion. The interpretation of what makes chemistry a common good determines what we define as good$_{adj}$ chemistry.

The reader may brainstorm for a minute on what comes to mind when thinking about what is *good chemistry*, or what is *chemistry done well*, or what is *a good chemist*. This could be done in the form of the following exercise that is not as childish as it may sound:

Box 1.1   Exercise: Draw/paint a Good Chemist

Take a pen or marker and draw an ideal chemist. Try to avoid the use of words, instead use symbols which will characterise a good chemist and her/his competences (knowledge, skills, attitudes, behaviour).

Experiences with this rather loose and open question answered by chemistry students (or students of other sciences) indicate that it is very likely that your ideas fit into one or more of the following three categories (see also for ref. 1):

1. *Methodology* – Good knowledge and competence: "*What is it that I am doing, and why am I doing it THIS way?*" Accordingly,

a good chemist is someone who understands the theoretical foundations of his/her profession, who knows how to apply the scientific method properly and adequately, and who is aware of the special position that scientific inquiry has in a society that affords such an expensive endeavour. This also includes practical hands-on competences like experimental and technical skills.

2. *Professional Integrity* – Good attitude and conduct: *"What does it mean to do my job well?"* This field covers questions of research ethics, scientific integrity, and what kind of behaviour is permissible within the guidelines of professional conduct. For other chemical professions outside academic science and research, analogous principles of professional ethics, work ethics, or economic and business ethics apply.

3. *Chemistry and Society* – Good impact and progress: *"How does my work impact the life world of society, and what is within my responsibility to do about it?"* In this understanding, chemistry is *good* when its impact on the environment and society is sustainable, when it supports a reduction of risks and a maximisation of benefits, and when chemists with their competence and expertise engage in science and technology (S&T) discourse and governance.

Table 1.1 shows some exemplary statements, sorted into these three categories. It also indicates how the mentioned aspects represent the chapters of this book (more about this in Section 1.5).

1.2.1   Research Methodology

Without a doubt, a good chemist is certainly someone who is good at chemistry, someone who *knows* chemistry. This includes *know-that* and *know-how*: textbook knowledge of all fields of chemistry, plus specific knowledge of the particular area of research or professional practice; and hands-on competence in laboratory work, experimental design, scientific writing and presenting, teamwork, and professional expertise. Usually, chemists – we will see in Section 1.3 below who is addressed with this term – have studied chemistry and obtained an academic degree related to the discipline, like Master and PhD degrees. Knowledge acquired in lectures, from books, in study groups, or in long overnight sessions in the university library in the days before important exams, gives the chemist the ability to make proper judgment in chemical contexts. Lab courses and the research work done for theses increase the chemistry student's

Table 1.1   Exemplary statements about good chemistry/chemists, illustrating aspects of methodology, good scientific practice, and social implications.

Research Methodology

"In the context of chemical research, good science means an advancement of existing knowledge."
Science theory, science history, epistemology   Chapter 2
"A good chemical researcher knows how to apply scientific methodology for the planning and conduct of experiments."
Scientific method(s), research design and experimentation   Chapter 3
"A chemist should be good at logic thinking and understand the differences between explanation and prediction, and between causation and correlation."
Logic, statistics, heuristic analysis, science and uncertainty   Chapter 4

Professional Integrity (Good Scientific Practice)

"A chemical researcher does a good job when complying with the rules of good science conduct!"
Good scientific practice, scientific integrity, research ethics   Chapter 5
"A chemist doing science should refrain from misconduct like faking data or stealing others' results!"
Scientific misconduct, fabrication and falsification of data, plagiarism   Chapter 6
"When it comes to publishing research results, a good chemist would not be motivated by fame or pride, but by truthfulness and fairness!"
Publishing, authorship, peer review, impact factors   Chapter 7
"Since chemists always work in teams, they better communicate properly and without bias!"
Mentorship, science-industry collaboration, conflicts of interest, funding, academic freedom, intellectual property   Chapter 8
"When chemists, for example as toxicologists, do animal experiments, they should follow regulations to limit the animals' suffering."
Animal testing, bioethics, 3R guidelines   Chapter 9

Chemistry and Society

"Chemistry is a common good and, thus, must not ignore the values that the society endorses!"
Neutrality thesis, chemistry as socio-techno-scientific practice, social construction of science and technology   Chapter 10
"Chemistry is good when it makes processes and practices more sustainable!"
Sustainable chemistry, definition and application of sustainability concepts   Chapter 11
"Chemical practitioners should know what they can be held responsible or accountable for!"
Responsibility, accountability, individual and collective   Chapter 12
"With their special expertise, chemists can contribute to reducing risks that go along with chemical processes in industry and business."
Risk, uncertainty, precaution   Chapter 13
"A good chemist is one who collaborates with non-chemists in order to make scientific and technological progress take a direction that everybody finds desirable."
S&T governance, assessment, interdisciplinary discourse   Chapter 14
"Chemists should be able to communicate with the public in order to be credible and create acceptance and understanding!"
Public communication, science journalism, education   Chapter 15

practical skills and the ability to apply the theoretical knowledge in activities that manifest ideas in something materially real. Chemistry, more than physics or biology, is a creative endeavour (while, admittedly, the disciplinary borders become more and more blurry). Even though most chemistry students pursue a career outside academia and outside laboratories, the study years determine and shape their thinking and their attitude when applying the gained chemical competences in their jobs.

In this view, in order to be (or become) a good chemist, one must study well and work hard. We may trust that university curricula in chemistry departments ensure that students have all the necessary resources and opportunities to acquire chemical knowledge and competence (if not, this may be an ethical issue itself). This book can't help much with that. Yet, there are methodological and theoretical aspects in doing chemistry and applying chemical knowledge that, reportedly, many chemistry students never learn formally. This concerns some philosophical issues like knowledge, truth, or logic, and judgments concerning the meaning and implications of chemical statements as the result of interpreting data. Here, we see that this category is not only of relevance for chemists as academic scientists or private sector researchers, but also for all those who make chemical judgments in the context of scientific testimony, for example in court, in public governance, in agencies like environmental bureaus or patent offices, or in other forms of public service. In these contexts, chemistry is applied as an interest-driven science in which the right application of chemical methodology and the right use and analysis of scientific insight is of crucial importance for the credibility of the chemical expert and for the fair and just treatment of those who are affected by the decisions that are made based on these judgments.

Case 1.1 provides an example of how science theory and the proper justification and application of its claims play an important role even outside of the academic scientific realm. Chemical expertise provides the means to inform decision-making and judgment in business, governance and society with evidence-based information that, in the right context, may be used as knowledge to persuade or convince people of the plausibility or meaningfulness of certain claims. Yet, in recent years, post-factual trends undermine the legitimacy and credibility of science as a reliable and most objective source of knowledge. Science as a social institution or sphere, represented by scientists (another blurry term), is under scrutiny and needs to stand its ground against opinionated dogma and ideology. Not everyone is familiar with the concepts of scientific inquiry and reasoning (like the defence lawyers in case 1.1), so that it may not be clear how empirical evidence is not mere opinion. At the same time, the explanatory power of scientific insights has clear limits. A chemist is able to enlighten our understanding of evermore sophisticated aspects of the material world, but may not be qualified to make judgments on the economic or environmental sustainability of a chemical process because that requires an assessment of non-naturalistic elements: human values and norms. In this respect, understanding science theory means finding the right position between epistemic confidence (defending the strengths of scientific inquiry) and humbleness (knowing what our methods are not able to inform about).

There is a huge amount of literature on science theory and research methodology. Some, or maybe most of that, is not targeted at practical chemists. In Part 1, this book attempts to cover all those aspects that are either of direct practical relevance for doing good chemistry, or deliver important insights for other chapters in Parts 2 and 3. For more insightful elaborations and a detailed description of strategies for practical application, the mindful chemical scientist is advised to read the excellent books by Pruzan[2] and Shrader-Frechette.[3]

1.2.2 Professional Integrity

Among the three different understandings of *good chemistry*, the second category is the best elaborated and most often discussed one. Case 1.2 is a good example that touches fraud, mentoring and, perhaps, publishing issues. Indeed, in view of countless cases of misconduct, fraud, betrayal and violation of codes of conduct that chemists are entitled to comply with, a demand for raising the awareness of research ethics may be identified. Fabrication and falsification of data, cases of plagiarism or other improper publishing practices, conflicts of interest and unscientific handling of intellectual property rights issues, academic freedom, that is at risk in view of contemporary funding and collaboration practices, all motivated by non-scientific goals and dispositions like careerism, financial benefits, greed for fame and power, but also systemic and organisational stress and pressure, are reported on a daily basis (see, for example, the online platform Pubpeer, or the blog forbetterscience.com). Insightful overviews with manifold researcher's possible real-life cases to practice one's scientific integrity are provided by Macrina,[4] Shamoo and Resnik,[5] Greer,[6] and particularly for chemists, by Kovac.[7]

As in Section 1.2.1, it needs to be pointed out that issues of professional conduct for chemists are not limited to academic scientific research and its ethos. Chemists working in the private sector have to fulfil organisational obligations and contracts while, at the same time, are part of a much wider network of various interest groups (or *stakeholders*), like business, management, marketing, trade, labour force, or clients and customers. Professional integrity, then, is not only a matter of research ethics and the virtues of good scientific practice, but also business and economic ethics, and perhaps environmental or even bioethics. Whereas chemists-by-training seldom work with human research subjects (like, for example, medical researchers, psychologists, or social scientists do), many do perform animal experiments, for example as toxicologists or analytic chemists. Yet, for all the issues in this section that may arise sooner or later in a chemist's daily professional practice, there is a connecting theme: the guidelines for good conduct are more or less clear, well-defined, and societally justified. Everybody would agree that stealing, cheating, lying or mistreating living beings are wrong. Thus, these guidelines as orientations for action are a matter of compliance rather than ethical evaluation.

1.2.3 Chemistry and Society

The connection of chemical activity with society and the environment – the third category in our list – is often overlooked in the context of academic chemistry and, thus, also in the education of young chemists (since that is done at academic institutions). Too complex are the various responsibility attributions; too uncertain are the causal trajectories of scientific and technological progress; and too far seem the actual impacts from the chemist's lab. Yet, there is an obvious impact of chemistry on society and culture; on the one hand, it facilitates a significant increase in quality of life through new products, processes and possibilities; on the other hand, at the same time, it contributes to environmental pollution, increased risk exposure (workers in factories, consumers through the food chain and global water cycles), creating challenges for the regulation of new compounds and chemical procedures.[8]

This inherent potential of *dual use* of the manifestations of chemical progress is, arguably, the most obvious ethical aspect in terms of the societal impact of chemistry.[9] A reflection on the role of chemists in S&T progress and its societal and environmental impact must be pragmatic and goal-oriented: in view of the duality of desirable and undesirable effects of chemical activity, *what is in our power to do about it?* Chemistry, from basic science to engineering, is not only part of the problem, but, above all, part of the solution. That is why it is not only a matter for engineers and the chemical industry, but – in specific ways – also an issue for chemical researchers and scientists. Emerging sub-disciplines like *green chemistry* or *sustainable chemistry* are devoted to the identification and development of compounds and processes that facilitate efficient and sustainable human activity, ranging from mobility, energy, healthcare, infrastructure, and communication to consumption and recycling of products. Chemists in industry are in touch with the environment, health and safety (EHS) regulations, the principles of *responsible research and innovation* (RRI), *value co-creation* in *industry 4.0*, or *corporate social responsibility*.

Many chemists with academic degrees occupy positions with high responsibility and power, making decisions and judgments that have direct and indirect implications for society or the environment. Case 1.3 illustrates the political and humanitarian dimension of a trade decision. Indeed, this case is very similar to a real case that we will revisit in Chapter 12: In the 1980s, chemist Frans van Anraat sold thiodiglycol and other chemicals to the Iraqi regime that fabricated chemical weapons from it and used them against their own people. Later he was convicted of complicity in war crimes and was sentenced to 17 years in prison because with his chemical background his claim that he didn't know what could be done with the chemicals that he traded was not convincing. In other words, he could be held responsible for what he did in his competence as a chemist.

Chemistry is $good_{adj}$ or a (common) $good_{noun}$ when its impact is positive (for example, beneficial, sustainable, desirable) and when it facilitates scientific and technological progress in the form of either enabling knowledge or useful innovation. But what about the *good chemist* question? What is expected from a socially and environmentally friendly chemist? Are researchers on green chemistry or sustainable materials better chemists than those who developed polymerisation reactions that enabled the large-scale industrial production of plastics that now pollute the water cycles? This definition would be misleading and unfair. Moreover, it would be wrong to shift all responsibility for S&T development to engineers and product designers, claiming that scientific research as such is always neutral. A *good chemist* in this category would be one who is aware of and concerned about the role that chemical progress plays in the wider network of technology, innovation, global trade and exchange, governance, consumption, and culture.[10]

Here, we close the circle with the first category; a mindful chemist is aware of the power that chemical knowledge and competence has, but humble enough to admit that this can't answer the question of how we want to live. A good chemist is the one who doesn't back off from this perhaps more difficult question, but who engages with the multitude of stakeholders to figure it out by sharing the strengths of chemistry – creating the material means to provide a higher life quality – and connecting it with the discourse on values, needs and demands.

1.3   Who Cares? Or, Who Should Care?

The headline question may be understood in two different ways:

- Who has an interest in chemistry being *good* in one or more of the abovementioned ways? Answer: Chemical practitioners and the society at large.
- Given the educational purpose of this textbook, who is addressed with the claim that chemists should try their best to be *good chemists*? Answer: Chemists in (i) academia, (ii) industry, and (iii) public service.

The first understanding concerns the call for methodological, ethical and socially responsible integrity. We may ask, of course, whether it is justified to claim that chemists should care about normative aspects of chemical activity. Isn't it enough when other experts and stakeholders (social scientists, ethicists, regulators, suppliers and appliers of chemistry, *etc.*) do that? It is important to see that knowledge and awareness of the normative dimensions of chemistry pays off, indeed, in various forms. The insights from Part 1 on methodology, hopefully, will increase the reader's professional skills and competences as a scientist and researcher. It may stimulate academic creativity and improve the quality of scientific output. It is also hoped, of course, that an awareness of the ethical pitfalls in research practice helps to increase professional integrity and compliance with the ethos of science. The benefits are a bit more difficult to realise in the case of Part 3 on the social implications of chemistry. An *ethical chemist*, generally spoken, has a higher credibility whenever it comes to chemistry-related discourses, be that community-internal (with other chemists, scientists, colleagues, *etc.*), with other stakeholders, or with the general public. Scientists' powerful position in factual knowledge-based discourses is challenged nowadays, so that a sense for normative aspects of science and the ability to articulate these in communication is an important skill of chemical practitioners. This will increase societal support and acceptance, which is of existential importance for the institution of chemistry. Moreover, with ethical chemists in important and influential positions in society, the goal of sustainable progress can be reached more easily. Last but not least, in the form of green and sustainable chemistry as an academic research discipline and economic strategy, ethical competence even has the potential for economic profitability.

Chemistry as such – justified or not – is under special scrutiny by the public. Take the Seveso and Bhopal disasters, for example: they are coined *chemical accidents*, whereas nobody would call an airplane crash a *physical accident*. Furthermore, the Contergan (thalidomide) case, chemical weapons, or the pollution of the oceans with plastic particles are often attributed to chemistry, often without any differentiation into academic chemistry, chemical engineering, chemical industry, manufacturing and trade. We can, of course, ignore public concerns as irrational, but as a matter of fact, this has a recurrent negative effect on chemistry as a science and as an industry. The future of the chemical profession stands and falls with societal acceptance and support. This goes far beyond the argument that a large part of public academic chemical research is funded by tax money and, thus, should pay off for the society in one or another way. It is a question of trust and public education.

Now that we have clarified the justification of the *should*-claim itself, we may figure out the details of *who exactly* should. Assuming that the reader of this text is a chemistry student, he or she may feel addressed directly, of course. Your interest is to finish a thesis successfully, to publish your first papers and to lay the basis for a professional career. Most likely, you have had your first experiences with lab research and, possibly, have encountered issues of research ethics and scientific integrity. You know chemistry as an academic scientific discipline. However, it is important to widen that scope. Where do we usually find chemists after completing their higher education and academic studies?

First of all, of course, we think of academic chemistry at universities and in other research institutes. In the former, besides basic and applied research, also the education of the future generation of chemists plays an important role. Chemists who pursue an academic career often do that for a particular reason: the academic freedom to ask interesting and creative questions for the main purpose of knowledge generation, perhaps out of pure curiosity, with an Einsteinian mastermind, or with the noble goal of doing something good for society. Some may also feel dedicated to teaching the next generation of chemists and form a team of motivated young Master and PhD students. The main activities are the design and conduct of experiments, interpreting the data to form scientific statements, and communicating these new insights to the larger scientific community and, perhaps, the wider public.

Then, there is the big field of chemical industries. Many companies have research and development (R&D) departments in which chemists do scientific research, but with motivations and goals that are different from those of academic institutions. More importantly, chemists working in industry deal with the production, storage, and transport of chemicals. Moreover, marketing and sales of chemicals may be part of the job activities in this field, too. For the purpose of the message of this book, we won't exclude chemical engineers from our list of 'chemists'. It is important, though, to keep in mind that the job profiles of chemistry majors and more technical chemical engineering graduates are overlapping more and more. As we will see in greater detail in Chapter 10, it is not the case that chemists populate science jobs and engineers occupy technology jobs, that chemists do basic research while engineers do applied research, or that chemists work in academia and engineers dominate the industry sector. Questions of good chemistry are not so much a matter of degree or title, but rather of professional work environment and particular tasks and responsibilities. Chemical scientists are as much a part of R&D and innovation as chemical engineers are.

An often-overlooked field of profession in which chemical competence is required is the public service sector. Examples are environmental protection agencies, food and drug administrations or other governance and regulatory bodies that deal with the assessment and regulation of chemicals and their impact, but also patent offices, science consultancy, auditing, science writing, or analytic services. Chemical R&D often plays only a minor role in these jobs. Yet, as explained in Section 1.2, chemical risk assessments, scientific testimonies, or any other publicly communicated chemical information that serves as a basis for decision-making require competences in making judgments concerning the value of chemical knowledge and progressive potentials. Some of the virtues of science that inform our concept of research ethics (like objectivity or truthfulness, see Chapter 5) also play an important role for agency staff, consultants or regulators with chemistry background, complemented further by questions of legal and social justice. Many chemical experts working in this realm work at the intersection of chemistry, industry and society on a daily basis. Therefore, those readers who consider a career in this direction should not put this book aside as it matters for their professional judgment and discourse skills as much as for basic and applied chemical researchers and scientists.

Table 1.2 summarises the considerations of Sections 1.2 and 1.3. It outlines, with broad examples, how all three categories of good chemistry have implications for all three fields of occupation in which chemistry graduates find their jobs.

1.4   Facts, Norms, and the Role of Ethics

After this initial reflection on the facets of chemistry and the people involved in it, it is now necessary to clarify some of the non-chemical terminology used in the previous two sections. Words like *normative*, *ethics*, *values*, or *discourse* may be unfamiliar to this book's chemical target group. Also, in this section, we will elaborate in further detail what kind of competence this book tries to equip the reader with. Hopefully, with the information from this section, the student in Case 1.4 would not have left the class.

The English word *ethics* has two meanings. As a plural term (*many ethics*), it is a synonym for *morals* and stands for the moral guidelines and rules that a society commits itself to. *Ethical*, in this respect, means *in*

Table 1.2   How the three categories of good chemistry matter for the three realms of the chemical profession.

		Chemists working in:		
		Academia	Industry	Public service
Categories of good chemistry	Methodology	Science theory, scientific method	Application of scientific knowledge in R&D	Scientific testimony, interest-driven science, post-factualism
	Good professional practice	Good scientific practice, research ethics, professional ethos	Research ethics, business ethics, legal and organisational compliance	Legal and social ethics, environmental ethics
	Social implications	Sustainable/ green chemistry, dual use of scientific insights	Responsible research and innovation, sustainability	Risk and precaution, governance and regulation of science and technology

accordance with the common-sense morality. This idea of ethics is applicable in the field of research ethics and good scientific practice: we don't need sophisticated philosophical reasoning to understand that cheating (fabrication and falsification of data), theft (plagiarism) or betrayal (hiding financial interests) are immoral. Here, ethics is a matter of compliance with established and unquestioned morals. As a singular term, *ethics* is the philosophical discipline that is dealing with a systematic and analytic approach towards what is *good* and/or *right* (to do). The result of ethics as a rational inquiry may be morals (the plural ethics), or a clarification of ethical dilemmas, conflicts, and clashing viewpoints. The impact of science and technology on society often results in such cases, in which compliance with morals is insufficient and ethical reasoning and argumentative discourse is required. In contrast to often expressed viewpoints, ethical assessment is not a matter of opinion and preference – thus, arbitrary and unscientific – but an academic field of expertise with established methodologies, tools and practical applications. Yet, it doesn't deal with factual knowledge as elaborated by the natural and engineering sciences, but with normative knowledge. Normativity, here, refers to all forms of evaluative judgments like good/bad, right/wrong, desirable/undesirable, beneficial/harmful, and so forth. Normative knowledge is not generated by observation and empiric investigation, but by argumentation and reasoning. This may best be explained by having a look at the general form of an ethical argument as shown in Figure 1.1.

Ethical arguments usually consist of two or more premises and a conclusion. We are interested in an acceptable and somehow correct prescription on what to do, which of several options to choose, or how to evaluate and judge a given or hypothetical situation – a should-conclusion (C_s).

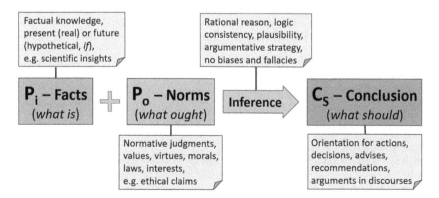

Figure 1.1 Elements of an argument.

A valid argument needs so-called *is-premises* (P_i) that describe the situation that is given (or hypothesised, "If...") and *ought-premises* (P_o) that introduce a value or normative orientation. Arguments that lack the ought-premise are called naturalistic fallacies as they base *what should* merely on *what is*. A standard example is the statement "*Smoking damages your lungs! You should stop smoking!*". In order to make this prescription valid, it needs to be introduced that an important value is at stake, for example health (with the additional is-premise that there is a relationship between smoking-affected lung capacity and one's health state, or increased cancer risks). Arguments may also lack necessary is-premises. Without the respective real-world case, ought-premises leading to should-conclusions may be accused of speculation and irrelevance. Respectively, if one wants to refute an argument, one can either attack the is-premise ("*What you assume here is not true/valid!*") or the ought-premise ("*Your value ascription is not tenable!*"), or the relationship between the two ("*The proposed value is not relevant for this case!*"). The smoker may argue that he or she values pleasure higher than health (refuting P_o), or point at studies that show that moderate smoking does not increase lung cancer risks to a larger extent than, for example, breathing polluted city air (refuting P_i).

In professional fields, the clarification and validation of both is- and ought-premises are necessary elements of assessments and discourses. Evaluating the impact of chemistry on society needs a thorough investigation of development pathways and trajectories (what is and will be), as well as a normative investigation of possible affected values and norms and the desirability of possible futures (what ought). The predominant role of chemists is, therefore, to feed the *is*-premises with input (knowledge, expert-based estimations). The discourse on the ought-premises involves other experts (ethicists, sociologists, regulators, *etc.*) and sometimes non-experts (political decision-makers, businessmen) or even laymen (citizen, general public). In Chapter 14, we will have a closer look at how such discourses are organised and put into practice. In any case, chemists are not expected to have a special competence in normative judgment (deciding on P_o). Yet, in interdisciplinary discourses like those on S&T progress, chemists need to be aware of the distinction between P_i and P_o and that both need careful analysis and validation. This may sound like a trivial statement. Yet, in practice, there are still many obstacles and methodological difficulties to overcome in order to make such assessments fruitful and efficient, one of them being scientific experts who refuse to participate in normative discourses with the argument that *ethics is not their competence*. Hopefully, this book can show that scientific expertise is a crucial element in tackling the big normative questions of our times.

These considerations show, once again, that *good chemistry* is not simply a matter of ethics, neither as compliance with morals nor as normative decision-making expertise or even philosophical skill. Many approaches to teaching research ethics understand it as the facilitated cultivation of a moral character: It is basically assumed that, after pointing out the virtues of good scientific practice, the young chemist will surely act in accordance with these virtues, either intrinsically motivated to be good or extrinsically incentivised by rewards and sanctions. If this was the case, the best choice would be case-based learning in order to acquire experience as orientation for action at hand whenever a similar situation arises in the context of one's own work. Unfortunately, empirical studies could show that training in research ethics doesn't prevent misconduct.[11] It cannot diminish a predisposition to being susceptible to committing fraud or being biased. Being a good chemist by making the right choices is, as is assumed throughout this book, predominantly a matter of attitude and discourse rather than character and knowledge of ethical rules and principles. Intra-community discourse is the most efficient mechanism to protect against misconduct. Inter-community (or interdisciplinary) discourse ensures that facts and norms (see Figure 1.1) are sufficiently enlightened. Extra-community discourse that reaches out to the public and various other non-expert stakeholders enables sustainable S&T progress. Therefore, rather than teaching ethical guidelines or preaching virtues, the main goal of Parts 2 and 3 of this book is to equip the reader with discourse skills and, directly linked to that, the ability of critical thinking.

Discourse is a crucial element of all sciences. Insights and knowledge are generated and verified by critical scrutiny in communicative exchange among different experts with respective competences. Discourse can, very broadly, be understood as communicative action. It may be characterised as a form of conversation that is more controversial than a *chat* or *small talk* but less aggressive than a discussion, debate or even quarrel. In a discourse, two or more participants exchange viewpoints on a topic, disagree about them, but try to figure out whose viewpoint is more plausible in view of parameters that need to be clarified as well. It is goal-oriented (finding a common ground or agreement) and content-based (in contrast to personal or emotional). However, we need to be more precise in order to classify the discourse type that is promoted by the approach of this book (see Figure 1.2).

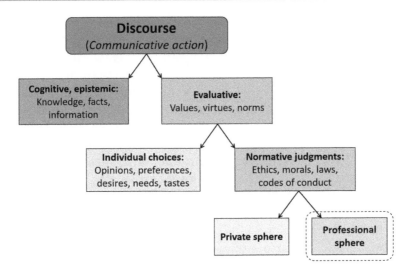

Figure 1.2 Discourse types.

In such discourses, different fields of inquiry need to be distinguished. When I tell you *"A whale is a fish!"*, you might disagree and inform me that whales belong to the group of mammals. When I try to support my argument with reasons like *"But it is swimming in the water, so it is a fish!"*, you will insist that this is an irrelevant factor, but instead we need to look at evolutionary pathways and the development of certain anatomic features to see that whales are mammals. This kind of discourse is a *cognitive* or *epistemic* one; we need factual knowledge (often delivered by certain branches of science) to find *correct* answers. In principle, we can solve our dispute by going to the library where we find what we need to know. In contrast to that, there are discourses on topics that no factual knowledge and no library can solve, namely those on norms and values. We may call them *evaluative* discourses. Ethical discourses fall into this category (being aware that descriptive ethics has a cognitive component which is not our concern here).

We still need to be more precise. There are many things we value, and not all of them have an ethical character. When I say, *"I prefer to buy a red car!"*, while you favour blue cars, it would be pointless to argue which of us is right. We need to distinguish preferences, opinions and personal desires from ethical argumentation and normative prescriptions. It is the latter that we are interested in! Moreover, we need to separate arguments that matter primarily in the private sphere from those that occur in a professional realm. You may argue with your partner about what would be the right thing to do, also in an ethical sense, but that is not necessarily anyone else's business. In this book, we will focus on those kinds of discourse that take place in a professional arena, in the public domain, or in a context that affects a larger circle of stakeholders.

Chemists see themselves and their professional impact in the epistemic discourse field. Clearly, chemists produce knowledge about matter and its exploitation. Yet, normative decisions and evaluative judgements are made in all stages of chemical inquiry, for example in grant proposals, choice of research topics, introduction sections of essays, statements in public communication, decisions in teams in industrial research, when making risk estimations, when choosing statistical models for inductive reasoning, and so forth. Moreover, the political and societal discourse on the impact of S&T progress – clearly an ethical question on how we want to live – is more and more entering the scientific realm itself.

The student depicted in Case 1.4 succumbs to a common misunderstanding: ethics is expected to serve as a tool with which we can figure out the right thing to do or to decide in case of a dilemma or a situation that challenges our normative judgment capacity. This is sometimes depicted as an ethical lens that helps us focus on the quintessential solution that gives us advice or a recommendation on what to decide and how to act (Figure 1.3A).

This image of ethics is misleading. Unlike many approaches in the natural sciences in which experimentation and investigation aims at identifying universalizable and generally acceptable

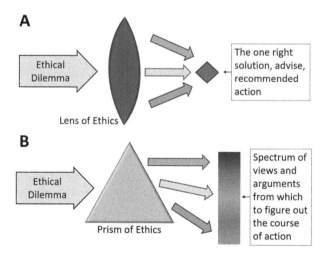

Figure 1.3 (A) Ethics as a focussing lens, and (B) ethics as a refracting prism.

claims, ethics as an applied methodology (not the philosophical discipline) is better understood as a strategy to disentangle argumentative and logic relationships between premises, clarify the consistency and plausibility of evaluations and judgments, and allow discourse participants to understand each other's arguments. Thus, the image of an ethical prism is more suitable: in the case of normative unclarity, ethical reflection allows one to refract the variety of views, identify overlaps and differences, or systematize and categorize types of arguments and their underlying worldviews. This may sound highly unsatisfying for empirically and analytically thinking natural scientists. Yet, if dogmatic or ideological attitudes concerning the truth of moral, ethical or legal claims are to be avoided, there is no other choice than regarding the decision on what is good or right to be the result of a discourse in which the participants express their views reasonably coherently, under scrutiny, and without outer influences that would undermine the *power of the best argument*. The example cases in the following chapters will provide many opportunities to understand this statement and put it into practice in the real-life situations that chemists typically face in their professional careers.

In very simple terms: this book won't tell you what *good chemistry* is. It will give you the practical competence to figure out in respective situations what you should do so that you deserve the label '*good chemist*'. This should be a good reason to keep reading rather than stand up and leave.

1.5   Book Structure and Learning Objectives

The three thematic fields elaborated in Section 1.2 – methodology, research ethics, social and environmental impact – serve well as an orientation for the structure of this book (see Table 1.1). We will first learn about the nature of scientific inquiry and its methods and strategies (Chapters 2–4). Then, we will discuss important aspects of good scientific practice and the daily pitfalls of research conduct and lab practice (Chapters 5–9). Chapters 10–15 deliver concepts and practices in managing the social implications of chemical activity, with a focus on the role of chemists themselves in this process. Chapter 16 summarises all these topics and illustrates, at the same time, how they overlap and connect in the daily life of chemical practitioners.

Every chapter starts with a summary of the content and an overview of the themes and learning objectives. This allows the reader to decide at one glance whether that chapter is relevant to their personal learning goals or not. Some may not be interested in animal experiments or public communication of chemistry; others might know that scientific publishing won't ever be an issue for them. Thus, it is intended to give every reader a chance to select the content efficiently.

After this abstract, the reader is confronted with introductory cases that provide an idea of what type of problem is addressed. It is advised to think of solutions or answers before continue reading. Comparing one's initial thoughts to that at the end of a lesson is always very insightful and a good way to identify the learning progress. Moreover, it gives the book a touch of inductive learning (from cases and observations to cognitive reflections to general principles and conclusions).

Throughout the book, the number of references is kept small. Wherever necessary, stated facts are backed up by reference to contemporary research and state-of-the-art knowledge. Yet, as this is not a research report, the main purpose of referencing should be to give the interested reader a chance to find further reading material to advance their understanding of the presented topics. The references are all provided as actual reading recommendations, not as back-up or proof of claims. Referencing papers and books that students wouldn't read anyway just makes the text more confusing and occupies space. Some of the listed books and essays include extensive reference lists that highlight specific topics for particular interests. Moreover, it is recommended to find material that is written in the reader's mother tongue. This is not only a language aspect, but also one of regional relevance. The mentioned books on scientific integrity,[4-7] for example, mostly refer to regulations and guidelines enacted in the USA, which might be irrelevant in some cases for chemists in the EU.

Every chapter ends with 20 exercise questions with multiple-choice answers (except for the shorter Chapters 5 and 6 that have 10 questions and the longer Chapter 8 that has 25). This gives readers a chance to check if basic concepts and ideas of each topic have been understood and are ready at hand for application in professional practice. Moreover, the questions and their suggested answers serve as examples that may clarify the addressed topics further.

The content presented in this book shall equip the readers or – if used as course material in a curricular course – the learners with competences and skills in basic research methodology and its philosophical foundations on the one hand, and in overseeing, understanding, evaluating and assessing contemporary ethical and social issues arising from chemical activity as part of scientific and technological progress on the other hand. The book is designed and planned in particular for chemistry students and their related fields, requiring no philosophical or ethical background knowledge. The content is strongly related to chemists' typical daily professional activity: science conduct, logic and theory of science, experimentation, writing publications, dealing with uncertainty, assessment of innovation and R&D, and the social and environmental impact of chemistry's creative potential. Applying the fundamentals in philosophy of science and research ethics to the particular conduct of chemical research and its internal and external domains of responsibility is expected to sharpen and solidify the learners' awareness of the theory of research practice, their knowledge of scientific integrity, and their ability to apply critical thinking for the assessment of the social sphere of *science and technology* as a field of human activity that impacts the quality of life of people all over the planet. As a major field in *applied ethics*, S&T ethics touches the domains of bioethics, medical ethics, environmental ethics, profession ethics and business ethics. With the help of countless examples from chemistry, science in general, research, engineering, R&D, and so forth in the history of societies worldwide, the reader will get a sense for the *ethos* of science conduct on the one hand, and for the ethical and social implications of S&T on the other hand. While the former is a matter of *internal responsibility* of individual researchers and their institutions, the latter topic on *external responsibility* will address risk issues, sustainability, multi-stakeholder discourses on S&T development, and the social construction of technology. The overall objective of this book is to contribute to a more *complete* education of young researchers and scientists as important enactors of progress and influential decision-makers in the future. It shall provide them with the skills to reflect on and deal with the major contemporary challenges in society and the environment with a higher degree of sustainability.

To summarise, this book is intended to support chemists and chemists-to-be in:

- Understanding basic science theory and applying it in daily research activity.
- Increasing knowledge on theory, conduct and communication of chemical science.
- Applying ethics to *scientific practice* and *science assessment.*
- Learning concepts of *responsibility* and *sustainability* in the context of chemistry.
- Acquiring skills for interdisciplinary normative discourse.

Exercise Questions

1. Which of the following types of evaluative statements, would NOT be considered normative in the sense of this book's terminology?

   A: Legal regulations.
   B: Ethical arguments.
   C: Personal preferences and opinions.
   D: Moral rules.

2. What should an ethical argument consist of?

   A: At least one is-premise and one ought-premise, resulting in a should-conclusion.
   B: A quote from a famous philosopher.
   C: An indication of a punishment or sanction in case of a violation of the suggested moral rule.
   D: A reference to a moral authority (for example the church or the constitution of one's country).

3. Which of the following is not considered an aspect of "Good Chemistry" in this book?

   A: Appropriate research methodology.
   B: Beneficial impact of chemistry on society and the environment.
   C: Well-paid job opportunities.
   D: Scientific integrity.

4. Which of the following accidents/problems are not commonly attributed to chemistry?

   A: Industrial accidents like those at Seveso or Bhopal.
   B: Harmful side-effects of drugs (like Thalidomide).
   C: Airplane crashes.
   D: Pollution of the ocean with plastics.

5. Who is addressed in this book?

   A: (Future) Academic chemists.
   B: (Future) Chemists working in the public service sector.
   C: (Future) Chemists working in industry.
   D: All of A, B and C.

6. Which of the following does NOT count as a legitimate motivation for considering ethical aspects of chemistry?

   A: It improves one's research skill.
   B: It protects against committing fraud and scientific misconduct.
   C: It supports sustainable development of science and technology.
   D: It proves one's integrity and, thus, protects against being accused of misconduct.

7. How is "discourse" defined in this book?

    A: A communicative action in which knowledge (factual or evaluative) is clarified by exchange of arguments among the discourse participants.
    B: A debate in which one tries to convince the other by any means.
    C: An emotional quarrel in which both parties express the sincerity of their viewpoints by facial expressions, gestures and body language.
    D: Any discussion that takes place in a professional realm (and not in one's private life).

8. "Good Chemistry", here, does NOT refer to:

    A: Research competence.
    B: Good scientific practice.
    C: Beneficial impact on society and the environment.
    D: Sympathy between two friends.

9. What is the relationship between ethics and morality?

    A: Ethical reasoning results in morals (moral rules for action).
    B: The two are the same (synonyms).
    C: Ethics is a matter of philosophy, whereas morality is a matter of religion.
    D: Ethics states what we should do; morality states what we should not do.

10. This book will teach:

    A: Current trends in moral philosophy.
    B: All breaches committed by chemists in the history of chemistry.
    C: Orientational knowledge for ethical conduct of chemistry.
    D: How to pass evaluations of the ethics board at one's institute/company.

11. Science ethics requires...

    A: ...to follow orders and guidelines.
    B: ...to think critically and evaluate appropriately what would be the best choice of action in particular situations.
    C: ...knowledge about moral philosophy.
    D: ...nothing but profound expertise and competence in one's scientific field.

12. Ethical competence as a scientist/researcher pays off in the form of:

    A: Public acceptance and credibility.
    B: Scientific integrity and good reputation in the chemical community.
    C: Economic profit (for example in "green/sustainable chemistry" business models).
    D: All of the above.

13. Which of the following is a matter for ethics?

    A: Preferences and feelings.
    B: Values and virtues.
    C: Scientific knowledge and factual truth.
    D: All of the above.

14. Ethics is a topic for chemists...

   A: ...only in terms of research ethics (good lab practice).
   B: ...only in their role as general citizens (committed to the commonly accepted moral codes).
   C: ...in various domains of their work (scientific practice, impact on society and environment) as an orientation for decision-making and professional conduct.
   D: ...because new education guidelines require that all future professionals study ethics before they are released into the job market.

15. Which of the following statements concerning the structure of an ethical argument is incorrect?

   A: Descriptive ("is-") premises give information about a given or hypothetical situation.
   B: Normative ("ought-") premises are randomly inserted because they are based on mere opinion or personal feelings.
   C: The premises need a logically consistent and plausible connection.
   D: The prescriptive ("should-") conclusion, as a result of the correct connection of premises, indicates what would count as "right" or "good" (to do).

16. [Preface question] Dr Jan Mehlich wrote this book because he...

   A: ...is a moral philosopher.
   B: ...once committed scientific fraud, was convicted, and can now share first-hand experiences.
   C: ...studied both chemistry and applied ethics, and worked in the field of "science and technology assessment", thus having the competences needed for this topic.
   D: ...is a member of the European Commission on Science Education that decided that such a course should be mandatory for chemistry students.

17. Ethical aspects of chemical activity...

   A: ...concern only chemistry students.
   B: ...are only important for chemists working in industry (private sector).
   C: ...are a topic for senior established professors in chemistry who can afford the luxury of spending time on it.
   D: ...concern all professional chemists at all stages of their career in all jobs and positions (in different ways, though).

18. Statement 1: "The whale is a fish." Statement 2: "Jazz is the most beautiful music!" Statement 3: "You should not cheat!" – Which of the following characterizations is correct?

   A: All three statements are opinions and, thus, wrong or at least debatable.
   B: 1 is a factual statement that is incorrect, 2 is a preference that can't be debated meaningfully, 3 is a normative statement that may be regarded as correct or incorrect in different contexts.
   C: All three statements can be verified or falsified (by encyclopaedia, poll, or sociological study) and, thus, are factual statements.
   D: Statements 1 and 2 are wrong because people have changed their views concerning these ideas over time. Only 3 is correct because this is knowledge that was possessed even by ancient cultures.

19. Which of the following understandings of ethics plays a role in this book?

    A: Ethics as moral philosophy.
    B: Ethics as applied/practical ethics.
    C: Ethics as the binding moral rules of a culture/society.
    D: Ethics as legal prescription.

20. Consider this statement: "Smoking damages the lung! Therefore, you should stop smoking!" Is this argument tenable?

    A: Yes, because it makes a scientifically correct claim.
    B: No, it commits a naturalistic fallacy by deriving a prescriptive conclusion without providing a normative premise.
    C: No, because smoking doesn't always damage the lung.
    D: It depends on whether the statement is made by an authority (a doctor, ethicist, parent, teacher, *etc.*) or not.

References

1. J. Mehlich, F. Moser, B. Van Tiggelen, L. Campanella and H. Hopf, On the Ethical and Social Dimensions of Chemistry: Reflections, Considerations, and Clarifications, *Chem. - Eur. J.*, 2017, **23**, 1210.
2. P. Pruzan, *Research Methodology. The Aims, Practices and Ethics of Science*, Springer, Switzerland, 2016.
3. K. Shrader-Frechette, *Tainted. How Philosophy of Science Can Expose Bad Science*, Oxford University Press, New York, 2014.
4. F. L. Macrina, *Scientific Integrity: An Introductory Text with Cases*, American Society for Microbiology Press, Washington, 4th edn, 2014.
5. E. Shamoo and D. B. Resnik, *Responsible Conduct of Research*, Oxford University Press, Oxford, 3rd edn, 2015.
6. S. Greer, *Elements of Ethics for Physical Scientists*, MIT Press, Cambridge, 2017.
7. J. Kovac, *The Ethical Chemist: Professionalism and Ethics in Science*, Oxford University Press, New York, 2nd edn, 2018.
8. J. Mehlich, Chemistry and Dual Use: From Scientific Integrity to Social Responsibility, *Helv. Chim. Acta*, 2018, **101**, e1800098.
9. *Innovation, Dual Use, and Security. Managing the Risks of Emerging Biological and Chemical Technologies*, ed. J. B. Tucker, MIT Press, Cambridge, 2012.
10. *The Ethics of Technology. Methods and Approaches*, ed. S. O. Hansson, Rowman & Littlefield Internationall., London, 2017.
11. M. D. Mumford, Assessing the Effectiveness of Responsible Conduct of Research Training: Key Findings and Viable Procedures, in *Fostering Integrity in Research (Report of the National Academies of Sciences, Engineering, and Medicine)*, The National Academies Press, Washington, 2017.

Part 1: Research Methodology

The three chapters in Part 1 introduce what makes chemistry *good* from the methodological and practical perspective. Chapter 2, the only philosophical chapter in this book, starts from very foundational considerations concerning epistemological presuppositions of scientific knowledge construction. Based on these insights on the strengths and limits of scientific enquiry, Chapter 3 illustrates, in a more technical way, how the success of chemistry as a science hinges on the application of appropriate scientific methodology including the design and conceptualization of research hypotheses, experimental planning and conduct for the acquisition of data, interpretation of this data, and subjecting it to scrutiny by communication of the results in order to refine and re-think the new insights. Chapter 4 discusses details of scientific logic and argumentation as an inevitable part of scientific reasoning.

2 Science Theory

Overview

Summary: Scientific researchers and practitioners, in the context of their profession, take many things for granted. Most scientists, for example, are naturalists and metaphysical realists. In a more hidden way, researchers often follow reductionist approaches in experimental designs and the interpretation of obtained data. Scientific inquiry is, moreover, often said to be a viable source for universal truth claims, based on facts, free from ideology and dogma, and even value-neutral. Some chemists (basic researchers, academic scientists, university scholars) regard their main job as generating knowledge of the material world, a mission of discovery (*What? When? Where?*) and sometimes explanation (*How? Why?*). Others (chemical engineers, applied researchers, private sector science) add a creative component of exploiting the chemical knowledge we have acquired over the centuries for the creation of something new, something that has not been part of the natural world before.

According to all we know about human civilisation, different cultures started reflecting on the nature of knowledge and truth about 2500 years ago. This knowledge of knowledge, or epistemology, is one of the major disciplines in academic philosophy. It is a central part of philosophy of science and science theory. Entire library shelves are filled with elaborations on this topic. Admittedly, it is not necessary for a scientist to study and know all this theory. Yet, a bit of it – the quintessential conclusion, perhaps – will surely make the chemical scientist and researcher a better practitioner. Therefore, this chapter attempts to introduce epistemology in a nutshell: what is knowledge, truth and meaning, what is science able to contribute, and when should the chemist be aware of limits and pitfalls of scientific knowledge? The practical relevance for the conduct of chemical science and its application in industry and innovation shall guide a short tour through this complex topic.

Good Chemistry: Methodological, Ethical, and Social Dimensions
By Jan Mehlich
© Jan Mehlich 2021
Published by the Royal Society of Chemistry, www.rsc.org

It is not a co-incidence or stylistic choice that the key themes below are questions. It is notoriously difficult to settle epistemological debates with definite answers. It may be unsatisfying for chemists, but this chapter's main goal is to raise awareness of questions that challenge beloved convictions and comfort zones. The result is more careful practice in scientific inquiry, and a reasonable balance between epistemic humbleness and confidence, that means knowing the strengths and limits of scientific knowledge.

Key Themes: What is scientific knowledge? Does science make truth claims? Do scientists have to be naturalists? Realists? Is science universal, or rather paradigmatic? What role does meaning play? What role does education play?

Learning Objectives:

In this chapter, you will learn:

- What it means to gain knowledge about the world, and how changes in our understanding of knowledge also change the way we characterise scientific inquiry.
- What presuppositions science is built upon.
- Why communication and discourse are crucial for the validity of scientific claims.
- That science is a powerful instrument for the generation of reliable knowledge that is threatened by contemporary developments towards post-factualism and political or religious ideology.
- What the limits of science are, and how a change of theoretical perspective can improve scientific inquiry.

2.1 Introductory Cases

Case 2.1: Alchemy versus Chemistry

The alchemists of Europe's late medieval age applied Aristotle's theories and principles in their investigations of Nature and its components. They intended to transmute cheap metals or even stones into gold and to find the formula for immortality. With the simplest means, polymaths like Albertus Magnus collected samples of minerals and organic materials, experimented with them, identified patterns and resemblances, catalogued, filed and indexed. But no matter how hard they tried, there is no reported case of successful gold fabrication or immortalisation. From today's perspective, it seems quite ridiculous to assume that it was possible. Yet, we may ask on what grounds the alchemists had the conviction that it was possible.

Case 2.2: The Politics of Molecules

Wilhelm Ostwald was a German chemist in the late 19th and early 20th century. Besides his chemical achievements, for example the famous Ostwald dilution law and his colorimetry research, he was a declared monist who believed that there are universal principles underlying everything from nature and its laws to human behaviour to society with its spheres. For example, he promoted his energetic conception of molecular forces as directly applicable to how politics should work: a constant harmonizing of attraction and repelling forces balancing each other out in the naturally most stable form. People in a society, if governed properly, find their most suitable position like the atoms in a stable molecule. Is this a plausible view?

Case 2.3: Chemical Dogma

The Russian biochemist Trofim Lysenko was in charge of advancing post-WWII Russia's agriculture program. He had the enthusiastic support of Joseph Stalin which he maintained by applying Stalin's voluntarist approach to the governance of social processes to chemical experimentation and its interpretation. Unfortunately for him, his proposed and realised measures to increase the output of Russia's agriculture failed utterly and almost caused a famine.

2.2   Why Does Chemistry Need Science Theory?

Obviously, science is chosen as a source of meaningful knowledge for certain reasons, that is, the method of knowledge generation is more reliable than other forms of insight. Science, as widely held, provides us with knowledge about the world that can be taken as *true*.[1] The goal of this and the next two chapters is to understand what makes science such a powerful institution, but also to see where its limits and dangers are.[2] These are questions for the *philosophy of science* and *epistemology*.[3] The *history of science* might be insightful, too: how did the scientific method change over the centuries, and what were paradigms of science (to use Thomas Kuhn's expression)?[4–6] We might have a look at the development from early alchemy to modern synthetic chemistry.[7] Differences to other strategies of knowledge creation (religion, spirituality, also ideology and dogmatism) have to be pointed out.[8]

We would have to talk about the impact of David Hume and Immanuel Kant on our scientific understanding of the world, and about Francis Bacon, Bertrand Russell, Karl Popper, Thomas Kuhn, and Willard Van Orman Quine, in order to get an overview of the various approaches to science theory.[9] Moreover, besides these rather philosophical "armchair" reflections, the contributions of more practical investigative minds like Galileo, Da Vinci, Einstein and Bohr to the development and advancement of scientific methodology are essential for an understanding of modern science. With this list getting longer and longer, and in view of several bookshelves in the university library on science theory, we begin to realise the danger of overloading our heads with too much information. Even those texts on science theory that are written for scientists and researchers easily reach book length.[10] Covering all this in one chapter requires a clear framing of the goal: we want to understand the strengths and limits of scientifically acquired knowledge;[11] and we want to be aware how this understanding impacts the ways we deal with and communicate chemistry-related matters.[12,13]

In order to perform this framing, we need to clarify what chemistry can gain from theoretical reflection on its foundations.[14] A first obvious reason to spend time and effort studying epistemology is the conviction that theory informs and improves practice. Chemical researchers and scientists in academia and private sector research and design (R&D) benefit from a solid understanding of scientific thinking when designing research questions, planning experiments, interpreting and analysing data, and convincing others of the rigidity of the chosen methods when communicating results. A second consideration concerns the credibility of representatives of the sciences – physicists, chemists, biologists, and so forth – in those realms in which scientific arguments collide with other forms of insight like spiritual convictions, political ideology, cultural tradition or common sense as *Zeitgeist*. Here, scientists equipped with the ability to explain the strengths and reliability of scientifically acquired knowledge have a big advantage in the discourse. At the same time, keeping in mind the limits of scientific inquiry may make the scientific practitioner more humble in his claims and more open-minded when listening to other non-scientific arguments. We will meet such situations in Chapters 4 (*scientific reasoning*), 8 (*chemistry as network activity*), 13 (*risk, uncertainty, precaution*), 14 (*science and technology (S&T) assessment and governance*), and 15 (*science communication*).

2.3 Epistemology: What Can We Know About the World?

We start from the very bottom; what, if anything at all, are we able to know and how? This old question has been asked across millennia and cultures, and has never been settled. Yet, there is something like a core tenet that is seldom challenged: knowledge is justified true belief. A wide variety of views on how we come to beliefs, how they are plausibly justified, and how to determine truth, makes it impossible, though, to present a simple formula for knowledge.[15] Figure 2.1 attempts to illustrate the following considerations on epistemology in one schematic overview.

A knower (an *epistemic agent* aiming at *cognitive success*) cannot know what he/she does not even believe. Belief, here, is understood as a kind of conscious state of awareness of something, even though some would disagree that consciousness is necessary for a belief. Belief, thus, ranges from mere opinions, intuitions, feelings, and guesses (educated, perhaps) to in some sense confirmed facts and states of affairs. Which of these count as knowledge? Those that are justified! This is where the trouble starts. Various viewpoints on methods and strategies of justification resulted in different paradigms, like the necessity of evidence for a belief, or the reliability of its cognitive verification method. Some claim that this justification is found inside a knower's mind (*internalists*), others locate it outside of it (*externalists*). Justification strongly depends on our concept of *truth*. As the third condition for knowledge (besides being believed and being justified), it introduces a normative component of knowing. First, the epistemic agent *should* aim at cognitive success by checking the truth of a belief so that it may count as knowledge, and, second, there are right and recommended ways of acquiring, justifying and evaluating knowledge. Yet, truth is itself a very tricky topic and an issue of global unsolvable controversy.

A first interesting field of inquiry – both as armchair philosophy and as empirical science, for example neuroscience, biology, or psychology – is the question of possible sources of knowledge. What forms our

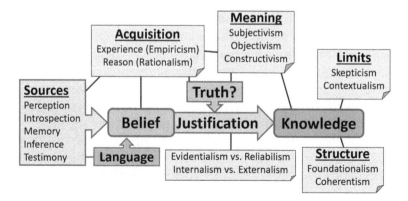

Figure 2.1 Aspects of epistemology that play a role in discussing what knowledge is.

beliefs and allows any judgement of their justification? The data (or the information) that we use may come to our awareness through perception, but also through introspection, as mnemonic recollection, through logical inference (from *a priori* knowledge, perhaps), or through testimony (*via* communicative acts from other sources, in the widest sense). Here, we begin to understand the role of language for knowledge. Can we only know what we have language for? To what extent does the validity of knowledge depend on the linguistic conventions that are at play in a communicative act of testimony? If you tell me A, do I know A, or do I know A*, a derivative of your knowledge of A, modified by my own epistemic background? Language does, of course, not only mean spoken words, but also logic, mathematics, symbols, signs, and even body language and, perhaps, molecular recognition.

Moreover, the question of knowledge sources is connected to more general views on knowledge acquisition. Classically, we distinguish between those who take experience as the major input (the *empiricists*) and those who regard reason as the real constitutor of knowledge (the *rationalists*). Even though not an entirely smooth analogy, rationalists are commonly endorsers of analytic *a priori* knowledge (that is inborn and must not be inferred, like logic and math), whereas the empiricists usually regard most valuable knowledge as *a posteriori* synthetic knowledge. A clarification of this dispute or even a decision for taking one of these sides has a strong impact on our understanding of belief and truth. Moreover, this debate cannot be solved (not even be conducted) without a reference to meaning ascription. Here, we meet the language aspect of truth again. Basically, we need to figure out where meaning is located. Common viewpoints are subjectivism (meaning is in the mind of the subject), objectivism (meaning is contained in the objects that are to be known), and constructivism (meaning is constructed from the social, cultural or environmental reality).

Now, we can talk about knowledge itself. Are there any ways to distinguish various knowledge forms (like know-how, know-what, tacit knowledge, or the lifeworld/microworld knowledge distinction)? Are there hierarchies of knowledge types? Foundationalists would claim so: there is a basis of knowledge, perhaps quite small, upon which all other knowledge is built by inference. Coherentists, on the other hand, would endorse a web-like structure of knowledge: knowledge is justified and true belief that fits into an overall scheme of reliably justified and true other beliefs. What are the limits of knowledge (if there are any)? Sceptics doubt that we can know anything for sure at all. One way of saving the value of knowledge is contextualism that allows different levels of sophistication and fallibility of knowledge in different contexts (for example ordinary daily life *vs.* a philosophy conference, or conventional *vs.* ultimate knowledge in Buddhism).

From this overview it becomes clearer that epistemology is not an isolated field of philosophical inquiry, but firmly connected to and informed by philosophy of language, philosophy of mind, philosophy of science, ontology, metaphysics, and in some sense even ethics (for example, are we ethically obliged to seek evidence in order to come from mere belief to knowledge *via* justification?). When all this is thoroughly explicated and spelled out, we can proceed to the methodology and theoretical foundation of scientific inquiry. We may figure out whether science *discovers*, or rather *constructs*, knowledge.[16] We may question scientists' realist and naturalist commitments on the basis of a solid theory of knowledge that calls realism into question.[17] We may define the criteria for critical scrutiny that make constructivism a cognitive theory (rather than a relativistic, psychologistic, or subjectivistic approach). We may even be in the position to identify the subtle differences between knowledge and wisdom and how to reconcile both to practical and pragmatic life skills.

Box 2.1 Exercise: Who knows?

Xhrk is a bronze age human and member of a clan in which he is assigned the role of a weapon smith. His father showed him where to find good raw material for the alloy fabrication and how to cast blades, but he found by himself that adding charcoal to the fire in which the ore is molten increases the stability of the weapon. John is a chemistry student who just received a good grade in the inorganic chemistry midterm exam. After studying the recommended textbook thoroughly, he could explain the chemical reactions involved in alloy manufacturing flawlessly. Who has knowledge about alloys? Xhrk, John, both, or none of them? Discuss with others!

In addition to this exercise, which is not meant to result in a definite answer, Case 2.1 can be reflected with these considerations, too: did the alchemists have any knowledge? Perhaps not, as most of their insights from trial-and-error may have been true rather by chance or luck than by proper justification of the method. However, from another perspective, perhaps yes, as given their possibilities of ascribing meaning to their observations and findings their conclusions may pragmatically count as true, especially if they resulted in successful applications, so that with sufficient justificatory evidence their insights count as knowledge at that time (but not now). Similar thoughts can be applied to the famous Phlogiston theory: proponents of this explanatory model were justified in their belief as observations and even experimental results, in their historical context, served as reliable sources of belief. But was it true? If we endorse a correspondence approach of truth – true is what corresponds to a real feature of the actual world – the Phlogiston theory proponents had no knowledge because it wasn't true belief, no matter how well justified. If we endorse other approaches to truth (coherence or pragmatic conceptions, for example), we may grant knowledge if the Phlogiston theory is sufficiently coherent with all other theories available at that time or if its application was successful.

What happens when a chemical researcher applies established chemical knowledge to generate new insights that, justified and true, becomes knowledge? In order to come to this situation, the chemist must have acquired knowledge by studying. In terms of Figure 2.1, we may say that large parts of the researcher's beliefs originate from testimony (textbook knowledge, teachers, peers, *etc.*), recalled from memory, thought further by means of logic inference, and aligned with observations (perception of the surrounding). Both theoretical and practical beliefs are acquired or formed by experience and cognitively processed by faculties of reason (rationality). The most interesting part is the ascription of truth labels *via* the construction of meaning.[18] What the chemist judges as true depends to a large extent on the context provided by experience (the textbook knowledge as testimony or the sense-making of perceptions in practice). If that is true, it would imply that truth is not an issue of the actual world, but one of our concept of reality. Many natural scientists reject this view because it shatters the comforting commitment to naturalism and scientific realism.

Box 2.2   Definitions

Naturalism: the view that the physical world follows natural laws. Thus, in principle, it is predictable, analysable, comprehendible, and explainable. Other entities like mental constructs, abstracts, ideas, deities, ghosts, and so forth, are either not real or the direct effect of physical processes (for example, a thought is a neural activity, or love is an emotion arising from neurotransmitters and memory).[19]

Realism: the view that statements are true if and only if what they state is really the case. It means that entities A, B, C and so on exist and have properties such as *A-ness*, *B-ness*, *C-ness* and so on, independent of anyone's beliefs, linguistic practices, conceptual schemes, and so forth.[20]

A scientific realist has no doubt that postulated particles like atoms, electrons, bosons, and so forth, really exist, and that properties like electronegativity, density, refraction indices, and so forth are the real

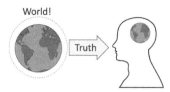

Figure 2.2 Naïve realist view of world perception.

inevitable measurable effects of the existence of such real matter. This sounds so familiar to most chemists that we hardly ever question it. But upon closer examination, is it really tenable?

With our cognitive tools we perceive the world we are living in. The most naïve view is that of a real world that presents itself to us. Our task, then, is to watch it with a clear mind (and clarifying the mind is a practice of philosophy) so that we are able to see as many facets of it as possible in order to increase the chances of a successful and fulfilled life in this world. This was the idea of the Ancient Greek philosophers, starting from Heraclitus and Parmenides up to Socrates, Plato and Aristotle. It was all about *the world*. It is out there and sends signals that stream into our mind through our sense faculties (see Figure 2.2). In order to get to the truth, we need to sharpen our senses so that there is a higher chance of perceiving the world correctly. In order to master the world and lead a successful life, thus, we focus on the world, because we are convinced that there lies the key to proper understanding and knowledge.

There are two dangers in this idea, and both are deeply entrenched in the further course of European-Western philosophy. The first is the dualistic division into *outside* and *inside*, into *outer world* and *inner me*. The second is the realist idea of *discovering* knowledge of real features of the world. This has immediate implications on our approach to scientific inquiry. It would mean that the observer (the subject) during the act of observing has no impact on the observed (the object). The result of a measurement, then, is an effect of the actual world and not a representation or manifestation of the subject's ideas and concepts.

Immanuel Kant is the most prominent philosopher who modified this image of world perception. His basic idea was that we can only become aware of those features of the world that we have a pre-formed image of. That means, we align the incoming signals with our previously made experiences (see Figure 2.3). He distinguished *things-in-themselves* (the features of the real world) from *appearances*, the things as they appear in our mind.

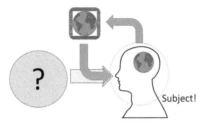

Figure 2.3 Reality as an image of the world.

As a consequence, we can never know for sure what the actual world is. It remains obscured. The world that is represented in our mind is fed by an image of the world, and at the same time it feeds this image (for example, by making new experiences that require a modification of the image). In this view, *world* is all about the subject (or: the observer). It still rests on realist accounts: there is this real world out there that needs to be comprehended by sharpening our senses and adding new experiences to the pool that forms the image. This is indicated by the direction of the arrow: signals stream in for us to process them. In order to increase the chance that our image of the world is identical to the actual world, we need to attempt to uncover the hidden features of the world. The scientific method, in this view, is then a discovery of the world and what is to be known about it. Yet, in contrast to the naïve realist's view, this knowledge is always embedded into a larger framework of meaning and dependent on it. A chemical experiment and its interpretation are only possible with previously acquired textbook knowledge and practical experience.

In order to operate with our world knowledge, and to question and refine it, we need to look at ourselves and our image rather than the world itself. What happens during an act of perception? It is plausibly suggested to separate it into a non-conceptual first and a conceptual second step. After the object of the perceptual cognition appears in the consciousness, the intervention of a conceptual construct enables the identification, classification, and naming of that object. A philosophical concept known as *phenomenology* looms large in this view.[21] It has an important consequence on our image. Not only is perception of objects aligned with our pre-formed image of the world, but we are also only able to perceive of objects that make sense within this framework. An act of perception, in this view, is not a mere *streaming-in* of stimuli, but an active *reaching-out* into the world (thus, the reversed direction of the arrow in Figure 2.4 compared to Figure 2.3). By nature, this is a highly selective process, illustrated by the lens in Figure 2.4.

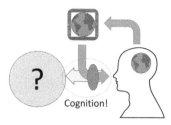

Figure 2.4 Phenomenological view of world construction.

It confines the cut of the world that we are able to pay attention to, and it also colours and shapes the incoming signals in the light of the existing image.

A famous experiment illustrates our selective perception; people were asked to watch the video of a volleyball match and count how often the ball was passed between players all dressed in white. A man in a black gorilla costume appeared in the middle of the scene during the match, beating his chest and making silly movements. The majority of watchers didn't see him, even though he was clearly visible among the white dressed players. We may say that it was unfair because the people were asked to concentrate on the ball, they cannot be blamed. But isn't daily life exactly like that? In order to function as self-determined agents, making decisions of which we believe will bring us forward (whatever that means), we focus our attention towards those environmental aspects that, according to our experiences, are significantly important for our agency (remember the abovementioned *cognitive success* considerations).

Phenomenology stresses the importance of *experience*. Every experience (extended to every act of cognition) involves the entire set of experiences made in the past. An experience is the manifestation of all experiences. As a simple example, when seeing the front of a house, we *know* that this is a three-dimensional building because we know the concept *house* from former experiences. In every perception of a part of the world, we are aware of the entire world, because only in this relation does the experience makes sense. This sense-making is the basis of all experience. Not only do we align all experiences with our worldview (constructed from previous experiences), we also can only experience what fits into our margin of *senseful-ness*. That is why we don't see the gorilla during the volleyball match, because a gorilla has no place in the microworld *volleyball*. The house front is automatically completed in our mind to an entire house. When walking around it we might find that it deviates from our imagination, for example the exact size, shape, and so forth, but these are just details. In the same way, we almost always succeed in identifying an item as a *table*, even when it is a very unusual modern art design, because its entire embedment into our world (including its functionality) is constantly present. Sometimes our imagination is fooled, misled, surprised, or puzzled. When we walk around the house front and find that it is only the decoration of a movie set, for example, then we either have to re-align the constructed reality (here: from the microworld *house as living space* to the microworld *movie making*), or we have to construct new meaning from the new experience.

What does that imply for chemical research and experimentation? Scientists often claim that good science is free from any human mind-dependent presuppositions, but a non-judgmental understanding of natural facts. But is this even possible, given the insights above? Every step of designing research hypotheses, conceptualising experimental setups to visualise effects (or make otherwise perceivable), interpreting data, and writing about the findings, necessarily apply human cognitive capacity and comprehensibility. The chemist can only notice and process what his expert microworld *chemistry*, as a realm of specific knowledge, allows him to. This threatens the scientists' dear truth claim because the referential system to judge whether a claim is true or not, the actual world, is lost. Moreover, it increases the risk of judgmental biases and fallacies, especially confirmation biases. Observations and experimental results can easily be interpreted on the background and in favour of established theories and paradigmatic models.

> **Box 2.3** Exercise: Shift work
>
> Thomas Kuhn is famous for his work on science theory, especially his view that science revolves around paradigms. A predominant paradigm is guiding the work and thought of scientists in a way that results are interpreted in favour of it, until a sufficiently large number of counter-evidences has been collected to shift the paradigm to a better alternative in an inevitable *scientific revolution*, as Kuhn calls it. That means, scientific progress is not a smooth increase or advancement of knowledge, but a step-like change of understanding that requires force and, possibly, power. An example is the Bohr–Sommerfeld atomic model that helped explain many observed effects until too many deviations and inconsistencies were detected so that it was given up by the scientific community and replaced by a better model. Can you find more examples that would support Kuhn's view? Are there examples that would suggest that Kuhn is wrong? Discuss with your peers!

If we accept that existing knowledge of chemistry shapes our experimental approaches to generating new insights, how can we be sure that the way we construct meaning from experience and claim universalizability is in any way justified? How do I know that what I see is the same thing as that what you see? There could be a simple answer: by talking about it!

World constructions do not represent the actual world sufficiently. Integrating two or more – almost necessarily deviating but also sufficiently overlapping – images into one, we have a higher chance to acquire viable knowledge. Mankind is a species that constitutes its environment through communication and collaboration. World construction is, therefore, always a process from the *inter*-space: inter-personal, inter-relational, inter-cultural. My world becomes my world by setting it into relation to yours. My experience is only valid (or not) in view of your experiences (and all others'). In case there are insurmountable differences, we need to engage in a conversation (or a *discourse*, to quote Chapter 1) in order to create new clarity (see Figure 2.5).

However, communication is not a trivial thing. Its most important tool is language. This includes our spoken language using words, but also numerical systems (mathematics) and symbolism, non-verbal interaction, body language, and so forth. Language itself is conditioned and constituted by experience, which means that we only have linguistic expressions for what is already part of our experience (made by us or any of our ancestors). Translatability of thoughts and other cognitive impressions is a difficult endeavour, not only between the different languages of different countries or cultures, but even on the very basic level of interpersonal conversation. With sufficient exchange of information, I might be able to anticipate your experience, but as my framework of experiences and their connection is different from yours, I will never be able to see the same thing in the same light. *World* can be defined

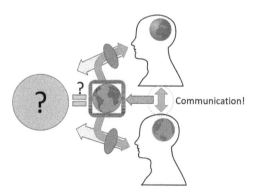

Figure 2.5 Communication as a means to construct the world from the inter-space.

as exactly this *framework of connected experiences*. Then, it makes sense to talk about *worlds* rather than *the world*, because what *world* is for you is more or less different from what *world* is for me. Identifying and becoming aware of the overlapping parts of our world constructions is what gives us an image that is more likely to resemble the actual world. This view is called *constructive realism*, a more recent epistemological concept of scientific inquiry.[22]

From these considerations we understand the importance of communication and discourse as a crucial part of the scientific method. It is impossible to conduct science on one's own in isolation. The viability of scientific claims is generated through critical intra-community scrutiny. That means that other competent scientists have to confirm or debunk what one scientist claims. Thus, the epistemic success of scientific inquiry as such depends on appropriate terminology and language usage, on proper writing and presenting approaches including graphical representations and imaging procedures, and on an attitude that features open-minded, impersonal (or, perhaps, selfless), unbiased objectivity.

2.4 Constructing Knowledge

When reflecting on the processes that are at play when we gain knowledge – that means – when we elaborate on justification for our beliefs and on what may deserve a *truth* label, we need to be aware of the distinction between how we *should* gain knowledge and how we *really do* gain knowledge. Classically, epistemologists focused more on the former. In more recent approaches, with insights from psychology and sociology, empirical observations of how people form and justify beliefs are not ignored.[23] For our practical purpose of chemistry as science and research, we are well advised to understand what chemical researchers and scientists actually do when they construct knowledge. We can do this on two levels, the individual and the communal level. The latter subject is discussed in Chapter 3 in greater detail. Here, the cognitive processes involved in meaning construction are enlightened further in order to understand the influences that accompany and shape our scientific thinking.

Why *meaning*? Didn't we discuss belief, justification and truth as the three crucial factors of knowledge? As Section 2.3 has shown, what we take as true depends to a large extent on what makes sense to us. Another way to say this is that we *construct meaning from our experiences*. Thus, enlightening this process will inform our concept of truth that is applied in scientific reasoning. Then, from there, we can also gain a clearer idea of other aspects of science theory.

Of course, the considerations in this section go far beyond science and chemistry. The mechanisms we are going to explore are at work in every moment of our lives. Often, examples from profane daily life, like the management of emotions or the choice to have dinner at a fast food chain restaurant, are easier to understand than the complexity of professional activity. As a 'chemical' guiding example, consider introductory Case 2.2: apparently, Ostwald transferred a scientific principle or paradigm to a non-scientific realm, politics. We may say that, in his case, meaning construction followed a pattern that is shaped and consolidated by his daily professional activity as a chemist. Logic and reason dominate empirical approaches to understanding the natural world. Ostwald applied the same channels in different social spheres, finding answers for his questions in a scientific way, being satisfied with the obvious plausibility of his conclusions that are in accordance with his scientific worldview.[24,25] This phenomenon that, as is claimed here, we all inherently follow may be described with the image of a *tree of knowledge* (Figure 2.6).

The roots constitute the sources of all our experiences. Everything we know about the world is constructed by our cognitive equipment: senses, central nervous system, brain. Parts of this system are

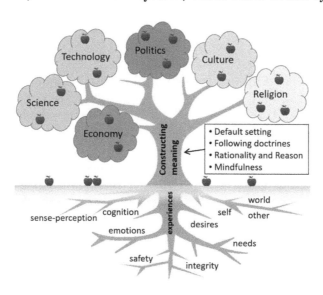

Figure 2.6 A tree of knowledge depicting the construction of meaning from experience.

memory, consciousness, emotions and other psychologically observable and explainable features. In simple terms: we observe, process, think, feel, recall and react. All experiences necessarily are made by us from the self-perspective. Nobody can make experiences *for someone else*. Same as a thought doesn't exist beyond its *being-thought*, experience doesn't exist beyond its *being-experienced*. The perception of a *self* (or an *ego*) inevitably goes along with the definition of everything else as *the other*. This illusion of separation creates the idea of *world* as something external. Within this world-space we experience desires and needs that feed our constant fear of *non-existence* and *ceasing-from-existence*. We experience many forms of suffering (in the literal form as pain, in the figurative form as non-satisfactoriness) and yearn for safety and integrity. This list of basic features is certainly incomplete, but I believe it is sufficiently precise to adumbrate the key point: all humans (as long as not physically or mentally impaired) share these features, and all humans build their decisions, viewpoints and their life on this foundation.

The trunk is the channel through which we process all these experiences in order to manifest them in our *being-in-the-world*. *Experiencing* is a process that only works in view of an experience background that is present in the experiencer, an active *sense-making* (see Section 2.3). This might be the biggest difference to Descartes' famous *tree of knowledge*: it is illusionary to believe that the act of sense-making for all humans is always only *scientific*, exploiting knowledge of *the real world* (nature). There are many more options. First, we all run on a kind of *default setting*. If not otherwise reflected or mindfully brought into our conscious awareness, the choices and decisions we make are controlled and determined by the cognitive and behavioural patterns acquired since we are born, under strong influence of our emotions, our education and other previous experiences that may be summarised as *the matrix*. In this default setting, we do not consciously reflect on the causes and triggers that underlie our choices and decisions. Then, there is dogmatism and indoctrination; someone tells us in one or the other form what certain experiences mean and what we have to conclude from them. In the light form, this includes the parental and institutional education at home and at schools. In the more drastic form we can find that in some organized religions, in political systems that employ manipulation as a tool to secure established power distributions, and in parts even in organizational hierarchies, business, and media (for example, in advertisements). In short, in all systems that have anything to do with power of some over others (in the widest possible sense). There are also more conscious and sceptical ways of sense-making: we can deal with observations and experiences empirically by setting them into perspective with other observations and experiences, we can contest them and refine our understanding of them. The most basic tool for this is logic. An important aspect of these strategies to *construct meaning from experience* is that they are more sustainable and stable the more mindful and aware a person is in the reflection and application of options.

Meaning construction results in decision-making and choices for particular actions. The branches in this image represent spheres of such manifestations, different modes of individual and collective practices and worldviews, patterns of thought and comprehension, and also the fields that serve as sources of insights for answering the questions that arise in our life, for explaining plausibly the new experiences that we make throughout our lives. Figure 2.6 shows six exemplary branches that represent such thinking modes:

- **Economy**: representing attitudes that are constituted by considerations of costs and benefits, efficiency, profitability, utility, maximization of achievements (pleasure, power, wealth, *etc.*);
- **Science**: symbolizing attitudes that regard knowledge and insight as the most valuable answers to any possible question;
- **Technology**: representing attitudes that assume that every problem can be solved with a suitable tool or procedure;
- **Politics**: including attitudes of diplomacy, discourse, power negotiations, and law, but also intrigue, rhetoric, manipulation and deception;
- **Culture**: representing attitudes that consult instances of tradition, custom, and social identity as orientation for action;
- **Religion**: representing attitudes that consult divine or spiritual authorities or their worldly representative institutions like church dogma, rites or rituals.

The following example may illustrate the differences between people with different attitudes or, in accordance with Figure 2.6, with different branch setups. When we experience difficulties with maintaining a joyful and healthy partnership, there are different ways of making sense of it:

- The person with a dominating **economy** branch will evaluate the partner and the partnership in terms of gained pleasure or utility: does my partner make me happier than any other ever could? Should I change to another partner with whom I have a higher gain?

- The person with a dominating **science** branch will ask what causes the problems, how they arise, what others did to solve the crisis, perhaps with insights from psychology, counselling, sociology, or human biology.
- The person with a dominating **technology** branch will understand the partnership as a kind of machine that needs maintenance and fine-tuning. There is always a way to repair it, and when it is really broken there are substitutes.
- The person with a dominating **politics** branch will seek conversation, refer to agreements and internal rules, enlighten power imbalances, and judge the health of the partnership in terms of justice and fairness.
- The person with a dominating **culture** branch will insist on traditions and customs, reject a divorce with reference to claimed societal values and virtues, worry about what the neighbours say, and look for guidance from ancestors and family members.
- The person with a dominating **religion** branch will rely on the personal spiritual practices (for example, meditation, prayer, mindfulness, mental and physical fasting, *etc.*) to get a clear mind, and seek explanation for the experiences and guidance for decision-making from spiritual or religious authorities (for example, the Bible, the Quran, Sutras, *etc.*, or contemporary teachers of such insights).

These are, of course, stereotypical patterns. There may be other branches representing other modes or patterns of sense-making. Moreover, there is no need to mention that no person is dominated clearly by only one type of cognitive approach to understanding the world and making decisions. What the illustration does depict is the phenomenon that modes of thought that are characteristic of one branch are applied in another realm of daily life or another social sphere, like the economic approach to partnership management, or technocratic governance, or the scientific explanation of morality or spirituality. Ostwald's idea of naturalistic politics (see Case 2.2) is a good example.

It is important to understand how the branches grow and why different people have very different setups of branches. Perhaps, our individual trees of knowledge are as unique as our fingerprints. Infants start with tiny sprouts when they make their first rudimentary experiences. Parents and their behavioural patterns have a big influence on the early development of their children's patterns, but also external events and patterns impact the formation of a young human's branches.

Here, we see the connection between the societal level and the individual level that the tree of knowledge depicts. Different societies express these branches in different fashions and to various extents, both regionally (an Asian society is different from a European one) and temporally (the Greek society of 500 BC differs from contemporary Greek society). From the historical perspective, some ancient branches disappeared while new ones flourished, others dried out or grew stronger. The societies of medieval Europe, for example, made sense of experiences under the influence of the church's dogmatism, fearing hell, praying for God's mercy and benevolence. An experience (for example, a disease) was interpreted in the framework of this sense-making (for example, a punishment by God for improper conduct). Traditions dominated the daily life and rituals of people (*"We do it because we always did it!"*). Scientific inquiry was mostly unknown or even suppressed, technology was not very advanced and didn't play a big role as a possible source of answers for the urgent questions of daily life that arose from the experiences the people made (see Figure 2.7).

Today, in our rather secular societies, we don't consult religion in the same way (to answer most of life's questions), but rather for specific private spheres. Instead, we seek solutions in the fields of science and technology, because the provided answers proved to be more reliable and sustainable in the sense that applying strategies from those realms more reliably lead to success than prayers or employing cultural traditions and customs (see Figure 2.8). This does not only mean that science and technology themselves are popular, important

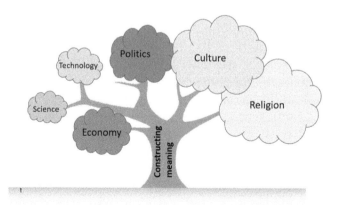

Figure 2.7　An exemplary tree for an ancient society in which meaningful answers are predominantly found in religious belief and cultural customs.

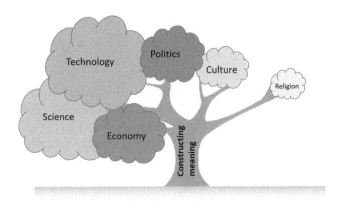

Figure 2.8   An exemplary tree of a contemporary Western European society in which meaningful answers are predominantly found in
scientific, technological or economic worldview patterns.

and/or dominant, but that any question, including matters of politics,
economy, culture, spirituality, and so forth, is treated like an issue
that can be enlightened with the means of science or technology.

There is a dialectic correlation between the society as a system and
its individual members. Each individual contributes to the characteristics of a society, but it is, at the same time, society that shapes
individuals and sets the margin for their self-expression. A religious
society will most likely produce religious members. A technocratic
society will support the flourishing of the technology branch of its
members. The process of social change and progress, therefore, is
usually very slow. This image shall not be misinterpreted in a way that
religious societies make people consider being a priest as a desirable
job while technocratic societies make people want to become engineers. This may or may not be a side effect. More importantly, the
characteristics of the branch configuration determines how the individual makes sense of the daily life experiences. This template constitutes the pattern along which a person is inclined to interpret new
experiences and find answers for occurring questions.

All parts – roots, trunk, and branches – are dynamic and subject to
change. Some roots grow deeper and stronger when a person puts a
focus on certain types of experiences or when outer conditions (for
example, the type of job, or the family situation) draw the person's
attention to particular aspects of life. Scientists, for example, deliberately induce new experiences by designing experiments that enable
new insights. The channels in the trunk are cultivated and expressed

to different extents, too. Children mostly follow their *default setting*, but during youth and adolescence they discover new strategies for constructing meaning. Some become open-minded empiricists, others indoctrinated religious fanatics (just to be sure: there are also open-minded religious people and dogmatic fanatic empiricists). Once a channel is formed and solidified, it is very difficult to change the setting, yet not impossible. Moreover, it is perfectly possible that many branches co-exist peacefully. A scientist can be outspokenly religious by separating the types of knowledge strictly – empiric physical knowledge into the science field, normative spiritual knowledge into the religion field. It would take an enormous amount of *active ignorance* to claim that *"there can only be scientific knowledge"* (as done by atheists) or *"there can only be religious belief"* (as done by religious fanatics).

In order to depict the impact of the branches and their dynamic manifestation, Figure 2.6 contains a fourth element, the fruits. When a branch flourishes, there are fruits growing that a person or a society has to harvest. A strong economy branch will support wealth and material well-being, but also greed and competition. A strong religion branch will increase the capacity of hope and identification with the community, but also fascism (separating their own beliefs from the others' beliefs) and dogmatism (for example promoting creationism and denying biological evolution). Strong science and technology branches will bring about knowledge, possibilities and better life quality, but also increased risks, environmental pollution and social injustice (if not taken care of well). Some fruits are sweet, others are poisonous or stink. It is these fruits that make people conclude that some branches are more valuable and viable than others, that some branches are better kept small or even cut off while other branches deserve more care and nourishment. Atheists often deny the legitimacy of the religion branch. Anti-capitalists see a social threat in the economy branch. Political reformists and anarchists would like to reshape the politics branch according to their political ideals. Reportedly, there are even *science-deniers* (but probably not among the readership of this book). However, same as it is impossible to change the roots (except for paying attention to the roots more mindfully), it is not possible to cut off or change branches (exception: revolutions, or tyrannies). The more effective strategy for an individual is to focus on the trunk, the channels of meaning construction, as this is where we have a choice.

Box 2.4   Exercise: Your Tree of Knowledge

Reflect on the thought patterns and habits that underlie your choices and decision-making strategies. Typical aspects of daily life are partner choice, study attitude, emotional competences, hobbies and spare time activities, political and religious convictions and commitments, phobias and manias, preferences and resistances. Can you identify how you delegate and channel your meaning construction into particular modes of thought? Are there situations in which you follow a default setting (not thinking much about it), or some in which you intentionally apply rationality and reason to understand what makes you form beliefs and choose answers? Draw your own tree of knowledge in which the size of the branches represents your preferred modes of thought. Indicate your predominant strategy of meaning construction by labelling the trunk accordingly. As this is a very personal illumination, you may choose not to share this with anybody. Yet, asking others for feedback may provide very insightful hints concerning your behavioural patterns.

How do we scientists construct meaning from experiences? We may be expected to employ rationality and reason. We challenge our experiences or induce new experiences, for example by exploring unknown features or by conducting experiments. According to the considerations above, however, we need to be aware of the fact that we don't look at isolated experiences. We cannot! We all have a pre-shaped margin of experiences within which new experiences (including the induced ones) have to make sense. This has been pointed out convincingly by W.V.O. Quine, who states that a belief (and he includes scientific statements as *beliefs with a high probability of being true or correct*) is always embedded in a network of other beliefs. A scientific statement only makes sense in view of all other scientific statements. This bears a danger as every scientist became a scientist by more or less formal education. We chemists all start our professional career with a backpack full of chemical knowledge as the state-of-the-art of chemistry. This knowledge serves as an orientation for us, unless we have good reasons (and the mind) to question this knowledge or parts of it. Richard Feynman gave an illustrative account of this fact by the example of the charge of an electron: After Millikan published his influential findings, over 100 years the value increased significantly, but very slowly, because scientists who did research on this matter and published it didn't dare to deviate too much from Millikan's value since that was the scientific fact accepted by the community. We can, also, see several eras in the history of science that are characterised by worldviews, like mechanistic concepts of matter, field theories, energetic models (like that of Wilhelm Ostwald), that influence the way we understand and interpret experimental data, and even our decisions on what counts as a suitable or insightful experiment. Now we understand what Kuhn means with scientific revolutions: for some time, scientific findings are interpreted following a certain way of meaning construction that is pre-shaped by the particular fashion of the science branch, until the evidences are powerful enough to "regrow" the branch in light of the new predominant understanding.

As this image wants to illustrate, a scientific approach to constructing meaning from experience does not only affect the science branch, but all the branches. Scientific knowledge is one (but not the only) important element in technological progress, it impacts production and consumption of goods and services, it influences political practices (for example it favours the development of deliberative democracy rather than a representative democracy); it pushes religion and its authorities into the private sphere (as religious teachings have no world-explanation authority), and it often (but not always) facilitates a cultural scepticism concerning questions of tradition and customs. In the positive sense, it leads to higher rationality and reason, in an exaggerated sense it leads to technocracy and scientism. It is this very danger that makes it important to see the limits of science as a strategy for meaning construction. The strong point of scientific inquiry – if employed well – is a high reliability and viability in questions about the natural features of our world. But it can't tell us how to live our lives or what to value. Yet, it is firmly embedded in this two-way matrix of experiences as the root of our inquiries and decisions, and the social spheres that are manifested by our convictions and practices. Throughout this book we will find countless examples in which the noble goal of scientific reason is corrupted by non-scientific interests and by misleading methodologies and reasoning strategies. That is how all science is intrinsically *normative*: the very act of scientific inquiry is a normative choice.

Whereas Case 2.2 (Ostwald) is an example of a scientific worldview that impacts the understanding of another sphere, Case 2.3 (Lysenko) illustrates the opposite: a socio-political concept is applied to scientific models.[26] It was not recorded exactly what Lysenko's intention was, but it is not far-fetched to assume that in the early stage of his career he understood that following the contemporary maxim or philosophy of the powerful leader would give him some advantages. His science was, thus, corrupted by non-scientific considerations. What might have worked as a model in social sciences could not be successfully applied to biochemistry. Is Lysenko's case unique or special, or are we chemists all at risk of blurring scientific rigidity with non-scientific meaning constructions, at worst with ideologies and dogmas? Current practices in science, research, innovation, and the funding of these activities puts the actors in these fields (scientist, researchers, innovators) in the position of having to decide what to perform research on and how best to acquire the means for it.

This makes it even harder to separate purely scientific considerations from non-scientific presuppositions concerning paradigms, values and norms of scientific research and its output.

To summarise the considerations so far, the goal of science is to create knowledge about the world and its elements. The method to acquire such knowledge is investigation and empirical reasoning by applying strategies of logic consistency. Systematic doubt is applied in order to refine and secure this knowledge. This is fundamentally different from religious or spiritual inquiry. The scientific method is a possible channel of world explanation and sense-making in the trunk of the tree of knowledge (Figure 2.6), processing experiences and leading to the flourishing of certain branches. It is even an important element of the scientific method to induce certain experiences – to make the invisible visible, or to draw something hidden into our awareness. These experiences are then processed with logic, rationality and empirical reason. With this kind of knowledge, we feed the spheres of our daily life (for example, the organisation of our society, the way we do politics, the economic system, the design and dissipation of technology, our understanding of the physical world and ourselves, *etc.*). We are not satisfied anymore with the dogmatic teachings of a religious elite (like the church) but want to be convinced by evidence that can withstand critical scrutiny. In this way, science shapes and influences our whole life and the way we choose to live it, but it does not decide it for us.

2.5 The Power and Limits of Science

There are many ways to summarise contemporary and historical science theory and philosophy of science for scientists. Why was it done in this way in this book? As practical or theoretical chemists, we are not so much interested in philosophical epistemology and even less in the metaphysical foundations of it. We want to be clear about two things: how DO we work and think in the lab, and how SHOULD we do it, that means, is there perhaps any way to do it better than we currently do it? The presented reflections may have the potential to challenge or even change some of scientists' dearest convictions and commitments.

2.5.1 Truth

There is a tendency to understand truth as correspondence between claims and actual facts about the world.[27] 'A' is true when it is really the case that *A* (truth as a *dehyphenation*). The metaphysical price for this conception of truth is too high: we have to make an *ontological commitment* (a non-scientific metaphysical claim about what the world is) that we are, at least principally, able to access the actual world as a reference to judge the validity of truth claims. This would be an infinite regress into even deeper reference systems in order to be able to represent reality with the claims we make by using language. A suggested alternative is a coherence theory of truth according to which the reference system for truth claims is the whole set of truth claims itself. *Coherence*, here, must not be confused with *consistency* since a consistent set of beliefs could still be mistaken. There has to be a link, a connection to the real world that allows a judgment concerning the validity or *force* of a claim. A metaphysically less challenging alternative to correspondence comes from a pragmatic truth conception: true is *what works*, what proves reliable in application, prediction and explanation, and what withstands the most critical scrutiny. The consequence is that truth can only be acquired in a communicative process. Moreover, it means that truth is not an eternal and universal feature of the world, but a subject of refinement, elaboration and change. Thus, rather than aiming at truth, scientific activity aims at *viability* (from Latin *vis* = force).

2.5.2 Presuppositions

The object of scientific investigation is nature and its elements. Whether things like emotions, laws, morality or art are part of the natural world shall not bother us here. The focus of chemical science and research is predominantly on material features of the world. Yet, what kind of forms these natural entities can take is a matter of scientific paradigm and as such part of a presupposition with which scientists elaborate new research ideas. Given the educational, social and cultural framework that shapes all chemical activity, there is no way to disconnect decision-making in a research context from values and worldviews that determine the means and ends we choose to focus on. The old claim that scientific activity is by definition value-free is not tenable under these circumstances. This *neutrality thesis* will be the topic of Chapter 10. It has to be kept in mind, however, that the impossibility of neutrality of science does not change the demand for scientific objectivity.

2.5.3 Worldview

Naturalism and scientific realism seem to be an integral part of the modern understanding of science. The scientific method, as such, would not work without a conception of strict causality and the postulation of laws that describe the existence and character-istics of natural entities. Natural sciences (but not only those) can only deal with those aspects of the world that can be expressed in numbers and analysed in terms of logic and empirical reason. The universality of logic and mathematics is necessary for the most crucial part of scientific inquiry: comparability, reproduc-ibility, and communicative discourse as peer review and scrutiny. Whereas this aspect of science is seldom questioned, others are more critical. Even the most central activity of scientific inquiry – doing experiments – is not a trivial matter. We understand that mere observation of the world is insufficient. In order to gain knowledge about the world and its components, we need to engage in de- and re-construction activities, which mostly means directed and strategic manipulation of the given by proper experimenta-tion. Yet, as we have seen, the relationship between the observer and the observed, between the manipulator and the manipulated, or generally, between the subject and the object, must be regarded through the lens of constructing meaning from experience rather than as a naïve perception of the actual world. Another important issue is the apparent complexity of the world. The human cogni-tive capacity is not sufficient to grasp this complexity. Therefore, for centuries, the method of choice was to cut the world into little pieces, reduce the degrees of freedom, and comprehend the little pieces. Wherever possible, we can set those well-understood pieces into perspective and, perhaps, understand the assembly of pieces by understanding the pieces separately. This reductionist approach has clear limitations.[28] Contemporary trends in scientific method-ology embrace complexity, conceptually supported by chaos the-ory and technically facilitated by machine learning and big data science.[29] According to these approaches, chemical science is sig-nificantly advanced by the move away from reductionism towards holistic system thinking.

2.5.4 Limits

Chemistry as scientific research applies a rigid methodology in order to produce viable knowledge. This allows an exploitation of this knowledge for many real-life applications in the form of tech-nologies, innovations, products, procedures, and so forth. We can rely on the validity and correctness of such knowledge because it was elaborated following such a methodology. At the same time, scientists are not the guardians of the truth. There are realms of meaning that play important roles in our lifeworld in which this rigid scientific approach would not yield satisfactory results. Most of these realms are constituted by forms of normative knowledge about values, needs, preferences, and laws (legal, not scientific ones), acquired by different methods of meaning construction and following different rationales. Understanding this helps the natural scientist to find the point of balance between epistemic confidence and humbleness. On the one hand, our methodology enables us to elaborate factual knowledge that withstands scrutiny. On the other hand, it is not the competence of scientific inquiry to enlighten value judgments and make decisions on what kind of life we would like to live.

Exercise Questions

Pick the one right/tenable answer, unless stated otherwise.

1. The social sphere "culture" as illustrated in the *tree of knowledge* in this chapter covers aspects of... (one *false* answer)

   A: ...language.
   B: ...art.
   C: ...traditional customs.
   D: ...weather and climate.

2. Science is... (one *untenable* answer)

   A: ...an accumulation of true facts.
   B: ...an inquiry about natural phenomena.
   C: ...a social sphere.
   D: ...a systematic work, based on research questions, assumptions and experiments.

3. Communication includes... (one *false* answer)

   A: ...numerical systems.
   B: ...telepathy.
   C: ...body language.
   D: ...spoken words.

4. Characteristics that serve as the identity-giving connecting fabric of an ancient society were... (one *false* answer)

   A: ...church dogmatism.
   B: ...fearing hell.
   C: ...traditional ways of doing things.
   D: ...new scientific insights.

5. Natural scientists...

   A: ...produce universal knowledge by investigating natural features of the real world.
   B: ...built their insights on a neutral basis that is free from any man-made presuppositions.
   C: ...refine the current state-of-the-art of scientific knowledge by investigating research questions at the borders of this knowledge.
   D: ...are the only reliable sources for any knowledge that deserves the label "truth".

6. Constructivism is the view that...

   A: ...reality is nothing but a mental construction.
   B: ...all knowledge is constructed and, thus, equally true.
   C: ...all knowledge we are able to have is shaped by our cognitive processes (perception, experience, sense-making).
   D: ...it is impossible to ever come to any reliable truth and, thus, scientific endeavour is condemned to fail.

7. According to W.V.O. Quine, all knowledge we have...

   A: ...results from a process of meaning construction of experiences in the context of all other previously made experiences.
   B: ...is relative and, therefore, there is no such thing as truth.
   C: ...must come from scientific inquiry, so that it can withstand critical scrutiny.
   D: ...will sooner or later be replaced by other knowledge that is closer to the real truth.

8. According to scientific realists...

   A: ...science is the only way to come to true statements.
   B: ...all aspects of the world can be described and understood by means of science.
   C: ...there is a reality "out there" that can be discovered by scientific means.
   D: ...scientists should not think of unrealistic future visions, but have to be pragmatically focused on their daily-life research practice.

9. Understood as a method to construct meaning from experience, scientific inquiry...

   A: ...follows doctrine and ideology.
   B: ...employs rationality and reason.
   C: ...depends on intuition and feeling.
   D: ...only works in an elevated mental state of mindful awareness.

10. Among the different spheres of social practices, science...

    A: ...competes with other spheres (politics, religion) for authoritative power.
    B: ...is an isolated field of practice, occupied by a rather small number of individuals (scientists).

C: ...serves as one possible source of answers for specific questions of daily life.

D: ...has the worst reputation, because the general public doesn't understand it.

11. Scientific activity is...

A: ...neutral (free from any value).

B: ...shaped by the ends and purposes of the society/culture where it is performed.

C: ...independent of the normative frameworks that dominate other societal or cultural spheres like politics or religion.

D: ...an exclusive endeavour of scientists, which works best by remaining detached from society and its irrational concerns.

12. Scientific knowledge...

A: ...impacts all other social spheres in one way or another (enabling technological progress, shaping policy paradigms, pushing religion into a private sphere, increasing economic power, etc.).

B: ...does not influence political practices.

C: ...is so far from religion in its characteristics that it hardly has any connection with it.

D: ...is better kept separate from cultural impact in order to secure its universal validity and ideological neutrality.

13. A scientific statement...

A: ...is validated in terms of the degree by which it is supported by real features of the surrounding world.

B: ...is validated by determining how well it represents the actual world.

C: ...is validated in terms of its match or mismatch with other scientific statements, requiring communication and collaboration.

D: ...does not require any validation.

14. According to constructivists...

A: ...the only way to arrive at pragmatically viable knowledge is communication and an exploration of the inter-space between different world constructs.

B: ...all statements about the world must be equally true because they are only mental constructions, anyway.

C: ...it is impossible to have any trustworthy knowledge about the world because different people construct different images of the world.

D: ...scientific inquiry is completely useless.

15. What was the connection that Wilhelm Ostwald saw between chemistry and politics?

A: Chemistry is corrupted by politics and, therefore, must be protected from it.

B: He subjected his energetic model of matter onto political practices, believing that it makes politics more natural and, thus, more viable.

C: Members of society should be arranged by politics in the same way that chemists arrange atoms in molecules.

D: He believed that all politicians should study chemistry before entering politics.

16. Nowadays, philosophers of science commonly suggest that scientific knowledge...

    A: ...should be evaluated along a correspondence theory of truth (how well it represents reality).
    B: ...can't be evaluated meaningfully.
    C: ...as history shows, is always proven wrong after some time.
    D: ...can only be evaluated along a coherence theory of truth (how much sense it makes in view of all scientific knowledge).

17. Which of the following changes in scientific paradigm is NOT supported by contemporary scientific theory?

    A: From reductionism to more holistic (system) thinking.
    B: From simplicity (parsimony) towards complexity.
    C: From strict causal determinism towards wider networks of conditionality.
    D: From naturalism towards spiritualism.

18. Interaction between scientists has the main aim to...

    A: ...increase productivity through competition.
    B: ...refine scientific knowledge through communication and discourse.
    C: ...form a collective voice against the power of economics and politics.
    D: ...provide a platform for their elitist theorising around their alienated worldviews.

19. In the "Tree of Knowledge" image presented in Section 2.4, the roots, trunk and branches represent...

    A: ...the sources of our experiences (roots), the strategies of meaning construction from those experiences (trunk), and the manifestations of such strategies in the social and individual lifeworld (branches).
    B: ...the hierarchical order of individual (roots), collective (trunk), and socio-cultural (branches) levels of knowledge, respectively.
    C: ...the epistemological (roots), metaphysical (trunk), and ethical (branches) aspects of humanity.
    D: ...the historical order of development in scientific theory from Ancient Greek (roots) to European Enlightenment (trunk) to postmodernism (branches).

20. Which of the following is NOT one of the historically meaningful views on what we know about the world?

    A: Naïve realism.
    B: Scientific realism.
    C: Optimistic realism.
    D: Constructive realism.

References

1. D. R. Trumble, *The Way of Science. Finding Truth and Meaning in a Scientific World-view*, Prometheus Books, New York, 2013.
2. B.-O. Küppers, *The Computability of the World. How Far Can Science Take Us?*, Springer, Cham, 2018.
3. *The Oxford Handbook of Philosophy of Science*, ed. P. Humphreys, Oxford University Press, New York, 2016.
4. *Philosophy, Science, and History: A Guide and Reader*, ed. L. Patton, Routledge, Abingdon, 2014.
5. F. Stadler, *Integrated History and Philosophy of Science: Problems, Perspectives, and Case Studies*, Springer, Cham, 2017.
6. R. DeWitt, *Worldviews: An Introduction to the History and Philosophy of Science*, Wiley Blackwell, Chichester, 3rd edn, 2018.
7. A. Ede and L. B. Cormack, *A History of Science in Society: From Philosophy to Utility*, University of Toronto Press, Toronto, 3rd edn, 2016.
8. T. Lewens, *The Meaning of Science. An Introduction to the Philosophy of Science*, Basic Books, New York, 2015.
9. M. Bunge, *Doing Science in the Light of Philosophy*, World Scientific, Singapore, 2017.
10. L.-G. Johansson, *Philosophy of Science for Scientists*, Springer, Cham, 2016.
11. J. Ziman, *Real Science. What it Is, and what it Means*, Cambridge University Press, Cambridge, 2000.
12. *Philosophy of Chemistry, Handbook of the Philosophy of Science*, ed. A. I. Woody, R. F. Hendry and P. Needham, Elsevier, Oxford, 2012, vol. 6.
13. *Philosophy of Chemistry. Synthesis of a New Discipline*, ed. D. Baird, E. Scerri and L. McIntyre, Springer, Dordrecht, 2006.
14. B. Bensaude-Vincent and J. Simon, *Chemistry. The Impure Science*, Imperial College Press, London, 2nd edn, 2012.
15. M. Steup and R. Neta, Epistemology, in, *Stanford Encyclopedia of Philosophy*, ed. E. N. Zalta, Fall edn, 2020, https://plato.stanford.edu/entries/epistemology/, accessed August 25th 2020.
16. J. Golinski, *Making Natural Knowledge. Constructivism and the History of Science*, Cambridge University Press, Cambridge, 1998.
17. J. Ritchie, *Understanding Naturalism*, Acumen Publishing, Stocksfield, 2008.
18. L. N. Cooper, *Science and Human Experience. Values, Culture, and the Mind*, Cambridge University Press, Cambridge, 2014.
19. D. Papineau, Naturalism, in *Stanford Encyclopedia of Philosophy*, ed. E. N. Zalta, Summer edn, 2020, https://plato.stanford.edu/entries/naturalism/, accessed August 25 2020.
20. A. Chakravartty, Scientific Realism, in *Stanford Encyclopedia of Philosophy*, ed. E. N. Zalta, Summer edn, 2017, https://plato.stanford.edu/entries/scientific-realism/, accessed August 25 2020.
21. D. W. Smith, Pheomenology, in *Stanford Encyclopedia of Philosophy*, ed. E. N. Zalta, Summer edn, 2018, https://plato.stanford.edu/entries/phenomenology/, accessed August 25 2020.
22. T. A. F. Kuipers, *From Instrumentalism to Constructive Realism. On Some Relations between Confirmation, Empirical Progress, and Truth Approximation*, Springer Science+Business Media, Dordrecht, 2000.
23. M. B. Fagan, Social Construction Revisited: Epistemology and Scientific Practice, *Philos. Sci.*, 2010, **77**, 92.
24. M. G. Kim, Wilhelm Ostwald (1853–1932), *HYLE*, 2006, **12**, 141.

25. P. Ziche, Monist Philosophy of Science: Between Worldview and Scientific Meta-Reflection, in *Monism. Science, Philosophy, Religion, and the History of a Worldview*, ed. T. H. Weir, Palgrave Macmillan, Basingstoke, 2012.
26. H. Sheehan, Lysenko case, in *Encyclopedia of Science, Technology, and Ethics*, ed. C. Mitcham, Thomson Gale, Famington Hills, 2005, vol. 3.
27. *The Oxford Handbook of Truth*, ed. M. Glanzberg, Oxford University Press, Oxford, 2018.
28. *The Problem of Reductionism in Science*, ed. E. Agazzi, Episteme, Springer, Dordrecht, 1991, vol. 18.
29. National Academies of Sciences, *Engineering, and Medicine, Data Science: Opportunities to Transform Chemical Sciences and Engineering: Proceedings of a Workshop—In Brief*, The National Academies Press, Washington, 2018.

3 The Scientific Method(s)

Overview

Summary: In Chapter 2 we have seen how scientific inquiry can be characterised, distinguished from other ways of knowledge construction, and that scientifically acquired knowledge has a high chance of being viable, reliable, and of withstanding critical scrutiny. It examines aspects of our natural world and enables evidence-based factual statements and judgments. But how can we make sure that a statement is *scientific* in a sense that it fulfils certain requirements of scientific knowledge generation? What is the method with which scientists arrive at insights that deserve the label *scientific*? This chapter and the next aim at describing all the features that make scientific research such a powerful way of gaining viable insights.

After identifying the basic steps of a scientific investigation and their characteristics in terms of typical activities of scientists and researchers, a systematic step-by-step guide through the elements of a chemical research project is presented in the form of Lee's scientific knowledge acquisition web. Key issues are the formulation of hypotheses, the analysis and interpretation of experimental results and data, appropriate strategies in case of errors and encountered difficulties, record keeping, and reporting and publishing considerations. While this chapter discusses conceptual and methodological issues, it defines the arena of good scientific conduct and helps to enlighten the standard of what counts as *good*. Thus, it lays a foundation for Chapters 5 to 8.

Key Themes: *One method* or *many methods*? Elements of scientific methodology; designing research inquiries; making hypotheses; choosing experimental setups; interpreting and analysing data; systematic and conceptual errors; record keeping; reporting and communicating results.

Good Chemistry: Methodological, Ethical, and Social Dimensions
By Jan Mehlich
© Jan Mehlich 2021
Published by the Royal Society of Chemistry, www.rsc.org

Learning Objectives:

After reading this chapter, you will be:

- Aware of the steps of scientific research, and the importance of each part of it.
- Able to identify where you are with your own research in the scientific knowledge acquisition web, and what this stage requires from you.
- Equipped with insights on the difference between scientific researchers and other personnel involved in research and development (R&D) (like lab technicians, editors and publishers, or engineers).
- Able to handle and apply scientific conceptual terminology like theory, model, hypothesis, observation, reproducibility, and so forth.

3.1 Introductory Cases

Case 3.1: Money for Nothing?

Gary started his first faculty position as an assistant professor recently. He just received a letter in which the Ministry of Science and Technology announced that his research grant application was rejected. He is extremely frustrated, because he thought he had a very creative research idea: a novel catalytic pathway to synthesise planar cyclobutane derivatives. After an institute meeting, an older colleague approaches him and offers help with drafting another grant proposal. His experience, as he claims, is that such grant proposals always need a clear statement on application potentials. *"You won't get money for nothing! But if it is not curing cancer, tackling climate change, or supporting sustainable innovation, the ministry guys think it is good for nothing!"*. Now, Gary is wondering if research ideas are really better when they have practical purposes as their goal.

Case 3.2: Dead End

Julia, a PhD student in a research group that does supramolecular chemistry, has been trying to functionalise cyclodextrin vesicles with short protein chains for more than a year. Even though her principal investigator (PI) mentioned that this project is more like an exploration of an unknown and uncertain territory, she wishes to have at least some successful experiments. But nothing seems to work as planned. She has tried hundreds of variations of synthesis conditions. Now she is at a point where she feels like throwing it all away and asking her PI for another topic for her thesis, not only because of her grade but also because it seems it is a waste of resources, material and money to continue with this hopeless project. Yet, there is always this slight bad feeling that the failure is due to her own incompetence.

Case 3.3: It Doesn't Work!

Ian is trying to synthesise an amine-based dendrimer that is equipped with a fluorescent marker. A comparable compound cannot be found in the literature. Short amines can be coupled with the fluorophore, but it seems not to work with common reaction conditions if the amine is a large dendritic molecule. The other option, building the dendrimer arms after coupling the core amine with the fluorophore, doesn't yield complete products, only irregular dendrimer products. After 4 months of trying several approaches, his best outcome is a yield of 15% of the final product. This is good enough for his purposes. Yet, he starts wondering whether he should include all the failed attempts in his research paper. Otherwise, someone else who is not satisfied with this yield will try all the other methods again that Ian knows are less promising. Isn't it necessary to publish negative results in order to avoid useless repetition of experiments that don't work and to make scientific research more efficient?

3.2 Elements of Scientific Research

When reflecting on what scientists do all day in their job, we have certain ideas, and the reader of this probably has some experience: we think of researchers in their laboratories, conducting experiments, performing measurements, analysing the data, drawing conclusions and visualising them. We also know that research usually starts with formulating a hypothesis that drives a scientist's research agenda. That means, it determines the choice of experiments and the type of information that has to be collected. Also, we know that the production of new insights is not the end point of scientific activity: it also must be communicated. In this chapter, we will have a more detailed look at this chain of research activities.[1] This backbone – research design, hypothesis formulation, experimentation, making scientific statements, producing knowledge, and communicating this knowledge – is labelled *the scientific method* (see Figure 3.1).[2-5] What exactly happens in each step?

Design Research: Every research project starts with an idea that turns into a plan that is then realized and carried out. The context matters significantly at this stage. Interest-driven research, industrial research and development (R&D), and academic basic research follow very different patterns in identifying, selecting and organizing research projects. Basically, research design has two components: the identification of a research question, and the acquisition of the means to conduct the project.[6,7] In academic settings, the choice of research is,

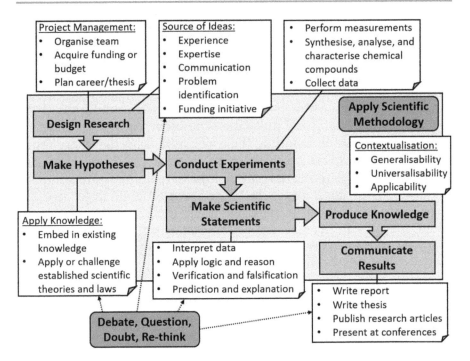

Figure 3.1 What scientists do.

usually, determined by the field of expertise of the PI. New ideas often arise from old ideas, extend former research to slightly different conditions, or are the logic consequence of previous findings. Another source is the many communication channels the academic scientist is using, like peer review, conference talks, or other correspondences within the scientific community. While academic scholars follow thematically consistent career plans, their students aim at completing research for a thesis. In both cases, different projects in the same research group, or by the same person, will not deviate too much. Furthermore, in many cases, the chosen research projects depend on available equipment and technical possibilities at the institute where they are carried out or where collaborations can be formed. The pragmatic consideration is: what are we able to do at this place at this time, given our competences and our infrastructure? In seldom cases, funding initiatives encourage scientists to pay more attention to specific fields that might be on the political agenda owing to their economic potential, such as the nanotechnology initiatives in the early 2000s. This can encourage related chemists (for example, in the field of colloidal matter) to design research projects that fit the portfolio of the offered funding. Researchers in the private sector face different conditions. The management provides the means for goal-oriented research that is strongly related to the company's objectives. The creativity of free academic research is traded for systematic troubleshooting and the development of novel materials or processes for clear end goals, often under economic or legal constraints. What both realms have in common is the requirement of chemical expertise for correct judgment of what is chemically possible. Fruitful research design has to be firmly rooted in state-of-the-art knowledge. *Bad* ideas are filtered out, ideally, before any cent or manpower is invested into experimentation.

Make Hypotheses: When this is settled – when the idea is turned into a solid plan – hypotheses (or sometimes only one) are formulated. These hypotheses are not coming out of thin air, but must be grounded in present literature and in commonly accepted theories and scientific laws. We will see later what determines a *good* hypothesis. However, it is obvious that the hypothesis paves the way for the course of experiments conducted and investigations initiated and carried out. The experimental setup and the data collected must be related to the hypothesis in a way that they allow insights into the validity of the hypothesis. In chemical research, the formulation of precise and clear hypotheses is surprisingly underrated compared to disciplines like psychology or sociology. Scientific inquiries in chemical contexts often ask *questions*: *How can the enantioselectivity of this reaction be increased by the proper choice of ligands on the organometallic catalyst? How much does this alloy need to be doped with a noble metal in order to change its colour?* Yet, as the name suggests, a hypothesis should be a statement or educated claim rather than a *question*: *Asymmetric organometallic catalysts reliably yield enantioselective reaction products. A doping of the alloy with 5-10 ppm noble metal ions shifts the absorption maxima by 50 nm.* These are more or less confident statements based on previous experiences and established knowledge while, themselves, are unexamined so far. They express expectations that are likely to be confirmed, but not proven, yet. With these, the researcher can compile a set of experiments that provide the information that is necessary to judge whether the claim is correct or not. In these examples, the chemist would suggest ten different promising organometallic compounds with different ligands, synthesize and characterize them, and apply them in the respective catalytic reaction to see the result; or, fabricate a series of alloys doped with different amounts of gold ions (5, 5.5, 6 ppm, *etc.*) and measure changes in the absorption spectra. The more background knowledge the researcher has the more intelligent the hypotheses he or she can formulate.

Conduct Experiments: The points made so far apply to all kinds of scientists, to everyone who follows a systematic approach to the generation of new knowledge, including natural scientists (physicists, chemists, biologists, geologists, *etc.*), social scientists, psychologists, medical researchers, but also historians, ethnologists and philosophers. They differ, however, in the methodologies of experimentation and data acquisition. The natural sciences perform measurements of certain factors that are relevant for the justification or the refutation of the hypothesis (for example, physical, chemical, biological or medical properties). Semi-empirical sciences like social sciences collect data in their own way (not measuring natural properties, but *soft* behavioural, cultural, or psychological factors). Normative sciences like philosophy and the humanities often don't collect numerical data but *arguments* that are then subjected to logic analysis with the same methodological rigidity as the other sciences. Experimental strategies and methods in chemistry are so manifold and diverse that a coverage of details is impossible in this book. The reader, assumingly a practical chemist, won't need a detailed introduction. Yet, a few remarks seem appropriate. In Chapter 2, we characterized experiments as inducing experiences in order to infer underlying principles from our observations. We deliberately produce and collect data. These data are the foundation of all scientific reasoning. Thus, good chemistry is inevitably and to a large extent about data quality, reliability, reproducibility and interpretative significance. Many forms of misconduct are about the mishandling of data (see Chapter 6). From an epistemic perspective, aspects of concern are the meaning and viability of images, the application and impact of technological means in data acquisition, the role of theoretical modelling in experimental designs, and the inference of meaning from numerical descriptions. Methodological and procedural issues arise in the context of record keeping and the appropriate representation of data.

Make Scientific Statements: The information collected (data, observations, arguments) must then be exploited to generate scientific statements. When are statements *scientific*? As the previous chapter has shown, we can distinguish scientific statements from other statements by referring to their following logic reasoning and being verified or falsified systematically. As this is arguably the most crucial step on the way from idea to knowledge, we will dedicate a whole chapter to this topic (Chapter 4). Basically, we may say that experimentation gives us *data*, which the scientific interpretation and analysis turns into *information*, subsequently contextualised in order to obtain *knowledge*. In this step, valid and plausible interpretation of data requires solid theories and models that serve as a convincing framework of meaning within which the data is exploited to extract its own distinct meaning. Logic and reason must be correctly applied and free from biases and fallacies. The conceptual distinctions between verification and falsification, and between prediction and explanation, need to be drawn carefully to withstand critical scrutiny.

Produce Knowledge: Scientific statements, then, allow us to know something that we did not know before. The difference between information obtained in the previous step (the interpretation of data) and knowledge is a successful contextualisation. That means, the scientific claim inferred from the data needs to prove viable when subjected to scrutiny. Remember Chapter 2: we use the term *viable* (which means something like having either explanatory force or application success) as a substitute for the metaphysically challenging *truth*. The bit of information that is acquired by analysing the experimental results has to pass a few tests: it must be generalizable, universalizable, and applicable. The first means that the pinpoint focus of the particular experiment must be extended to similar contexts in which the claim can unleash its applicability potential. The second means that the claims prove viable independent of the location and time of their generation and of the person (or generally, the human mind) who generated them. The third means that claims can be tested and exploited so that their content can be manifested in reality. Take the enantioselective catalysis example that was used to demonstrate hypotheses formulation: the organometallic chemist will synthesize the compound and obtain NMR, mass, and other spectra as the raw data. These are interpreted in order to infer the success of the planned synthesis. The compounds are then applied in a specific context (the enantioselective catalysis) which again yields data that is interpreted so that, finally, the scientific claim is demonstrated viable by yielding a chemical compound with specific properties (a high *ee*-factor). Obtaining respective results under exclusion of other factors is a good hint that the scientific claims are correct and, thus, viable knowledge.

Communicate Results: This knowledge is only valuable when it is communicated. By communicating it – for example in the form of lab reports, theses, research papers, book publications, or at conferences – the scientist also confronts it with feedback and possible criticism from colleagues, fellow scientists, peers or even the entire public. This is a crucial element of science: debating, questioning, doubting and re-thinking, not in an individual isolated fashion, but as a collective and systematic process of the scientific community and those who collaborate with it (for example industry or science policy). This discourse is such an important aspect in the research methodology that we can consider it part of the scientific method. Communication is covered in this book in Chapters 7 (community-internal), 14 (community-external), and 15 (public). The critical stance of doubting and re-thinking concerns almost all aspects of *good chemistry* and should be considered the major overall theme of this book.

Box 3.1   Exercise: Your Research Project

Assuming that you are an experimental chemist conducting research of one or the other kind, please relate your daily activities to these six elements of the scientific method. Answer the guiding questions as mindfully, thoroughly and honestly as possible!

- Who made the research plan, you or a supervisor? How? Is there a clearly communicated purpose or application?
- Is there a hypothesis in the form of a statement (not a question) that guides the experimental investigation?
- What kind of data do you generate? How do you record, store and process it?
- What exactly happens when you analyse the data and make scientific claims? Are you always aware of the logic inferences you make?
- How concerned are you about the transition from information (the scientific claims) to knowledge (the contextualisation)? How much proof is enough to satisfy your doubtful and sceptical attitude?
- Which channels of communication do you use? How do you deal with critical feedback? Do you give other chemical scientists feedback, and how harsh and strict are you with them?

The translation of these conceptual reflections into particular strategies and practices in research design, experimentation, analysis and communication shall not be part of this section. There is a large variety of handbooks and guides in research methodology for specific fields of research. The reader is encouraged to look for books that discuss methods and approaches for his or her special interest. Even with the confidence that the knowledge of methodology acquired in chemical training is sufficient, it will be insightful to gain a larger perspective. It is the author's experience, confirmed by many peers and colleagues, that in chemical training at universities, or at the beginning of a new job, the main approach of learning methodology is *'this is how we do it here'*. If not studied formally, yet, it will be enlightening how fruitful it can be to systematize one's approaches to methodology of chemical science and research![8]

3.3 The Scientific Knowledge Acquisition Web

In the following, we will have a look at the elements of conducting a research project in much greater detail than the overview compiled in the previous section. We will do that with the visual support of *the scientific knowledge acquisition web* (Figure 3.2) as proposed by Lee.[9] However, simply going through that scheme step by step will be tiring and boring. Therefore, we will fill it with life by designing and running our own project in our imagination. As readers have different chemical backgrounds and might not all be familiar with the same field of chemical research, an exemplary research project that can be understood by everyone is suggested: we will investigate the fluffiness of bread (inspired by an exercise in Pruzan's book on research methodology[3]).

3.3.1 Identify the Problem Area

Every research project starts with the observation of *missing knowledge*. Something needs to be known that is not known, yet. Scientists usually become aware of a lack of certain knowledge from reports, conversations with peers, or communication with non-scientists (for example with other stakeholders or the public) who have a problem that can be solved by scientific inquiry. Even though there is a trend

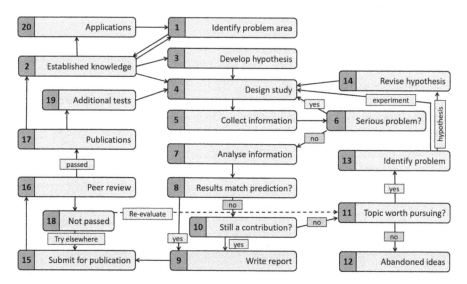

Figure 3.2 The Scientific Knowledge Acquisition Web.[9] Adapted from ref. 3 with permission from Macmillan Publishers Ltd, Copyright 2016.

towards all research endeavours being *meaningful* and *purposeful* in
one way or another, there are still projects that have no obvious benefit besides gaining knowledge for the sake of knowing it. Introductory
Case 3.1 illustrates the dilemma that some academic basic researchers find themselves in. Research ideas may be driven exclusively
by pure scientific curiosity, but related instances (public, funders,
etc.) ask *what it is good for*. The ethical dimension of this phenomenon is discussed in Chapter 8 (Section 8.6 on *Academic Freedom*).
From a methodological perspective, the important consideration is
where the research takes place. We would locate purpose-driven and
application-oriented research in the private sector and technical universities, whereas academic research institutes, as we find them at
universities or in public research organisations, focus at least in part
on basic purpose-free research. A general judgment on which is better
is impossible as both have their justification.

> *Fluffiness of Bread Example, Step 1:*
>
> *Bakers report: The fluffiness of a loaf of bread appears to turn out differ-*
> *ent on humid and dry days when all other procedures of the baking pro-*
> *cess are kept the same. Some customers complained that they can't trust*
> *that their favourite bread is the same every time they buy it. Bakers*
> *want to know how to control the fluffiness of bread because it impacts*
> *their business.*

3.3.2 Checking Established Knowledge

First, it has to be made sure that the issue in question is not already
answered by someone else in a study that is analogous or even similar to the intended one. If nothing can be found, the body of existing
knowledge (in the form of available literature) is scanned for useful
information that is related to the project idea. This can provide hints
for how to specify the research questions (the hypotheses) and for
where to start with investigating. Furthermore, it embeds the project
in the state-of-the-art of science which will be of utmost importance
in peer review and external evaluation. In Chapter 7, which discusses
issues related to publishing chemical research, we will discuss the
ethical dimensions of citing other researchers' works. Besides giving proper credit and avoiding plagiarism, providing references to
other scientific work serves the important purpose of peer review. We
would, usually, cite only good research that inspires us or informs our
research fruitfully. Ethical aspects aside, there are methodological

issues, too. The ideal image of studying the literature before everything else starts deviates from many researchers' lab reality, especially those in university research groups. Most teams work with established procedures and pass experiences and protocols down to new generations of group members. Not every research plan or experiment requires extensive literature review. It is not uncommon to perform the literature search when writing the report, thesis or paper draft in order to prove with these references that one knows the field. However, this Step 2, grounding one's research idea in the established knowledge, is an ongoing process that accompanies all steps in this web in one or the other way.

Fluffiness of Bread Example, Step 2:

The literature review reveals that obviously nobody studied the relationship between air humidity and yeast systematically, yet. However, we find papers that suggest a link between yeast activity and the presence of humidity.

3.3.3   Develop a hypothesis

The researcher sets the scene for his investigations by formulating a hypothesis. Usually, this is a statement that expresses a certain expectation in the form of *"Condition A always and reliably leads to the effect E."*. This guides the researcher in her experimentation (for example, inducing condition A and showing that effect E can be observed with a sufficient degree of reliability and reproducibility). There are criteria for *good* hypotheses:[3,10]

- *Fruitfulness*: Elaborating on these hypotheses will result in useful and applicable knowledge and insights for further progress, for academic and scientific purposes, or for a clear purpose in industry, policy, society, and so on.
- *Clarity, precision, and testability*: The hypotheses suggest a clear experimental or investigative strategy for verifying or falsifying them.
- *Framework for organising the analysis*: It becomes clear from the hypotheses what sort of knowledge from what kind of sources is necessary to generate viable insights.
- *Relationship to existing knowledge*: It becomes clear from the hypotheses formulation how to set the findings into perspective to experiences and approaches in other parts of the world

and in other fields of expertise and application. Sometimes, the major goal is to connect and draw relationships between fields of knowledge that haven't been connected so far.

- *Resources*: The hypotheses should indicate and determine a workload that is feasible given the available resources (funding, lab equipment, manpower, time frame, *etc.*).
- *Interest*: The hypotheses should indicate a field of research and a scientific question that the investigators are academically and also personally interested in.

Hypotheses are often too vague, for example assuming a *relation* between two factors without specifying whether a *correlation* or a *causal relation* is sought for. Most hypotheses are written in a positive wording that suggests what a scientist is looking for and expects to prove. It must be noted, however, that a scientist should have a neutral stance towards his hypotheses, rather interested in falsifying than verifying them! There are also many reported cases in which researchers change the hypotheses at the end of a study in order to make them match with the results they obtained. This may be classified as *questionable* when the researcher is not able to provide proper *scientific* reasons for doing so.

Fluffiness of Bread Example, Step 3:

Hypothesis: There is a positive correlation between the fluffiness of bread and the surrounding humidity.

3.3.4 Study Design

While the hypothesis outlines what shall be enlightened as *knowledge* at the end, the study design is sketched in order to have a clear plan on how to achieve that goal. It states in particular what kind of measures are required to generate useful insights, and how these measures can be operationalised. It must convincingly explain why a certain experiment is intended to be carried out and in which way the acquired data has any relation to the question posed in the hypothesis. A clear proposal also describes the contribution of additional competences, external expertise, special equipment and interdisciplinary collaborations. In this step, a profound understanding of scientific models and heuristic analysis frameworks is required. A good chemical researcher is able to anticipate the weaknesses or insufficiencies of experimental designs, aware of critical questions that will arise if the scientific logic is not sound and complete. How can predictions be employed to inform possible explanations? How can observed effects be attributed with plausible certainty to distinct causes? Details of logic in science will be discussed in the next chapter.

Fluffiness of Bread Example, Step 4:

The measure "fluffiness" must be operationalised, that means a way must be found to measure it reliably and reproducibly. It would be insufficient to simply evaluate the resulting bread as "very fluffy", "medium fluffy" and "only slightly fluffy". One way could be to measure the volume of a bread based on 500 g flour after baking, as the volume is proportional to the air entrapped in the dough which is the major determinant of "fluffiness". Measuring the humidity is easily done with a hygrometer, but a way must be found to control the humidity. Furthermore, all kinds of extraneous factors have to be anticipated (e.g. pressure of kneading and kneading time) and kept constant over a series of test bakes.

3.3.5 Collect Information

In this step, the researcher performs experiments, records observations, conducts measurements, in short: collects data. Experimental apparatuses are set up, calibrated and used to perform measurements. Protocols are followed, refined, adapted, or tested. Protocols of observations and measurements are written, recorded or compiled. Some experiments yield spectra, photographic images, scans, or other visual images, others result in electronic data that need to be subjected to various processing operations. As mentioned above, the correct documentation of data is crucial for proving the integrity of one's experimental and analytical approaches and actions.[11] When experimental data and its interpretation and analysis are communicated and discussed, it is important to be able to present the whole process from experimental design, conduct of measurements and experiments, data collection, and processing of the raw data. This helps not only the researcher, but also the peers in arguing critically on the meaningfulness and correctness of the scientific claim made on the basis of this data. The key for reliable and credible science is writing research protocols, keeping track of acquired data, and recording and storing observations, measurements, spectral analyses, images, and so forth. Many cases of fraud accusation arise from poor record keeping. In other words, many cases of suspicion could have been avoided if the defendants had been able to provide a properly written lab notebook.[12] Most institutions require their staff (and, in the case of universities, their students) to keep track of everything they do, and there might even be courses or seminars on how to do that. Complete and well-organised lab notebooks can easily solve disputes or misunderstandings. Moreover, they also help the researcher, of course, to maintain an overview of activities, the state of a study, the necessity of further experiments, and so forth, so that writing a research report becomes much easier. Especially for Masters and PhD students, it is therefore highly recommended to put efforts into writing proper lab notebooks.

What are the features that make a record a *good* one?[13] A lab notebook should be a bound book, consecutively numbered so that references to entries on other pages can be made. Each entry is labelled with the date and a headline. The researcher introduces the plan for that day with his/her idea and thoughts that explain the choice of experiment. This experiment plan is shortly summarised, giving a citation from which the protocol is taken. Then, the actual conduct of the experiment is listed, with the exact values for used substances, solvents, for time measures, and so forth. The result section includes a physical output (for example, spectra, elementary analysis, gas chromatography print-outs, the visualisation of an agarose gel electrophoresis after a PCR, *etc.*), and a short evaluation or interpretation of it. Concluding remarks may give an outlook for further experimenting (change in protocol, or other factors tested, *etc.*). If possible, all obtained result items (spectra, images, data sheets, *etc.*) should be taped or glued into the notebook so that they can't get lost.

There are certain challenges for record keeping, such as electronic data,[14] the computer-aided processing of data that is not trackable in a lab notebook, or the sometimes interdisciplinary character of collaborative studies in which data must be shared.[15] It must also be noted that data including the lab notebooks are usually the property of the institute (faculty, university), not that of the researcher! Scientists are not allowed to take their notebooks home or keep them when they finish a project, leave the institution or change to another one. We will address this issue in the section on (intellectual) property rights and its conflicts (Chapter 8).

Fluffiness of Bread Example, Step 5:

A strict baking protocol is followed to bake test-bread. All procedural factors are kept constant except the humidity. The humidity is varied in a controlled fashion and the volume of the resulting bread is measured. The humidity dependence of the volume is graphically illustrated in a curve chart.

3.3.6 Serious Problem?

It is very seldom that a research plan yields useful results upon the first attempt. Experimental setups might have flaws, procedural difficulties occur, delegation of (sub-)tasks are unclear or inconsistent, factors that the researcher has been unaware of become visible. Sometimes the acquired data give a hint that something must be wrong with the way the data is collected (statistical insignificance, inconsistency, *etc.*), so that procedural and systematic errors in the experimentation, measurement or data compilation method can be identified. Problems of this kind usually don't put the entire project into question, but require a revision and improvement of the study design (going back to Step 3).

> *Fluffiness of Bread Example, Step 6:*
>
> *According to the experimental protocols, it seems the results obtained on Mondays differ from those yielded on Tuesdays. A look into the lab organisation reveals that different technicians were in charge of kneading the bread dough on different weekdays. A skinny dainty girl kneaded on Mondays, a tall muscular guy kneaded on Tuesdays. It may be assumed that this makes a significant difference for the baking protocol. Indeed, after delegating the task of all kneading work to the guy, the results are more consistent.*

3.3.7 Analysing Information

Datasets and their visualisations have to be analysed. This usually does not happen after all data has been collected, but in parallel to further data acquisition. Raw data is subjected to calculations with certain equations, statistically checked, graphically illustrated, and explained in regard of the possible meaning of its content. A few things have to be ensured:

- *Validity*: The obtained data indeed allows insight into a certain factor that was intended to be analysed. The measures and values represent the *real* occurrence and not just co-incidentally appear at the same time (as far as it is possible to confirm that).
- *Reliability*: The applied methods result in consistent output and don't vary owing to procedural or systematic errors. Here is an example of how validity and reliability are related: imagine a brain scanner that measures the size of brain tumours. Option 1: If in several measurements it always gives the same value and this

value resembles the actual size of the tumour, it has high reliability and high validity. Option 2: If all measurements give the same value, but that value is 40% smaller than the actual size, it has high reliability, but insufficient validity. Option 3: If the variance of the values is large, but their average is very close to the actual size of the tumour, it has low reliability and high validity (but, maybe, rather by chance). Option 4: If the measured values vary largely and neither any of the values nor their average is close to the actual size, the scanner has a low reliability and its output a low validity. Only case 1 is acceptable. The other scanners need to be repaired or discarded.

- *Reproducibility/replicability*: A measurement made on one day should turn out the same on the other day (given that all conditions are kept the same). Values once obtained should occur again when the experiment is repeated. Reproducibility refers to the clarity of the experiments and methodologies in their descriptions according to your laboratory notebook or the report you write: someone else should be able to recapitulate what you did and should obtain the same data when repeating the experiment. There are, of course, experiments and data acquisitions that can't be repeated (*e.g.* observations on astrophysical events or seismic data during earthquakes for geoscientific research). In this case, even though the measurements are not replicable, the conduct of the experiment (setup, data processing, *etc.*) can be reproduced and reconstructed theoretically.

- *Relations*: One of the most crucial points in scientific analysis and debate is the relationship between measures. Some experiments (sometimes mere observations) are certainly *only* descriptive: They describe phenomena without stating anything about relations. In other cases, researchers attempt to make statements about correlations between variables, or even causal relations. Example: researchers in Seoul (Korea) found that the number of suicide cases is always significantly higher two days after a day with exceptionally high air pollution. Referring to the collected statistical data, that is a simple description of a phenomenon. In the next step, they claimed a correlation between the two, because toxicologists could show that ozone radicals (whose concentration is higher in polluted air) affect a certain centre in the brain that is believed to play a role in depression. Whether this is even a causal relation (*"ozone radicals make people die by suicide"*) remains questionable though, as the social, psychological and pathological mechanisms that lead to the decision to attempt suicide are much too complex to draw it down to a simple cause-effect-relationship.

Fluffiness of Bread Example, Step 7:

When depicting the breads' volume (our measure for fluffiness) versus the humidity (our controlled variable), the graph shows an increasing fluffiness for higher values of humidity, but from a certain humidity on, the fluffiness decreases again. If desired, it can also be attempted to explain this observation: maybe too much humidity makes the dough "too heavy", so that the CO_2 produced by the yeast is not able to make the dough loaf grow bigger.

3.3.8 Do the Results Match with the Prediction?

If the results match with the prediction – that means, they are in accordance to the theoretical and conceptual framework that was construed based on available knowledge – and, therefore, confirm the hypothesis, we can proceed to writing a report and communicating the findings (Step 9). If they don't, it must be decided how to proceed (Step 10).

Fluffiness of Bread Example, Step 8:

The observation described in Step 7 – environments with different humidity result in different fluffiness, maybe even quantitatively determinable – matches with the hypothesis made in Step 3. This can be communicated in the form of a research report and, later on, in a publication.

3.3.9 Write a Report

A report consists of a description of the problem, the necessary background knowledge that determines the procedural and theoretical framework of the project, the hypothesis, methodologies and experimental design, the collected data, its interpretation, and a conclusion. All related data, ranging from electronic data files, microscopic or other images, to spectra of all kinds to physical items (blotting gels, vials with substances, substrates with surface modifications, *etc.*), should be considered part of the report. Sometimes the data is so voluminous that it is attached in an appendix or as *supplementary information* in an online database. In any case, it should be made publicly available together with the report in Step 15. Here, Case 3.3 raises an interesting question: should a research report always necessarily describe only the successful part of a study? Indeed, many scientific papers are written in a way that idealises the actual lab work and shapes it into a narrative form. Failures and unsuccessful trials are not mentioned unless they are part of the study. The protagonist of Case 3.3 is concerned that the low yield of reaction product might encourage other chemists to try alternative synthesis routes, just to repeat the same failed attempts. In such a case, an overview of experimental setups, conditions and varieties may be presented in an appendix or supplementary information.

Fluffiness of Bread Example, Step 9:

We write a report on the influence of humidity on the fluffiness of bread with all the necessary information as listed above.

3.3.10 "Negative" Result: Still Worth a Contribution?

Sometimes, a mismatch between data and the expectation is not necessarily a *bad* outcome. Maybe the results explain something important in a different way than expected. Or, the non-relation is an important insight, too. It is bad style, however, to simply re-formulate the hypothesis in a way that the results confirm our expectations. In case the results are still exploitable and insightful for a report, we can proceed with Step 9. If the results pose major problems and confusion, we proceed with Step 11.

Fluffiness of Bread Example, Step 10:

If the result is that there is no obvious relationship between humidity and fluffiness, this might also be an important information for bakers – then they don't have to care about it anymore. However, if the results are not clear at all and leave too many open questions, it is better we don't write anything, yet, but try to analyse the source of the mismatch.

3.3.11 Is it Worth Continuing the Project?

When the results are not as expected it must be checked thoroughly what the reason could be. There is, of course, the possibility that the entire project is misconceptualised or based on improper assumptions.

However, maybe something has been forgotten or overlooked. Maybe the unexpected results occur because of false initial assumptions, a flawed theoretical or conceptual framework, or an improper formulation of hypotheses. In many cases, especially in interdisciplinary collaborations and those with industrial partners, *giving up* is not an option. Then, it is even the researcher's duty to go on with the study (Step 13) rather than announcing its end (Step 12). In some cases, it is very difficult to figure out whether the failure is resolvable or not. The protagonist of Case 3.2 is in such a dilemma. Conceptual errors can be identified and eliminated, and lack of experience can be compensated, by the learning effect that the research activities provide. To admit that an idea does not work at all is challenging, both in terms of research practice and psychologically. Communication with peers and studying the scientific literature are essential, here. The decision to end a project (going to Step 12) is rather seldom. Researchers' experience is, usually, potent enough to design projects and experimental studies that have a good chance of success. Yet, when reaching the point at which it would be unreasonable and stubborn to insist on spending efforts on a fruitless project, choosing that option would be the wiser idea. As for all the steps in this web, the decision should be made on scientific grounds rather than on the basis of personal, economic, financial or other non-scientific interests.

Fluffiness of Bread Example, Step 11:

Giving up the study on the fluffiness of bread should only be the last resort – when we can't find any clue on what influences the fluffiness and would have to search entirely in the dark. Most likely, however, it will be possible to identify a strategy for the efficient improvement of the study. Maybe our study is even financed by the bakers' guild and it is written in the contract that we have to yield results.

3.3.12 Abandon the Idea

A researcher must know when it is time to bury a project in the graveyard of discarded ideas! It would not be the first one! When it can be expected that a project would "burn" more resources and efforts than reasonably acceptable compared to the outcome, it might be advisable or even ethically demanded to abandon the idea and focus on other projects. This can save resources, both financially and in terms of manpower and time. However, this decision is especially difficult for projects that PhD and Masters students work on, as their theses might depend on it.

Fluffiness of Bread Example, Step 12:

The idea of a simple relationship between fluffiness and humidity is stupid! The experiments have shown that the issue is far too complex to be studied like this. We will tell the bakers that they better focus on their craftsmanship's expertise and experience and that science can't help them with their competence.

3.3.13 Problem Identification

In case we decide to continue the project – which, by the way, happens every day as there is almost no research project that runs smoothly from idea to publication – we need to analyse what could be the problem. One option is that the obtained confusing, unexpected, or unexplainable results are simply wrong in a way that their validity must be questioned. Another option is that those results are, actually, correct but either don't represent a significant factor of importance (as we mistakenly believed) or indicate that the focus of our study has been on the wrong direction. Here, however, we must distinguish between errors that occur as a result of improper experimenting (as mentioned in Step 6) and errors that occur as a result of a flawed research concept. As a result, we can't simply solve the problem by re-thinking the study design (which is its experimental details). We must revise the entire study, which means we might need to re-evaluate and eventually change our hypotheses.

Fluffiness of Bread Example, Step 13:

Other factors besides humidity, for example the kneading time and pressure, or the surrounding temperature during dough preparation, are more significant. A deeper literature research, for example into the behaviour of biopolymers (like starch and flour), reveals that external pressure can force them to uncoil and intertwine with neighbouring polymer chains. This indicates that kneading time as a fluffiness factor should be investigated, too.

3.3.14 Revise Hypothesis

If the problem analysis in Step 13 results in deeper insights into the matter, the initial hypotheses can be revised and adapted to the new framework of the background knowledge. This, of course, might influence the entire study design and its course of experimentation, so that all the Steps 4 to 9 have to be gone through again, with the risk of ending up at Step 11 again.

Fluffiness of Bread Example, Step 14:

Our new hypothesis: the fluffiness of bread is significantly dependent on the kneading time and pressure applied to the raw dough.

3.3.15 Submit for Publication

After the study has finally resulted in an acceptable and meaningful output that is summarised in a study report (Step 9), it should be communicated in the next step. The most common way to communicate research findings is to publish an essay in an appropriate journal. We will talk about publishing and its implications in greater detail in Chapter 7. Important discussions that precede the submission of a research paper manuscript are the authorship (who is eligible of being mentioned as an author of a paper), the decision to publish one longer article or several shorter papers, or possible conflicts with other forms of publishing like patent applications, report obligations to public organisations (for example, a project report to the European Commission as a project funder) or intra-organisational communication in private R&D. The choice of journal has, in addition to an ethical dimension (see Chapter 7), a methodological dimension: the main purpose of publishing research as part of the scientific method is the scrutiny and critical feedback from peers. Recent trends in open access publishing and increasing numbers of profit-oriented journals that offer uncomplicated publishing procedures without peer review undermine this purpose. *Good chemistry*, in this respect, does not shy away from rigorous and professional channels of critique. This starts with the choice of journal for one's research paper.

Fluffiness of Bread Example, Step 15:

We write an essay entitled "Environmental factors impacting the fluffiness of wheat flour baking products" and submit it to the "International Journal of Advanced Breadology".

3.3.16 Peer Review

As pointed out before, the internal feedback and control system of the scientific community is a crucial aspect for the institutional and societal justification and implementation of science in its whole. Peer review is one of the tools that has been established to ensure a sufficiently high quality level and to supervise the compliance with common standards. The editor of a serious and professional scientific journal sends the draft to two or three reviewers for them to evaluate the essay. Their comments and recommendations are considered in the editor's decision to reject or accept the article, or to ask the author for revisions before the manuscript can be accepted for publication. In this step of the web, fairness and freedom from bias are of extraordinary importance. Ethical issues arising in the context of peer review are discussed in Chapter 7.

Fluffiness of Bread Example, Step 16:

Our editor sends the article to three other baking experts and fluffiness researchers in order to get their opinion on our essay. In a "double blind" review process, they don't know our names, and we won't know theirs. This helps reduce the risk of bias in the review process.

3.3.17 Passed: Publication

When a manuscript has passed the review process successfully, it is published in the next issue of the journal, often earlier online. The new insight becomes publicly available knowledge, as such it is part of Step 2 and may serve as background knowledge for other research project designers.

Fluffiness of Bread Example, Step 17:

Now, there is a little bit more knowledge for baking theory.

3.3.18 Not Passed

There can be several reasons for rejecting a submission. In some cases, the editor rejects a submission because the paper may not be suitable for this particular journal. In that case we may submit it to another journal. It may be that the reviewers come to the conclusion that the quality of the study (or better: the quality of the description of the study) is too low, that important possible experiments are missing, that the results are insignificant or that the interpretation of data is unsatisfactory. In that case it is advised to re-evaluate the study and discuss how to proceed (back to Step 11).

Fluffiness of Bread Example, Step 18:

Option 1: We slightly change our article and submit it to another journal, the "Journal of Baking Theory and Practice". Option 2: One reviewer remarked that our claim of a causal effect of humidity on the fluffiness needs more experimental support. We decide to design a more sophisticated quantitative study in cooperation with the baking engineering department, hopefully yielding more comprehensive results that justify a new, longer, more convincing research article with more authors.

3.3.19 Additional Tests

In most of the cases, a research project doesn't result in only one publication. In other words: most research projects are not finished after the publication of a related research article. Despite its acceptance for publication we might still not be entirely satisfied with our study outcome. Then, we continue the investigation, perform more experiments, gather more data and solidify our insights with more evidence.

Fluffiness of Bread Example, Step 19:

Our study on humidity and fluffiness feels incomplete. Some of the experiments suggested that surrounding air humidity is less significant than the amount of water added to the dough as part of the recipe. We assign a PhD student in our group with an additional investigation on this matter (start again with Step 4).

3.3.20 Applications

Sometimes, the knowledge that is revealed in the form of published research articles finds its way into particular applications. Other researchers cite your articles, attention is paid to your essay, maybe engineers obtain important insights from your experimental results and use it for the improvement or invention of a technical artefact.

Fluffiness of Bread Example, Step 20:

Bakers decide to control the humidity in their bakeries in order to obtain ideal baking results. Upon their request and with our article as a convincing argument, engineers invent an oven that has an in-built humidity control unit.

Certainly, this overview lacks depth. Details of most of the steps depend strongly on the actual condition and topic of the research project, and we can't cover all of them here. For example, as mentioned earlier, in non-academic research like industrial R&D, Step 9 (writing a report) is not necessarily followed by Step 15 (publishing a paper). The details of Steps 5 and 7, for example how many times a measurement series is repeated in order to achieve statistical significance, are not elaborated here. Important for the purpose of this book is the connection between the elements of scientific methodology (Section 3.2) and the scientific rationale of following this scheme with its crucial steps. Upon a closer look, we identify many elements in this scheme that require normative judgments that can't be resolved by the research itself or by scientific methods. The decisions in Steps 10, 11 or 18, for example, require evaluations of circumstances, of interests, of resources, and chances of success. Indeed, the topics in Chapters 5 to 8 can be related to some particular steps in this scheme: Common forms of scientific fraud (fabrication or falsification of data) are committed in Steps 5 or 7; publishing issues arise in Steps 15 to 18; collaboration issues (conflicts of interests, intellectual property, *etc.*) arise in Steps 3 and 4, sometimes in 11 or even Step 1, the initial decision on what to do research on. That this decision is a normative choice can be illustrated with the example of *green chemistry*: it is the goal of research in this field to gain insights into chemical processes and procedures in order to make them more sustainable, to increase safety aspects, or to protect the environment. A chemist in this field might regard it as a duty to spend public resources on research that serves the purpose of increased quality of life or environmental integrity.

We may also use this scheme to reflect on our particular goal that we follow with our research activities. In many academic setups, it is publications (Step 17) as these are regarded as the most important merit of scientists. As a Masters or PhD student, your primary goal might be your thesis, which would be Step 9 of the web. You may also state that all your scientific research is purely curiosity-driven and only serves the noble goal of increasing knowledge (Step 2). In many cases, our goal – directly intended, or secretly hoped – is in some way an application (Step 20). This must not necessarily be a technical artefact or an innovative invention, but at least we hope that someone reads our articles, gains useful information for his or her own research and then cites our work in articles or books. We want to be visible and meaningful.

In the next chapter, we will have a closer look at two crucial steps of this scheme: Step 3 (how to come to good hypotheses) and Step 7 (how to apply logic and scientific thinking to generate scientific statements).

BOX 3.2 Exercise: Umbrellaology

The following is a letter which was received by the editor of a science journal.

'Dear Sir:

I am taking the liberty of calling upon you to be the judge in a dispute between me and an acquaintance who is no longer a friend. The question at issue is this: Is my creation, umbrellaology, a science? Allow me to explain. For the past 18 years assisted by a few faithful disciples, I have been collecting materials on a subject hitherto almost wholly neglected by scientists, the umbrella. The results of my investigations to date are embodied in the nine volumes which I am sending to you under separate cover. Pending their receipt, let me describe to you briefly the nature of their contents and the method I pursued in compiling them. I began on the Island of Manhattan. Proceeding block by block, house by house, family by family, and individual by individual, I ascertained (1) the number of umbrellas possessed, (2) their size, (3) their weight, (4) their colour. Having covered Manhattan for many years, I eventually extended the survey to other boroughs of the city of New York, and at length completed the entire city. Thus I was ready to carry forward the work to the rest of the state and indeed the rest of the United States and the whole known world.

It was at this point that I approached my erstwhile friend. I am a modest man, but I felt I had the right to be recognized as the creator of a new science. He, on the other hand, claimed that umbrellaology was not a science at all. First, he said, it was silly to investigate umbrellas. Now this argument is false, because science scorns not to deal with any object, however humble and lowly, even to the 'hind leg of a flea.' Then why not umbrellas? Next, he said that umbrellaology could not be recognized as a science because it was of no use or benefit to mankind. But is not truth the most precious thing in life? Are not my nine volumes filled with the truth about my subject? Every word in them is true. Every sentence contains a hard, cold fact. When he asked me what was the object of umbrellaology I was proud to say. "To seek and discover the truth is object enough for me." I am a pure scientist; I have no ulterior motives. Hence it follows that I am satisfied with truth alone. Next, he said my truths were dated and that any one of my findings might cease to be true tomorrow. But this, I pointed out, is not an argument against umbrellaology, but rather an argument for keeping it up to date, which exactly is what I propose. Let us have surveys monthly, weekly, or even daily, to keep our knowledge abreast of the changing facts. His next contention was that umbrellaology had entertained no hypotheses and had developed no theories or laws. This is a great error. In the course of my investigations, I employed innumerable hypotheses. Before entering each new block and each new section of the city, I entertained a hypothesis as regards the number and characteristics of the umbrellas that would be found there, which hypotheses were either verified or nullified by my subsequent observations, in accordance with proper scientific procedure, as explained in authoritative texts. (In fact, it is of interest to note that I can substantiate and document every one of my replies to these objections by numerous quotations from standard works, leading journals, public speeches of eminent scientists, and the like.) As for theories and laws, my work represents an abundance of them. I will here mention only a few, by way of illustration.

There is the *Law of Colour Variation Relative to Ownership by Sex*. (Umbrellas owned by women tend to a great variety of colour, whereas those owned by men are almost all black.) To this law I have given exact statistical formulation (See vol. 6, Appendix 1, Table 3, p. 582.) There are curiously interrelated *Laws of Individual*

(continued)

BOX 3.2 (*continued*)

Ownership of Plurality of Umbrellas and *Plurality of Ownership of Individual Umbrellas*. The interrelationship assumes the form, in the first law, of almost direct ratio to annual income, and in the second, in almost inverse relationship to annual income. (For an exact statement of the modifying circumstances, see vol. 8, p. 350.) There is also the *Law of Tendency Toward Acquisition of Umbrellas in Rainy Weather*. To this law I have given experimental verification in Chap. 3 of Volume 3. In the same way I have performed numerous other experiments in connection with my generalizations.

Thus I feel that my creation is in all respects a general science, and I appeal to you for substantiation of my opinion ....'

As an exercise, please compose a reply to this letter.

[Taken from Klemke *et al.*[16]]

Exercise Questions

1. Why is it important to develop a research hypothesis before commencing experimental work?

   A: It is required by the historical rules of scientific conduct.
   B: Subsequent development of a hypothesis based only on the experimental results obtained carries a high risk of bias and opportunistic interpretation.
   C: Supervisors and colleagues have to be informed about the hypothesis before experimental work starts.
   D: In the case of a laboratory accident, the insurance doesn't cover the damage when the experiments have been conducted without guidance from a hypothesis.

2. A hypothesis is...

   A: ...an idea.
   B: ...an educated guess based on prior knowledge and experience.
   C: ...an explanation of observations that must be logically consistent, is tested by experiments, and can change after more observations are made.
   D: ...information that is collected.

3. A special kind of claim that states how one variable in an experiment will affect another is called...

   A: ...hypothesis.
   B: ...scientific theory.
   C: ...scientific method.
   D: ...educated guess.

4. What skill is a student using when (s)he listens to the sounds of burning hydrogen?

   A: Interpreting data.
   B: Making observations.
   C: Drawing conclusions.
   D: Making a hypothesis.

5. What is the important characteristic of the knowledge produced by means of a scientific method?

   A: It is true.
   B: It confirms the researcher's hypothesis.
   C: It is generalizable and reproducible by other researchers.
   D: It enables applications.

6. What makes a statement "scientific"?

   A: It cites a renowned independent scientist or a publication in a high-impact journal.
   B: It is true, independent of time, space, and who is stating it.
   C: It contains references to data as a proof of its validity.
   D: It is the consistent interpretation of observation(s) using the logic of a hypothesis.

7. Which of the following would NOT count as conducting experiments?

   A: Analysing a substance using a spectroscopic method.
   B: Counting events and making marks on a checklist.
   C: Using software to plot a graph from raw data.
   D: Using various derivatives of a compound in chemical syntheses and recording the reaction yields.

8. Which of the following is not an appropriate approach to generating research ideas?

   A: Discussions with one's supervisor or colleagues.
   B: Reading the literature in one's research field and identifying unanswered questions.
   C: Listening to what your peers are doing and trying to do the same faster than them.
   D: Presenting one's own research work at a conference and receiving constructive feedback from peers in the audience.

9. Which of the following would not count as peer review?

   A: Feedback from a reviewer on a manuscript submitted to a publisher.
   B: Critical comments in the Q&A section after a presentation at a conference.
   C: Remarks from a committee that evaluates a grant proposal.
   D: Letters from readers commenting on an interview that was printed in a newspaper.

10. Which of the following would NOT be regarded as a function of a literature search?

    A: To reveal whether one's own research question has already been answered or not.
    B: To provide useful hints on how to conduct one's own experiments.
    C: To help the researcher predict the outcome of his/her experiments and align the results with the published ones.
    D: To be able to give references and cite well-known publications that are relevant to one's own work.

11. Which of the following is the most important factor in deciding whether one's research has been successful or not?

    A: The interpretation of experimental data yields useful insights concerning the hypothesis.
    B: It has resulted in a useful application.
    C: The researcher was able to go through the web of scientific knowledge acquisition in one smooth run without repeating any steps.
    D: The supervisor/boss is satisfied and provides more funding for future projects.

12. Hypotheses don't necessarily have to be...

    A: ...clear.
    B: ...simple.
    C: ...precise.
    D: ...testable.

13. Which of the following could be considered a good reason for giving up a research project after encountering difficulties?

    A: After going through the steps of the scientific knowledge acquisition web several times, the source of the problem could still not be identified. The problem seems to be much more complicated than assumed.
    B: Time is running out.
    C: Major inconsistencies in the research design, the hypothesis, or the experimentation have been identified with the help of senior peers. Adopting their suggestions would involve too much work!
    D: Another project with a larger research grant is currently available.

14. What should be the first reaction when a manuscript that was sent to a publisher has been rejected, with the argument that the presented experiments are not sufficiently convincing to justify the drawn conclusions?

    A: Send the same manuscript to another journal.
    B: Add an influential scientist with a big name to the list of authors and submit it again.
    C: Identify how the scientific claims presented can be substantiated with additional experiments that allow a more convincing interpretation.
    D: Give it up and focus on other more promising research projects.

15. Ideally, who should decide when it is time for you to publish your research results? (Select the best answer.)
    A: You as the researcher have the best overview of the progress of the project and the validity of the scientific claims. You should know when the results are ready for publishing!
    B: The supervisor/boss/PI, because he/she has the most experience and knows when "enough experiments" have been conducted.
    C: It is determined by the funding period. When the funding ceases, the experimental work stops and any achieved results are then published.
    D: There is no "perfect time" for publishing. You just keep trying to get manuscripts accepted by journals. Where this is unsuccessful, you just continue doing more experiments until your submission is successful.

16. Which of the following statements is correct?

    A: After some historical refinements and improvements, the current idea of the scientific method is the best we ever had.
    B: An academic discipline that doesn't follow the commonly accepted standard method as applied in physics or chemistry can't be called "science".
    C: There is no THE scientific method. Different disciplines of science employ different standards and methodologies, so that it is better to talk about "scientific methods" in the plural.
    D: They key element in any scientific method is the collection of experimental data. Disciplines that don't collect numerical data (like some Humanities or social sciences), therefore, can hardly be considered as "science".

17. Why is the work on a research project not finished after publishing a journal article? (Choose the best answer.)

    A: It is possible that other researchers send an email with a question to the author(s) after reading the article.
    B: No research question is ever completely answered. There will always be follow-up questions, related experiments to provide deeper/further insights, new research ideas enabled by the previous findings, and so on.
    C: It is necessary to deal with the remnants of the experimental work. Chemicals have to be disposed of or properly stored, the laboratory has to be cleaned and prepared for the next project, and samples have to be catalogued, filed, or safely disposed of.
    D: There is still a lot of administrative work to do after publication: bills have to be paid, taxes have to be filed, lab books and other data have to be organised and stored in case they are needed for legal processes, and so on.

18. Which of the following parameters is least important (or possibly of no importance) when analysing experimental data?

    A: Validity
    B: Reliability
    C: Reproducibility
    D: Simplicity

19. Lab books and other records of research work usually belong to...

    A: ...the researcher who writes/records them.
    B: ...the supervisor/boss/PI of the work group.
    C: ...the institution (research unit, university, academy, *etc.*).
    D: ...nobody – they are public property.

20. Researchers spend time and effort on record keeping for many reasons. Which of the following is NOT one of them?

    A: It makes it easier for the researcher to keep track of experiments, organise research work, and remember/recapitulate experiments and their outcome.
    B: In case of accusations of fraud, a properly written lab book can provide evidence that data have not been falsified/fabricated.

C: In the case of a researcher leaving an institute and the project work being passed on to a successor, the records help the new researcher familiarise quickly with the state-of-the-art of the research project.

D: Well-written lab books can be used at job interviews and other career-relevant occasions to prove one's professional competence when carrying out lab work.

References

1. I. Valiela, *Doing Science: Design, Analysis, and Communication of Scientific Research*, Oxford University Press, New York, 2001.
2. *Research Methodology. A Practical and Scientific Approach*, ed. V. Bairagi and M. V. Munot, CRC Press, Boca Raton, 2019.
3. P. Pruzan, *Research Methodology. The Aims, Practices and Ethics of Science*, Springer, Cham, 2016.
4. S. Gimbel, *Exploring the Scientific Method*, University of Chicago Press, Chicago, 2011.
5. A. Chalmers, *What is This Thing Called Science? An Assessment of the Nature and Status of Science and its Methods*, University of Queensland Press, Queensland, 4th edn, 2013.
6. A. M. Novikov and D. A. Novikov, *Research Methodology. From Philosophy of Science to Research Design*, CRC Press, Boca Raton, USA, 2013.
7. L. B. Christensen, R. B. Johnson and L. A. Turner, *Research Methods, Design, and Analysis*, Pearson, Harlow, 2014.
8. *Research Methodology in Chemical Sciences. Experimental and Theoretical Approach*, ed. T. Chakraborty and L. Ledwani, Apple Academic Press, Oakville, 2016.
9. J. A. Lee, *The Scientific Endeavor: A Primer on Scientific Principles and Practice*, Benjamin Cummings, San Francisco, 1999.
10. E. Shamoo and D. B. Resnik, Data Acquisition and Management, in *Responsible Conduct of Research*, Oxford University Press, Oxford, 3rd edn, 2015.
11. R. Grant, Recordkeeping and research data management: a review of perspectives, *Rec. Manage. J.*, 2017, **27**, 159.
12. K. Shankar, Order from chaos: The poetics and pragmatics of scientific record-keeping, *J. Am. Soc. Inf. Sci. Technol.*, 2007, **58**, 1457.
13. A. A. Schreier, K. Wilson and D. Resnik, Academic Research Record-Keeping: Best Practices for Individuals, Group Leaders, and Institutions, *Acad. Med.*, 2006, **81**, 42.
14. C. L. Bird, C. Willoughby and J. G. Frey, Laboratory notebooks in the digital era: the role of ELNs in record keeping for chemistry and other sciences, *Chem. Soc. Rev.*, 2013, **42**, 8157.
15. W. M. Wang, T. Göpfert and R. Stark, Data Management in Collaborative Interdisciplinary Research Projects—Conclusions from the Digitalization of Research in Sustainable Manufacturing, *ISPRS Int. J. Geo-Inf.*, 2016, **5**, 41.
16. *Introductory Readings in the Philosophy of Science*, ed. E. D. Klemke, R. Hollinger and D. W. Rudge, Prometheus Books, Amherst, 3rd edn, 1998.

4 Scientific Reasoning

Overview

Summary: In addition to the technical and experimental skills in daily lab work and a profound knowledge of one's professional field, chemical researchers need competence in analysing and interpreting their acquired experimental data in view of the claims they made in their research hypotheses. Both making proper hypotheses and interpreting data in a scientific manner are topics in this chapter.

First, it will introduce the most important concepts of logic that play a role in scientific thinking and reasoning: deduction, induction, and abduction. More important than the correct application of logic concepts is the awareness of pitfalls, biases and fallacies. Then, we will learn how logic informs the heart of the scientific method, the predictive power of proper hypotheses and the explanatory power of data analysis and interpretation. Furthermore, a subsection is dedicated to heuristic and methodological analysis. Last but not least, it is worthwhile to have a closer look at statistical analysis and the differences between frequentist and Bayesian approaches.

The goal of such a chapter in a book on good chemistry is, of course, a higher awareness of factors that determine the consistency and plausibility of scientific claims. In intra-community discourse among scientific experts, but also in extra-community communication with non-expert stakeholders or the public, argumentation and the logically correct positioning of evidence-based facts decide over the success of the proposed claims. This is not only essential to come to viable – that means, pragmatically true – knowledge, but also to maintain credibility and trust in the force of scientific methodology. Furthermore, it gives a foretaste of Chapters 5 and 6 in which we will elaborate in greater detail what is good scientific practice and what is scientific misconduct. Many types of misconduct and fraud are, in one way or another, the intentionally fallacious or undermined application of scientific reasoning.

Good Chemistry: Methodological, Ethical, and Social Dimensions
By Jan Mehlich
© Jan Mehlich 2021
Published by the Royal Society of Chemistry, www.rsc.org

Learning Objectives:

This chapter is intended to equip you with:

- Basic logic skills for scientific thinking and reasoning.
- The competence to identify and avoid biases and fallacies.
- Strategies for heuristic and conceptual analyses of hypotheses and research questions.
- Awareness of the importance of statistical analysis.
- The ability to think and argue clearly with scientific concepts.

4.1   Introductory Cases

Case 4.1: Certainly Uncertain

An analytic chemist is conducting toxicological studies on functionalized nanoparticles (NPs). These are designed to serve as imaging agents for the detection of joint inflammation (arthritis). None of the tests, so far, indicate any detrimental physiological effects. Moreover, a critical component, a short protein receptor on the surface coating of the NPs, can't be detected in free form, which means that the composition of the NPs is stable. When communicating the results to the medical partner who will conduct clinical tests, her formulations are questioned. The doctors wish to have certainty that the chemical composition and stability of the NPs is safe for patients. Is the toxicologist able to claim this degree of safety with sufficient certainty?

Case 4.2: Spectral Proof

At a group seminar, a PhD candidate announces the successful synthesis of a new compound. The principle investigator (PI), though, is not satisfied as the student's only analytical result is an H-NMR spectrum that shows at least two uninterpretable peaks ("dirt") and in which the assignment of one hydrogen atom in the molecule deviates from the expected value. The student argues that the mass spectrometry of this compound is difficult owing to decomposition reactions in the spectrometer, and that a CHN analysis is ambiguous. Yet, as most of the group members agree, having only one indication of the successful synthesis of the product, it is very likely to interpret it too benevolently and with a bias.

Case 4.3: Just Call it 'Proof of Principle'!

Ryan used a chemical deposition technique to fabricate a field effect transistor. The precise and flawless formation of the different layers of semiconducting, isolating, and conducting materials is notoriously difficult. Finally, on one of his specimens, he is able to perform electronic measurements that show the field effect. When he tries to repeat the measurement, he is not able to locate a suitable spot for the attachment of the circuit wires. His PI is satisfied, anyway. It was successful once, that means it works! In the paper, it will be declared a *proof of principle*!

4.2   Logic in Scientific Reasoning

Scientific rationality is devoted to logic.[1] A large part of the success of scientific research hinges on the correct application of logic in argumentative strategies. As seen in Chapter 3, the goal of constructing and following a sophisticated research methodology is to create explanatory frameworks in which logic inferences from generally accepted starting points yield plausible and universalizable insights. But is that even possible? Doesn't scientific activity aim at elaborating exactly these *generally accepted starting points* (scientific theories and laws), and wouldn't that mean that we are lost in an infinite circular regress? First, we need to understand basic strategies of logic inference. Then we can understand how they can go wrong, and how they can be exploited for generating scientific insights.[2,3]

4.2.1   Deduction, Induction, and Abduction

The following situation (Box 4.1) happens more or less exactly like this in every class on scientific logic that the author has taught. The reader is encouraged to accompany the reading of the case with own thinking. What would you have said? What would be your suggestion?

What happened here? If, upon reading the first three digits, you immediately looked for a pattern, found one, and continued the sequence accordingly, you did what most people would do. You applied a principle that most of us follow in ordinary life and that researchers use for their experimental designs and interpretative strategies as well: a positive confirmation, or *verification*, of the rule in your head. Let's have a look at the most basic forms of logical reasoning to make this clearer. We distinguish deductive, inductive and abductive logic (Table 4.1).

Deductive logic follows a path from a *rule*, a *law* or any known (in mathematics: axiomatic) starting point *via* a *case* or the occurrence of a certain condition to a conclusion that is drawn by relating the condition

Box 4.1 A Logic Game

In a class on logic thinking, the teacher challenges the students with this task: " *I will give you the first three numbers of a sequence of numbers that follows a certain rule. The rule is in my head. You have to find out the rule. Please make a suggestion for the next number in the sequence. I will tell you if that is a valid continuation of the sequence or not. Then, tell a possible rule. Here are the first three numbers.*" He writes:

2, 4, 6

Student: "*8!*"
Teacher: "*Yes, that is a possible next number. What could be the rule?*"
Student: "*It is all the even numbers.*" – "*No, that is not the rule in my head.*"

2, 4, 6, 8 – "*What could be the next number?*"

Student: "*10!*" – "*Yes, possible. Rule?*" – "*The next number is the previous number plus 2.*" – "*No, not my rule, sorry.*"

2, 4, 6, 8, 10 – "*Any more suggestions?*"

Student: "*17?*" – "*Yes, that is a possible next number!*" – Student: "*A random sequence of numbers?*" – "*No, that is not the rule!*"

2, 4, 6, 8, 10, 17 – "*What could be next?*"

Student: "*3?*" – "*No, that is not a possible number!*" – Student: "*A number must be higher than the previous number?*" – "*Yes, that is my rule!*"

Table 4.1 Basic logic strategies.

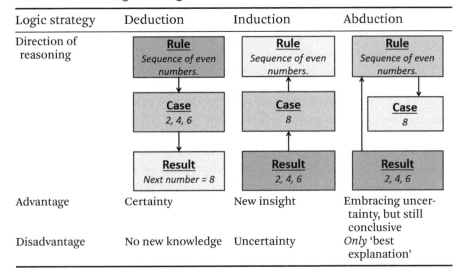

Logic strategy	Deduction	Induction	Abduction
Direction of reasoning	**Rule** / *Sequence of even numbers.* → **Case** / *2, 4, 6* → **Result** / *Next number = 8*	**Rule** / *Sequence of even numbers.* ← **Case** / *8* ← **Result** / *2, 4, 6*	**Rule** / *Sequence of even numbers.* → **Case** / *8* ← **Result** / *2, 4, 6*
Advantage	Certainty	New insight	Embracing uncertainty, but still conclusive
Disadvantage	No new knowledge	Uncertainty	*Only* 'best explanation'

to the rule. A common example is the statement that *"All humans are mortal!"* (which can reasonably be taken as true) and the case of Peter being human (the condition of Peter), so that it can be concluded (deducted) that Peter is mortal. Of course, it is possible to make mistakes here! We could, for example, claim that all humans are mortal, observe that Peter dies (which means he is mortal), and deduct from this that Peter must be human. This is clearly wrong, because Peter could be member of a different species which is also mortal (but not mentioned in our rule). However, if we apply deductive logic correctly, it has the advantage of a high certainty in its predictive power. As long as rule and condition are *true*, the (correctly deducted) conclusion is also true. Yet, we learn nothing new by deductive logic, because the result is inherently contained in rule and condition (we only need to *pull it out*). Deductive logic is not applicable in the logic game in Box 4.1, because we don't know the rule. But if we had known it, the case that we observe (the sequence 2, 4, 6) will lead us to the inference that the next number must be 8.

In science, however, we are interested in increasing our knowledge of rules, laws and principles. They are the end point of our inquiry, not the starting point. We start from observations that we consider the *results* of rules that are at work. In Box 4.1, the provided first three numbers were the initial observation. In order to find out the rule you *made up* cases to extend your observation. This is similar to a scientist conducting experiments in order to go beyond what is naturally observable. This manipulation and directed observation are expected to enlighten the researcher on the underlying mechanisms that constitute the rule that he or she intends to find out. This is the advantage of the inductive method: It reveals new insights. However, the downside is that we can never be sure if we are correct with our conclusion – there is always uncertainty. Indeed, in Box 4.1, students *observed* something (number 8 was declared a valid continuation of the sequence by the teacher) but the concluded rule was still wrong. Another illustrative example is the case of drawing candy from a bag of unknown composition of flavours. If the first three pieces are strawberry candy, you may conclude that this is a bag of strawberry candy. If the next 10 are still strawberry candy, you may feel confirmed that your conclusion is correct. Yet, you can't reach a stage of 100% certainty. The 14th draw could be a lemon candy. Case 4.1 illustrates a more chemical example. The inability of inductive reasoning to come to definite conclusions poses a great challenge to chemists! Toxicological tests, for example, when showing no toxic effects, allow exactly that and only that statement: *We observe no toxic effects*. With appropriate methodology, the probability that this means that the tested substance is, indeed, non-toxic can be reasonably high, but there is simply no 100% certainty.[4]

At a closer look, the students in the logic game applied a third strategy of logic reasoning. Abductive logic, like induction, starts from observations (the *results* as effects of a rule or principle), but formulates a rule first, before making up a case that confirms or falsifies that rule. You saw the number sequence (2, 4, 6), thought that *it must be the even numbers*, and made the case "8". You even found "8" when the teacher as the omniscient oracle told you it is *correct*. Therefore, you concluded that your rule was confirmed. However, it was not! The game was solved when one student had the courage to say something completely beyond expectation. Abduction is useful when no other form of knowledge acquisition is possible, but it can never result in (secure) new knowledge, but only in *the best explanation*. There is a fine line to be drawn between induction and abduction. As we found earlier, a scientist does not conduct experiments into the void, without any orientation. There will always be pre-formed ideas and directions, manifested in the formulated hypotheses. We can say that this is a form of abduction, because there is already an assumption concerning the rule even before we investigate our *cases*. The main difference is how easily we are satisfied with the results of our case investigation (the experiments). Good scientists do not look for confirmation of their hypotheses or a proof for their endorsed theories or explanatory frameworks (*rules*). They try to challenge them by the method of falsification! If you want your theory to be confirmed, do a negative test! Ask something that would be excluded by your theory! If you still find it, your theory must be wrong! Historically and contemporarily significant scientific theories that sustained over centuries have been solidified by negative confirmation (falsification), not by positive confirmation (verification), most notably Karl Popper's epochal work.[5] The most prominent example is the biological evolution of this planet and its life forms that nobody seriously doubts today. Darwin did not publish his insights before years of experiments and investigations with falsification experiments to make sure his interpretations of observations were correct.

The scientific method tries to combine both the advantages of deduction and induction. The overall logic is induction, but this strategy of inquiry only works efficiently with iterative deductive confirmations. In other words, science is *macro-induction by micro-deductions*.[6] The experiments we design to extend our observations towards insights about rules and theories have to make sense in a deductive logic sense. In this way, we can have both certainty in reasoning and progress into new fields of knowledge. However, this way is full of pitfalls and fallacies, as we will see.

4.2.2 Logic Mistakes, Biases, and Fallacies

To err is human. Being aware of this increases the chance to identify errors in reasoning and to do better science. Deductive logic can go wrong when premises are connected in a fallacious way or when inferences misinterpret key words in the premises such as *all*, *only*, *every*, and so forth. These mistakes are termed *formal fallacies*. Inductive logic suffers from inadequate reasons for acceptance of premises rather than their employment in inferences. These fallacies are termed *informal fallacies*.

4.2.2.1 Formal (Deductive) Fallacies

- *Affirming the Consequent*: If A then B; $B \rightarrow A$!
  This is a fallacy because B could have other causes than A, unless the rule states explicitly that *only* if A then B. In all other cases, it is invalid to apply the backwards reasoning that observing an effect B is a proof that A occurred.
  Example: When the chemical reaction successfully yields product XY, the solution turns yellowish. The solution turns yellowish. Therefore, we must have obtained XY.
- *Denying the Antecedent*: If A then B; Not $A \rightarrow$ Not B!
  Something else can be or cause B, too! Again, the word *only* changes everything.
  Example: All esters have a strong smell. This compound is not an ester. Therefore, it doesn't have a strong smell.
- *Undistributed Middle*: All Zs are B; Y is $B \rightarrow Y$ is a Z.
  This is only true if only Zs are B!
  Example: Radicals are very reactive. Compound X is very reactive. It must be a radical.
- *Appeal to Probability*: Taking something for granted because it is probable or possible.
  Example: Cyclodextrin is nontoxic and can be washed down the drain. Your functionalised cyclodextrin derivatives are most likely nontoxic, too. You can wash them down the drain, too!
- *Bad Reasons Fallacy* (*Argumentum ad Logicam*): Drawing a conclusion based on *bad* (implausible, questionable) reasons.
  Example: There is a black residue in your reaction flask. Black indicates a failed reaction. You should discard it and start all over again.
- *Masked Man Fallacy*: Committing a mistaken substitution of parties. If the two things that are interchanged are identical, then the argument is assumed to be valid.

Example: We observe that our carbon deposition technique results in a material with semiconductive properties. According to the literature, a particular configuration of single-walled carbon nanotubes is semiconductive. Thus, we must have produced that type of carbon nanotube.

- *Non Sequitur*: Asserting a conclusion that does not follow from the propositions.
  Example: Azo-dye solutions are very colourful. Compound X is not an azo-dye. Thus, its solution is not colourful.

4.2.2.2 Informal (Inductive) Fallacies

Fallacies of Presumption:

- *Complex Question Fallacy*: Inflicting questionable assumptions. The way the question is asked presumes an intended claim either way it is answered.
  Example: "Are you going to admit that you're wrong?"
- *Hasty Generalization Fallacy*: One (often abnormal) situation leads to drawing a much broader conclusion.
  Example: Our novel catalytic system proved efficient in a specific azide-alkyne coupling reaction. Therefore, this catalyst is powerful in all kinds of click reactions.
- *Non Causa Pro Causa* (False cause): Postulating a cause that, more or less obviously, cannot be the cause of something.
  Example: We should really repair our gas chromatograph. I use that often, and never get the expected result out of it!
- *Post Hoc, Ergo Propter Hoc* (*after this, therefore because of this*):– mistaken assumption of cause and effect: A happened, then B happened, so A must have caused B.
  Example: I observed a colour change in my reaction flask. A minute later it burst. The colour change must have caused an increase of pressure.
- *Cum Hoc, Ergo Propter Hoc* (*with this, therefore because of this*): Connecting two events which happen simultaneously and assuming that one caused the other.
  Example: Hospitals are full of sick people. Therefore, hospitals make people sick.
- *Slippery Slope Fallacy*: Falsely assuming drastic consequences of actions.

Example: Once we fabricate self-replicating nanobots, we will have to face a global grey-goo scenario!

- *Sweeping Generalization Fallacy*: Applying a premise too broadly.
  Example: We found a way to use cellulose as an alternative material substituting industrial plastics. Now, all environmental problems are solved!

- *Tu Quoque ("you, too") Fallacy*: Turning criticism against the other person.
  Example: Colleague XYZ told me that my proposed reaction mechanism has a flaw. But didn't he publish an implausible paper on another reaction mechanism just last year?!

- *Appeal to Ignorance*: Arguing that a proposition is true because it has not yet been proven false.
  Example: There is nothing in the toxicological testimonial that disproves company A's lacquer additive's toxicity. Thus, it must be toxic and banned!
  Circular Argument: What is to be proven is part of a premise.

- *Example*: Established scientific theory suggests to interpret our data in the way that *A*. This result (*A*) is in accordance with scientific theory.

- *False Dilemma*: Presenting an argument in such a way that there are only two possible options.
  Example: In the review of our manuscript, you express the doubt that our proposed reaction mechanism features a nucleophilic substitution in the 4th step. That means that you must suggest a radical substitution which is even less likely!

Fallacies of Ambiguity:

- *Equivocation Fallacies*: Using words multiple times with different meanings.
  Example: Risks are of concerns for insurance companies and regulators, so I as a chemical researcher don't need to care about chemical risk assessment of my compounds.

- *Straw Man Fallacies*: Misrepresenting someone's claims or positions to make an argument look weak.
  Example: Chemist A: "The Western blot image you presented in your talk is not very clear, which makes its interpretation a bit ambiguous." Chemist B: "You underestimate the explanatory power of the Western blotting technique! You can't seriously doubt that this tool that millions of biochemists use around the world is very useful for the comparison of proteomic samples!"

Fallacies of Relevance:

- *Appeal to Authority*: Attaching an argument to a person of authority in order to give credence to it.
  Example: Well, Isaac Newton believed in alchemy! Do you think you know more than Isaac Newton?
- *Attacking the Person (Ad Hominem)*: Responding to an opponent with a personal attack, insult or defamation rather than with reference to the actual argument.
  Example: I don't need Prof. Miller's advice on my research paper manuscript. He didn't publish any paper in a high-impact journal last year!
- *Bandwagon Fallacy*: Making claims that are only appealing because of current trends and growing popularity.
  Example: [From a research grant application] Homeopathy enjoys increasing popularity in recent years. Therefore, its healing effects deserve a more thorough scientific investigation.
- *Gambler's Fallacy*: Assuming that short-term deviations will correct themselves along the lines of probability.
  Example: This coin has landed heads-up nine times in a row. So, it will probably land tails-up next time it is tossed.
- *Red Herring Fallacy*: Using irrelevant information or other techniques to distract from the argument at hand.
  Example: Yes, the mass spectrometry may have yielded ambiguous results, but I have applied my compound in a surface modification process and the results look very promising!
- *Weak Analogy*: Employing analogies between things that are not really alike.
  Example: Persistent organic pollutants (POPs) are bad for the environment, just like oil spills. But if you are not going to ban heavy oil in marine traffic you can't ban the application of POPs in agriculture.
- *Two Wrongs Make A Right*: Defending something done wrong by citing another incident of wrong-doing.
  Example: This little bit of photoshopping of my fluorescence microscopy image is not so bad! Look at what some Indian and Chinese scientists are doing!

Strictly speaking, not all of these fallacies concern the scientific reasoning as part of the hypothesis formulation and analysis based on acquired data. Some concern other parts of the scientific method, for example the communication of results or the discourse on the social and environmental impact of scientific progress. Certainly, many if not all of these fallacies deserve our attention and need to be treated with clear and critical thinking in many realms outside scientific research.[7–10] However, as part of scientific reasoning, avoiding these fallacies has a particular urgency: committing these fallacies makes our scientific output attackable and weak. This does not only have a negative drawback on us and our career but, even worse, may cause harm to those who rely on the validity and legitimacy of the chemical knowledge we produce.[11]

The following list summarises the most important biases that the chemical researcher is well advised to avoid as best as he/she can:

- There is a tendency to conclude from the inability to prove or show *A* that *non-A*. There are two ways to make sense of it, and they are distinguished by the expectation of what scientific logic is able to achieve. In a deductive logic approach, we can understand this problem as a form of *denying the antecedent*: A positive test indicates property xyz (for example, a toxic effect); the test is negative; the tested substance is non-xyz (for example, non-toxic). In an inductive logic approach, this line of argumentation commits an *appeal to ignorance* fallacy. The burden of proof is put on an opponent to disprove a proposition which, as seen above, inductive reasoning can't achieve with 100% certainty. This is especially problematic in interest-guided forms of research like policy-relevant toxicology. In particular, companies that aim at profitable business of their products tend to conclude (sometimes aggressively) easily and hastily from *lack of evidence* that there is *an evidence that there is no* risk, harm, negative side effect, and so forth.
- The following fallacy is, somehow, the opposite of the previous one: from a few positive results confirming A, we conclude that A is true/valid/proven. This is a form of the *affirming the consequent* fallacy. It lacks scientific scrutiny and reasonable scepticism. It is a form of confirmation bias (we tend to accept hints for the correctness of our beliefs much easier than hints that they are wrong or at least still unproven), or availability bias (we take a few example cases as a proof for our claim, regardless of cases that point in another direction). The student in Case 4.2 is an example for this case: eager to be successful, he hastily interprets his results benevolently and takes one analytic test result as the proof of yielding the desired substance.
- When making logically valid arguments, in principle, we combine a few premises to yield a conclusion (see, for example, the form

of an ethical argument in Chapter 1, or the deductive or inductive arguments in this chapter). It is very important for the validity of the claim that the premises are plausible and tenable without the claim being made in the conclusion. Or, in other words, what is to be shown or proven must never be assumed in any of the premises! Arguments that do so are *begging the question*. Famous historical examples are attempts to prove the existence of God in circular arguments like *"God must exist, because we have an idea of him planted by him into our minds!"*. In chemical research, this fallacy occurs whenever a claim is investigated by experiments that are designed on the premise of what should be shown, for example when models or technical equipment are employed for the analysis that are built on theoretical considerations that presuppose what the experiments aim at showing.

4.2.3 Predictions and Explanations

During his or her research, a scientist asks many questions:

- What? Description of observations and phenomena.
- How? Causality and other relational mechanisms of processes.
- When? Predicting when an effect occurs, both temporally (at what time) and conditionally (under what circumstances).
- Why? Explanation of certain phenomena.

Some fields of science ask almost exclusively *what*-questions, while others focus on *how*-questions. *Why*-questions are addressed by surprisingly few scientific disciplines. This might be – as you maybe agree – due to the fact that *why*-questions are much harder to answer than *what*- and *how*-questions. Eliciting causality – or more precisely: aspects of relations, including correlation and causal relations – with scientific investigations necessarily follows an inductive reasoning approach which, as seen above, can't achieve full certainty. Moreover, pathways of causal relations are not always linear, but can be more complex than we anticipate. Imagine the following scenario: today I have a headache. I believe it is because I did not sleep enough last night. My wife implies that it is because I stared too much at the PC screen while writing this text. Moreover, I have been very stressed lately, which could also be a cause of my headache. There are four options of how these things are related (Figure 4.1).

The first is that one of the mentioned possible causes (lack of sleep, staring too much at the PC screen, too much stress) leads to the effect

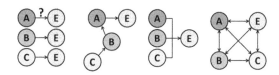

Figure 4.1 Possible cause-effect configurations.

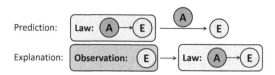

Figure 4.2 The logic of predictions (deductive) and explanation (inductive).

(headache). The second possibility is that all these causes lead to the effect in a chain-like fashion: first, I was stressed; so I forced myself to finish the book and looked at the screen for too long; this decreased my sleeping quality and time; and this, ultimately, gave me a headache. Option 3 is that it needs all three factors together to cause me a headache. If only one was missing, I wouldn't have a headache. The last option is that the cause-effect-relation-network is far more complicated. Maybe we can't even say what is cause and what is effect. It is very likely that all are somehow cause and effect for everything else, not symmetrically though, but in an imbalanced way. Looking at the screen too long is caused by my inner mental stress level, both together cause a slight headache (maybe unnoticed, yet), which decreased my sleeping quality, which in return made my headache stronger and also increased my stress level additionally.

Remember what we said about deductive and inductive logic. A prediction follows a deductive reasoning approach: when we know a law that links a certain effect E to the occurrence of a condition A, we can predict that E occurs as soon as condition A is established/happening/induced. When we have no clue of the law (or rule, principle, *etc.*), in order to give an explanation for the observation of an effect E, we need to gain insight into the causal mechanisms of a condition A leading to effect E. This resembles an inductive reasoning approach (Figure 4.2). Again, we can see why *why*-questions are harder to answer than *when*-questions, and that both, however, are necessary to deliver reliable insights in a scientific way: only by confirmation through correct deductive predictions is it possible to verify or falsify possible inductive explanations.

As mentioned above, knowing logical reasoning strategies doesn't prevent us from committing mistakes sometimes. Fallacious premises and/or their invalid connection to come to conclusions happen often and easily. Yet, we should try to avoid logical fallacies and cognitive biases as much as we can. Here lies the connection between critical thinking and scientific methodology. Only when the tools of logic and reasoning are applied correctly can scientific research play out its advantages and power.

4.3 Heuristic and Methodological Analysis

Logic thinking is applied for both *making* (or formulating) and *checking* hypotheses. In other words: first, we need to engage in a *heuristic* analysis of hypotheses, and then in a *methodological* one. Heuristic, here, means refining the possible claims into a rather confined set of options in a certain direction, ruling out all very unlikely situations. What counts as a reasonable or plausible hypothesis is often defined by established knowledge. We almost never conduct scientific research *into the blue*, digging in the dark for new insights. We make analogies to comparable phenomena, to experiences with similar substances or reaction types (for example), or to our knowledge about particular elements, molecules, materials, and so forth. In many cases, we think through experiments and their setup before we start the actual work in the laboratory: we conduct thought experiments. With sufficient experience and reasonable assumptions, these theoretical considerations often help refining hypotheses and increasing the efficiency and success rate of experimental designs and conduct.[11-13]

Scientific plausibility is often derived from experiences, intuition, expertise, competence and a profound knowledge base. However, these aspects are sometimes difficult to capture in scientific terms (especially in numbers) and to communicate with non-scientists. Several strategies have been elaborated to express heuristic considerations in scientific terms, ultimately in calculations and numbers, for example statistical significance, algorithms, or other rules. Very often, these assumptions cause situations that are counter-intuitive or implausible. In toxicology or pathogen research, statistical significance based on a threshold that is randomly chosen sometimes hinders progress in regulation and the production of reasonable handling instructions for better safety. Insisting on empirically derived values is, then, corrupting scientific inquiry rather than supporting it or ensuring its epistemic and explanatory power.

The same can be said about the methodological analysis. As mentioned above, scientific knowledge acquisition is widely taken as the attempt to harvest the advantages of both induction (new insight) and deduction (certainty). This method – a careful exploitation of successful predictions and explanations – is commonly referred to as the *hypothetico-deductive method*. Other scientists and philosophers of science argued for alternatives, pointing at the impossibility to ever gain certainty in the sense of scientific truths resembling reality. The pragmatic focus on viability led to the idea that hypotheses are acceptable whenever their ability to solve a problem or to explain observations and phenomena is greater than any other known claim. This *historical-comparativist method* compares hypotheses to assess their relative historical problem-solving ability. In return, this concept has been the subject of criticism from both scientists and science theorists.

In cases in which scientific certainty can't be achieved, but where a fact- or evidence-based decision has to be made – for example in regulation, toxicology, medical research, and so forth – *inferences to the best explanation* might be more goal-oriented, helpful and fruitful than insisting on continuing the scientific investigation until full certainty is reached. In those cases (interest-driven science), abduction is still better than claiming certainty where there is none. Here, too, scientists can reach a maximum gain of insights by feeding inferences with their competence and experience without strictly following empirical rules and guidelines.

A special topic, here, might be the conduct of case studies. Well established in sociology and related fields of research, they suffer from a generally rather bad reputation in chemistry, physics and other so called hard sciences. Case studies, so the claim goes, are arbitrary, random, cherry-picking, and explain nothing. Yet, if chosen carefully, following plausible strategies and reasonable considerations on conduct, analogy-drawing, and explanatory power, a good and sophisticated case study design can be as convincing as other study designs while, at the same time, saving time and resources, and circumventing methodological and conceptual difficulties.

A crucial question in scientific reasoning is the amount of data that is necessary for drawing viable conclusions. It has an ethical dimension in the sense of '*how many resources and efforts need to be invested so that we establish the right balance between certainty (and, in applied contexts, safety) of claims and investment of money, materials, and manpower?*'. However, there is also a purely methodological dimension: how many times must a measurement series be repeated so that we can perform a significant and valid statistical analysis? How many different spectroscopic analysis departments do we need to consult so that our claim of a successful synthesis is properly reasoned? Answers must be found case by case. These answers should always be *scientific* in the sense that epistemic reasons and logic considerations should outweigh personal interests like saving time or budget. Case 4.3 offers an opportunity to compare scientific and biased judgments. Usually, we would claim that experiments and measurements must be reproducible and replicable. When researchers don't do this and publish results based on single events or experiments performed only once, it looks suspicious. At best, they are lazy; at worst, they can't reproduce their own experiments. The fabrication of a nanomaterial-based field-effect transistor requires extremely precise and flawless deposition methods and sophisticated technical devices for the measurements. Yet, the only way the measurement shows the field effect is that the layers of the transistor are formed properly. The PI is correct to say that *in principle* the fabrication was successful, even if not over the complete surface area of the substrate. Finding the right spot for the measurement may have been luck or the result of time-consuming effort. In other cases, in which only one successful measurement series could be achieved, the judgment whether this counts as a proof of principle or not may be different, dependent on the conditions for applying logic reasoning and biases from personal interest considerations.

4.4   Statistical Analysis

There exists a complex and intimate relationship between observing/ testing, probabilistic reasoning, statistical analyses, and the interpretation of the results of one's observations/experiments in the sciences.[14] Probability is the deductive approach to uncertainty: simple rules together with mathematically coherent and physically sensible axioms lead to the deductive derivation of theorems. Probability serves science in the same way that a single unified theory of arithmetic does. Statistics is the inductive approach to uncertainty. Statistics (as applied inductive logic) can contribute in two major ways: in designing efficient experiments for gathering data, or in formulating inference procedures for analysing data. Of significance for us are Bayesian and frequentist concepts.[6,15]

The much older approach to statistics is a strategy elaborated by the British mathematician William Bayes. In an essay on chance (Bayes 1763) that was sent to the Royal Society of London, two years after Bayes' death, by a colleague who found the essay in Bayes' papers, he developed a theorem now named after him, the *Bayes theorem* (Figure 4.3). It is considered by many to be the chief rule involved in the process of learning from experience as to how to decide between hypotheses.

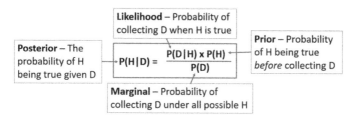

Figure 4.3 Bayes Theorem and its components.

For ease of presentation we interpret H as a hypothesis and D as data that has been collected to test the hypothesis. As P(D) is unconditional and therefore does not affect the relative probabilities of different hypotheses we obtain the simpler form:

$$P(H|D) \propto P(D|H) \times P(H), \text{ or: Posterior} \propto \text{Likelihood} \times \text{Prior}.$$

The posterior, P(H|D), is the probability of hypothesis H being true given the new evidence/data D; the prior, P(H), is the previous/initial/old belief/knowledge as to the probability of H being true; the likelihood, P(D|H), summarizes the impact that the new data has on the probability of the hypothesis H being true. Confusion about conditional probabilities can lead scientists to confuse P(H|D) and P(D|H). In simple terms, Bayes' theorem prescribes how probabilities are to be modified in the light of new data or evidence. In the focus of controversies stands the contribution and role of the prior. Obviously, the input here may be previously collected data, but also experiences, beliefs, qualitative extrapolations and estimations. Proponents claim that this puts statistical inferences on more pragmatic and reasonable grounds, rooted in daily life experiences and not disconnected from intuition and ratio by *cold* numerical calculation (see the discussion in the previous section). Thus, it often makes statistical analyses simpler, more plausible, and practical.

For historical (but in no way rational or pragmatic) reasons, the *frequentist* approach to inference is the one that is predominantly taught at universities. According to the frequentist interpretation, the probability of the occurrence of an event is defined as the limit of the relative frequency of the occurrence of the event (such as the outcome of an experiment or the occurrence of heads when tossing a coin) in an infinite sequence of trials (experiments/tosses of a coin). Thus, according to this frequentist interpretation of probability, when we say that the probability of getting heads with a fair coin is ½, we mean that in the potentially infinite sequence of tosses with the coin, the relative frequency of heads converges to or has the limit ½. More generally: the relative frequency will converge to the true probability.

In frequentist statistics, one of the hypotheses under consideration is designated the *null hypothesis*, H_0. Ordinarily, the null hypothesis is to be compared to the alternative or research hypothesis, H_1, which is the hypothesis the researcher really wants to test. The null hypothesis is used to predict the results which would be obtained from an investigation if the alternative hypothesis, the one the researcher is really investigating, is not true. For example, if the alternative hypothesis is that variable A affects variable B, the null hypothesis is that A does not affect B, and this can then be tested by manipulating the independent variable A and measuring the corresponding variations in the dependent variable B. If statistical tests do not reject the null hypothesis that A does not affect B, then the research hypothesis H_1 should be rejected. If on the other hand the null hypothesis that B is independent of A is rejected, then more confidence can be placed in the alternative hypothesis that A affects B and it is accepted as being true. Thus, there are four possibilities when testing a hypothesis, and these include two types of errors:

- Type I error: This error occurs when one rejects a null hypothesis that is in fact true.
- Type II error: This error occurs when one accepts – or perhaps more precisely, fails to reject – a null hypothesis that is in fact false. Note that this *fails to reject* wording is more in line with a falsification understanding of how to corroborate/justify scientific statements.

As low Type I and II error rates permit reliable learning from experiments, the basic consideration of a frequentist approach to hypothesis testing should be to minimize the likelihood of both types of errors. The basic idea of frequentist inference is that procedures with low error rates (in principle both Type I and II, but in practice only Type I) are suitable for sorting true/acceptable from false/not acceptable hypotheses.

Proponents of the frequentist approach claim that this strategy is purely objective and, thus, acceptable for scientific means. They criticise that the Bayesian approach with its possibility to include beliefs (non-scientific knowledge, experience, estimations) in the calculations is subjective and, therefore, inacceptable for scientific reasoning.

Yet, scientists often interpret frequentist analyses as though they are Bayesian: the latter provides data-based evidence on hypotheses, while the former does not. We will return to this topic in Chapter 13 on risk and uncertainty, in which we will see that we might face situations in which we have to reflect on the acceptability of Type I and II errors with regard to their impact (causing damage or harm). In this way, the frequentist approach is also not purely objective but requires careful ethical reasoning of the choice of thresholds for error rates.

Exercise Questions

1. Which of the following is NOT a common approach to logical reasoning?

   A: Abduction
   B: Deduction
   C: Production
   D: Induction

2. What is the advantage of deductive reasoning?

   A: It reliably leads to correct conclusions (if no mistake or fallacy has been committed).
   B: It leads to new knowledge (if premises are correctly combined).
   C: Everybody can do it (even with limited knowledge).
   D: It is in-line with the philosophy of science (and, thus, counts as "good reasoning").

3. Which of the following statements about inductive reasoning is correct?

   A: Inductive reasoning is the only acceptable strategy for scientific inquiry.
   B: Inductive reasoning can lead to new knowledge, even though conclusions from inductive inference can never be 100% certain.
   C: Inductive logic is historically outdated and not used in modern science.
   D: Induction can only give a "best explanation" of a phenomenon.

4. You make an observation. Based on this, you postulate a principle or rule for how this observed phenomenon occurs. In order to test if your suggested rule is correct, you make a prediction and confirm that you have been correct by appropriate experiments. Which approach to logic have you followed here?

   A: Inductive
   B: Hypothetico-deductive
   C: Historical-comparativist
   D: Abductive

5. Which of the following is a danger of abductive reasoning?

   A: Too much bias can make a scientist accept confirming experimental results too easily and form an over-confident conviction in a postulated rule or principle.
   B: It focuses too narrowly on data and doesn't see the larger picture.

C: Rhetorically skilled arguers can attack it.

D: Reviewers of research manuscripts often insist on inductive reasoning as the scientific standard and, therefore, might reject the paper.

6. Which of the following statements on uncertainty in scientific reasoning is incorrect?

   A: Probability is the deductive approach for dealing with uncertainty.
   B: Statistics is the inductive approach for dealing with uncertainty.
   C: Precaution is the abductive approach for dealing with uncertainty.
   D: Frequentist and Bayesian approaches are the most common statistical models for dealing with uncertainty.

7. Why do scientists engage in heuristic analysis of hypotheses?

   A: It excludes possibilities that are inconsistent with the scientist's intuition and ensures that the research direction matches his/her research portfolio.
   B: They do it because they are lazy. Actually, it is not even scientific and they would be better off avoiding it.
   C: It shows that they are literate people, who care about society and the ethical implications of their research.
   D: It is always necessary to narrow down possible explanations to a set of plausible options. Scientists usually don't approach a research question from zero, but with reasonable pre-knowledge and valuable experience.

8. "In our tests, compound xyz did not show any toxic effects. Therefore, it can be concluded that it is non-toxic and totally harmless." – Is this proper reasoning?

   A: Yes, it is a simple deductive inference and, thus, true.
   B: No, it is a form of "begging the question" because other people will not be convinced and will keep doubting it.
   C: No, this is a case of "appeal to ignorance" because it attempts to take a lack of evidence of A (toxicity of xyz) as an evidence that non-A (xyz is not toxic).
   D: Yes, it is proper because scientific reasoning can't result in any claims that are "stronger" than this one.

9. Which of the following groups of words are of minor importance when checking the validity of deductive logic arguments?

   A: All, only, every.
   B: Therefore, thus, it follows that.
   C: I, you, they, us.
   D: Not, none, no.

10. When the student in Case 4.2 ignores unassignable peaks in his NMR spectrum and interprets the other peaks allowing significant deviation from expected values, what bias does he have?

    A: Appeal to authority.
    B: Confirmation bias.
    C: Narrative bias.
    D: Appeal to ignorance.

11. The most important parameter in scientific reasoning is:

    A: Consistency with existing theories.
    B: Consistency with the written research proposal of the project within which the analysis is being performed.
    C: Consistency of the logic employed.
    D: Consistency with respect to societal and cultural values and norms.

12. What constitutes a bias in scientific reasoning?

    A: Using fallacious logic in argumentation.
    B: Being too careful/shy when drawing conclusions and interpreting data.
    C: Doing research for a Masters or PhD thesis.
    D: When chemists do research on a topic that is actually in the field of physics or biology.

13. How should you respond to being asked, "Show me the evidence for your scientific claim?"

    A: Present your raw data.
    B: Compile a list of relevant literature references.
    C: Perform a statistical analysis with a software package.
    D: Explain how the data and its plausible interpretation reveal insights for a well-reasoned and consistent research hypothesis.

14. Which of the following comments about scientific theories is true?

    A: They serve as the standard by which my results have to be evaluated.
    B: They describe natural laws. All hypotheses have to be informed by these rules. Without these, no deductive reasoning is possible.
    C: They are developed by carefully setting all available scientific insights into perspective and making sense of them. Thus, they can change over time and describe the development of scientific knowledge at a particular point in time.
    D: They have to be understood as larger hypotheses and guiding paradigms in science. Scientists work to find evidence to support these theories.

15. If chemistry is an "inductive science", and inductive logic can never achieve 100% certainty, how can chemical research produce any useful insights?

    A: Chemists actually employ a scientific methodology on the basis of predictions and explanations, proceeding in iterative steps of deduction and induction (exploiting their respective advantages).
    B: The insights are useful. We just have to live with the uncertainty. That is why we keep doing research.
    C: Although it is often labelled "inductive", chemistry actually usually follows abductive approaches.
    D: Chemistry doesn't produce useful insights in the sense of true knowledge. It should be clear that what we know today will be refuted tomorrow. Therefore, any attempt to reach 100% certainty is doomed to failure!

16. Which of the following plays no part in a scientific investigation?

    A: Prediction
    B: Speculation
    C: Observation
    D: Explanation

17. Which of the following paths corresponds to an inductive approach?

    A: From rule/law/principle *via* case (particular situation) to result (conclusion).
    B: From result (observation) *via* case (experiment) to rule/law/principle.
    C: From case (experiment) *via* result (observation) to rule/law/principle.
    D: From result (observation) *via* rule/law/principle to case (experiment).

18. What is the basic question that Bayesian statistical analysis asks?

    A: Does the gathered data have the explanatory power to change/refine the initial hypothesis?
    B: How can I make sense of my previous beliefs in view of the obtained data?
    C: How can I prove mathematically that my hypothesis is true?
    D: How can I reduce error rates to the smallest possible level?

19. What do frequentists accuse Bayesians of?

    A: Having plagiarised frequentist models and presented them as Bayesian ideas.
    B: Simplifying complex mathematical issues into student-friendly but imprecise equations.
    C: Employing a "prior" that could, potentially, be built on beliefs and intuitions, so that Bayesian statistics is subjective and, therefore, unscientific.
    D: Encouraging scientific misconduct by enabling researchers to camouflage fabricated and falsified data.

20. "Researchers see in their data what they wish to see in it!". What is this bias called?

    A: Survivorship bias.
    B: Authority bias.
    C: Outcome bias.
    D: Confirmation bias.

References

1. M. Black, *Critical Thinking. An Introduction to Logic and Scientific Method*, Prentice-Hall, New York, 1946.
2. T. Nickles, *Scientific Discovery, Logic, and Rationality*, D. Reidel Publishing, Dordrecht, 1980.
3. J. Trusted, *The Logic of Scientific Inference: An Introduction*, Macmillan, London, 1979.

4. *The Cognitive Basis of Science*, ed. P. Carruthers, S. Stich and M. Siegal, Cambridge University Press, 2002.
5. K. Popper, *The Logic of Scientific Discovery*, Routledge Classics, Routledge, London, 1959, 2004.
6. P. Pruzan, *Research Methodology. The Aims, Practices and Ethics of Science*, Springer, Cham, 2016.
7. F. MacRitchie, *The Need for Critical Thinking and the Scientific Method*, CRC Press, Boca Raton, 2018.
8. G. A. Foresman, P. S. Fosl and J. C. Watson, *The Critical Thinking Toolkit*, Wiley Blackwell, Oxford, 2017.
9. D. Jackson and P. Newberry, *Critical Thinking. A User's Manual*, Cengage Learning, Stamford, 2nd edn, 2016.
10. R. Dobelli, *The Art of Thinking Clearly*, Spectre, UK, 2013.
11. K. Shrader-Frechette, *Tainted. How Philosophy of Science Can Expose Bad Science*, Oxford University Press, New York, 2014.
12. *Model-based Reasoning in Science and Technology: Theoretical and Cognitive Issues*, ed. L. Magnani, Springer, Berlin, 2014.
13. *Model-based Reasoning in Science and Technology: Logical, Epistemological, and Cognitive Issues*, ed. L. Magnani and C. Casadio, Springer, Cham, 2016.
14. J. L. Myers and A. D. Well, *Research Design and Statistical Analysis*, Lawrence Erlbaum Associates, Mahwah, 2nd edn, 2003.
15. J. Vallverdú, *Bayesians Versus Frequentists. A Philosophical Debate on Statistical Reasoning*, Springer, Heidelberg, 2016.

Part 2: Research Ethics

Part 1 expanded our understanding of how to perform chemical research well from a methodological and science-theoretical perspective. Research ethics and scientific integrity from a profession ethics point of view must base its judgments directly in this code of professional conduct. Scientific methodology is what it is *because* this methodology *makes* it scientific and, thus, viable and societally meaningful. With this orientation in mind, it is possible to extract attitudes and codes of behaviour that meet the requirements for good research conduct (Chapter 5). We may call these attitudes *scientific virtues*. If a researcher has not yet incorporated and manifested them as character disposition, they can be trained or cultivated. In daily professional practice, they help in solving disputes and dilemmas, serve as argumentative tools in communication of conflicts, and can provide answers to the question of what counts as *the right thing to do*. Experiences have shown that chemical research as it is typically performed today bears a few common dangers of ethical conflicts or misconduct. These are the intentional mishandling of data (Chapter 6), ethical issues arising in the context of publishing (Chapter 7), conflicts among various actors and interest groups in chemical research including mentorship, collaboration between different experts or non-experts, funding, and intellectual property (Chapter 8), and – at least for some chemists – experimentation that involves animals (Chapter 9).

5 The Virtues of Science

Overview

Summary: Part 1 depicted chemical scientific research as an endeavour based on epistemic and cognitive agency (that means; applying and advancing knowledge by thinking and acting in rational and reasonable ways) as well as on methodological rigidity and logical plausibility. This allows the formulation of behavioural rules for researchers and expectations on their attitude towards doing their job. These guidelines – we may call it *the ethos of science* – gain their justification directly from the methodological concepts that define *good chemistry*. This is especially important in view of science being a societal institution that is built on social acceptance and support. No researcher or scientist is working outside of this normative framework that the social conventions and expectations span up.

In order to reach the goals of scientific inquiry, a set of behavioural attitudes – *virtues*, as we may call them – is demanded. Truthfulness, objectivity, honesty, self-control and fairness are certainly among them. In addition to these individual character attributes, there are also communal virtues as proposed by Robert Merton: universalism, communalism, scepticism. All these virtues are clearly framed by respective vices as either a lack of that virtue or an excess of that virtue. In this chapter, we will discuss all these considerations in detail.

This chapter serves as an introduction and orientation for all the chapters in Part 2. In research ethics, both as guidelines and as discourse tool, the virtue approach serves a good purpose. Here, we are not very concerned about ethics as a philosophical method of figuring out solutions for dilemmas and problems, but about compliance with moral conventions as part of a social contract. Some regard the ethical rules as trivial: everybody knows these rules! Yet, as practical experiences show, knowing the rules doesn't protect from acting against them. Thus, the main goal of this chapter is to present a basis for argumentation and goal-oriented discourse. We learn these virtues of science not primarily in order to develop our virtuous character, but in order to have a discourse strategy at hand that we can apply when we encounter situations of misconduct and violation of scientific integrity.

Key Themes: Virtues of science (honesty, truthfulness, objectivity, disinterestedness, systematised doubt, disciplined self-control, fairness); communal virtues (universalism, communalism, scepticism); virtues for chemists other than scientific researchers.

Good Chemistry: Methodological, Ethical, and Social Dimensions
By Jan Mehlich
© Jan Mehlich 2021
Published by the Royal Society of Chemistry, www.rsc.org

> **Learning Objectives:**
>
> If you take this chapter serious, you will:
>
> - Know the virtues that constitute the ethos of good scientific practice and relate them to your daily work as a chemist.
> - Be able to make arguments in favour of or against certain practices – your own or observed on others – on the basis of these virtues, thus giving them justificatory power and yourself more confidence and, subsequently, positive influence.
> - Understand the ethical guidelines for chemical professions as the result of communal and social agreements and expectations, built on rational considerations and derived from the methodological foundations of scientific activity.

5.1 Introductory Cases

Case 5.1: Swear the Baconian Oath!

Jenny and Peter discuss a newspaper report about recent fraud cases in scientific research. They come to the agreement that scientists, indeed, have quite a huge impact on society and, thus, a high responsibility for doing their job well. Peter thinks that the current institutional guidelines for good scientific practice are enough. When people have the 'criminal energy' to violate them, anyway, it is rather a structural or systemic error of how we organise science institutions and their individuals. Jenny, on the contrary, thinks that researchers' social impact makes them almost like medical doctors. Thus, she insists, after their education and before they start their jobs, they should prove their ethical integrity by swearing an oath in analogy to the Hippocratic Oath of doctors. With Francis Bacon as the forefather of the modern approach to science, we could formulate a Baconian Oath for scientists!

Case 5.2: Special Motivation

Gerry and Olivia, both academic chemists at two different universities, are a happily married couple. Yet, one evening after dinner, they get into an argument. Gerry's research has been influential and groundbreaking over the past two decades. He is extraordinarily ambitious and even talks about the Nobel award, recently. Rumours that he might have a chance to receive it spread across his institute, and his disappointment when he wasn't chosen last year was quite obvious. He is now pushing harder than ever to have papers published in high-impact journals. Olivia is worried that his greed for fame might blur his scientific integrity. Yet, when she addressed this issue after said dinner, he becomes upset and claims that success is one of the main drivers of scientific achievements. Yet, she has a bad feeling and looks for a good argument that explains why fame should not be the motivation for doing good science.

Case 5.3: Our Own Biggest Critics

The PhD candidate of Case 4.2 (the one with the H-NMR spectrum as the only analysis of his compound) has a personal talk with his principal investigator (PI) after the group seminar. They discuss how much evidence is necessary to back up scientific claims. *"NMR spectra don't lie!"*, says the PhD candidate. *"For complex organic molecules like yours the interpretative margin of NMR signals is still too high!"*, claims the professor. The PI has the feeling that her student is desperately trying to defend himself. This is understandable as the student works hard for his grade and wishes to have success with his research. At the same time, this can't be a legitimation for sloppy conclusions. She tries to convince the student with an argument on what scientific research is about: *"You have to understand that other people will be very critical with our claims. They want to see more evidence than just one NMR spectrum. In order to be a respected member of the chemical community, we have to be the biggest sceptics of our own work in the first place, and our own biggest critics!"*. The student leaves with mixed feelings. Did she make a proper point?

5.2   Virtues of Scientific Practice

The professional activities of scientists, engineers and researchers go along with high responsibilities for intended and unintended effects, ranging from intra-communal impact in the form of knowledge and competence to environmental and social implications. Therefore, it is important to follow guidelines of *good scientific practice* and refrain from misconduct.[1-12] This means, similar to medical professions, scientist should comply with a professional *ethos* of science conduct.[13-15] 'Ethos' is a term used for a set of virtues that members of a professional community should cultivate and manifest in their actions and choices. The idea of virtues as a source for knowing what is *good* and *right* (to do) is the oldest form of ethics (the singular term, as in *moral philosophy*). The most prominent advocates of virtue ethics are the ancient Greek Philosophers, especially Plato, Aristotle and Socrates. The Asian schools of thought from that time (~500 BC) – Confucian, Daoist and Buddhist philosophy – revolve around the idea of *cultivating virtues* as a source for good life conduct.[16]

What was the idea these wise men had? If you want to find out what you should do in a particular role, imagine the *ideal* person and what he/she would do, and that is what you should do, too. For example: when you are a soldier and wonder how you should act in a certain situation, you imagine the ideal soldier and will certainly find that he would have virtues like bravery, courage, and persistence, but would certainly not have vices (the opposite of virtues) like daredevilry or cowardice. Therefore, you – as a soldier – should act brave and wise. Here, you can see that virtues are thought of as a golden mean, the middle way between two extremes (the vices) of *a lack of it* and *too much of it* (see Table 5.1). A lack of courage is cowardice, an excess of courage (or *an over-ambition of showing courage*) is daredevilry. Another example could be the virtue generosity, with its vices of stinginess and wastefulness, for example for a businessman.

In all kinds of life situations and the roles people can find themselves in – politician, citizen, marriage partner, parent, colleague, and so forth – this consideration can be applied. Some virtues are valid for all and therefore considered *universal virtues*, like wisdom or benevolence. An important aspect of virtues is that they are expressions of personality and not utilised strategies for certain goals in particular situations. Virtues are *cultivated* by the willingness to act according to them and the repeated application so that they become part of one's habits and personality. The approach sounds simple and, actually, it has been criticised for being vague and arbitrary. However, in many cases it serves the purpose of defining codes of conduct very well and is easy to understand for those who are obliged to follow the rules. Based on the considerations of the previous chapters concerning what

Table 5.1 Examples of virtues and their respective vices.

Vice (lack of...)	Virtue (mean)	Vice (excess of...)
Cowardice	Courage	Daredevilry
Stinginess	Generosity	Wastefulness
Foolishness	Wisdom	Astuteness

scientific conduct means and implies, the *virtues of science* can be elaborated by imagining *the ideal scientist*. Same as above, we regard the virtues as the mean between two vices, the *lack vice* and the *excess vice* (see Table 5.2).[17]

Intellectual Honesty: This virtue is directly derived from the main purpose of scientific activity to elaborate viable knowledge from appropriately acquired data. It addresses the scientist's duty to handle and communicate data and all the means employed for their interpretation with honesty, sincerity and righteousness. The vice as a lack of honesty is more than simply dishonesty. More important than the act of being dishonest is the motivation for keeping data and methods secret or hiding failure. Arguably, in the majority of cases, lack of honesty originates from the fear of having to suffer from disadvantages like poor grades, delayed research paper publication, bad reputation, or damaged pride. Thus, we find the term *opportunism* in Table 5.2: rather than being honest, a researcher with this vice will say what she believes others like to hear or what brings her closer towards her goal. At the other end of the scale, is it possible to be *too honest*? Speaking too openly about one's research projects puts the scientist at risk of foolishly revealing good research ideas or making promises and prospects that are not well grounded and quite likely disappointed in the end. This form of *babbling* undermines the maxim of honesty that endorses a focus on what really happened in the course of experimenting and interpreting acquired data.

Truthfulness: In order to maintain credibility and trust, a scientist must stick to *truth* in the sense that he can and should only state what the case is and what is in accordance with the insights that are elaborated with applied scientific methodology. A lack of truthfulness is secretiveness. This must not be confused with the discretion or secrecy to which a researcher may be sworn in a collaborative project with industrial partners on organisational and

Table 5.2 The virtues of scientific practice.

Vice (lack of...)	Scientific virtue	Vice (excess of...)
Opportunism	Intellectual honesty	Babbling
Secretiveness	Truthfulness	Injudiciousness
Bias	Mindful truth-seeking	Relativism
Ideology	Objectivity	Aloofness
Advantage-seeking	Dedicated disinterestedness	Indifference
Credulity	Systematised doubt	Categorical doubt
Self-adulation	Disciplined self-control	Self-humiliation
Unfairness	Fairness	Self-abjection

legal grounds. Secretiveness, rather, refers to deliberately hiding or miscommunicating scientific findings (data, the methods of their acquisition, and the rationale of their interpretation) if they don't fulfil own or others' expectations. On the other side, again, we may wonder how someone can be *too truthful*. It may be understood as injudiciousness, a kind of rashness or indiscretion, which makes the scientist communicate insights from his activities without the necessary reflections on interpretation of data or the implications of the elaborated meanings.

Mindful Truth-seeking: Whereas the former virtue rather refers to communicative actions in the science community or in public spheres, objective and mindful truth-seeking aiming at truth assurance is of procedural importance. When conducting experiments, the scientist is obliged to free himself from biases and other factors that might influence his objective devotion to truth. The quest for truth is *mindful* when streams of thought, logic inferences, influences of interests and character dispositions, but also the epistemic conditions for cognitive success (as explicated in Chapter 2) are present in the scientist's awareness when conducting scientific research. A lack of this attitude puts the scientist at risk of being blurred by dispositions that aim at achievements other than truth (for example, power, money, fame, success). Whereas *truth* in the context of the *truthfulness* virtue refers to what the researcher really does and finds out in the course of his research – a low-stakes case of truth that is comparably simple to determine – *truth* in the context of *mindful truth-seeking* refers to the knowledge that the scientific inquiry elaborates – a high-stakes case of ontological truth. Recall from Chapter 2, it should be clear that *truth*, here, refers to the viability of scientific claims rather than their metaphysically demanding correspondence with features of the actual world. In order to understand the choice of *relativism* as the excess vice of mindful truth-seeking, the reader shall be reminded of the *tree of knowledge* image from Chapter 2 (Figure 2.6). In order to make sense of experiences, we tend to colour and shape them according to our previous experiences that constitute our world-explaining framework. Mindful truth-seeking, in this image, means to be aware of the mechanisms that are at work when we interpret experiences such as the data of a measurement series. Following patterns and default-settings without much reflection is a lack of mindfulness and, thus, the lack-vice that we called bias. Stripping off all these mechanisms and influences from one's meaning construction, the excess-vice, would result in an entirely non-judgmental *looking at things*. Anything goes. This attitude is not helpful for science as a community endeavour because it makes results incommunicable, and because it amounts to a position in which a clear universalizable claim as a viable result of the scientific inquiry is never reached.

Objectivity: This virtue is expressed by the scientist when actions, decisions and judgments are motivated and informed solely by scientific considerations concerning methods, models, mechanisms or worldviews. Objectivity – a focus on the *object* of one's scientific investigation rather than the subjective stance of the interpreting and judgmental scientist – is required in all six elements of the scientific method (Chapter 3, Figure 3.1). A lack of objectivity is shown, for example, when the scientist sticks to ideologies and non-scientific presuppositions like Lysenko in Case 2.3 (Chapter 2), but also in the lighter but daily case of approaching a scientific insight with expectations on what this insight should be (like the student in Case 4.2). On the other end of the scale, objectivity can amount to a very distant relation to one's own research. This aloof attitude resembles the relativistic stance of too much *mindful truth-seeking* ambition.

Dedicated Disinterestedness: Gerry in Case 5.2 shows an attitude that makes his wife worry for a good reason. His motivation seems to arise from the prospect of fame and, perhaps, money. This must not necessarily result in improper science conduct but, as the scenario suggests, may create a bias in Gerry's scientific judgments, for example, trying to place a mediocre paper in a high-impact journal. A scientist should have no other interest but the generation of insight and knowledge, and especially no interest in the kind and type of result he obtains. The selfless devotion to the ambitious goal of constructing knowledge and gaining deeper comprehension should not be blurred by selfish careerism, the interests of any sponsors, or other forms of seeking advantage. This is slightly different from the objectivity virtue that concerns prejudgments and expectations about the meaning of our observations. Here, we talk about external motivations that conflict with purely scientific attitudes. Of course, this doesn't mean that we show no interest in our results at all. Too much disinterestedness might amount to indifference. With this attitude, it would be impossible to write any convincing research paper or grant proposal. Scientific results are defined by their meaning (not purpose!), and showing no dedication whatsoever to one's own research results is neither credible nor scientific.

Systematised Doubt: Doubt (next to curiosity, perhaps) is the main driver of scientific inquiry, but it must be systematised and reasonably approached with the scientific method. When it turns into categorical scepticism, the scientific endeavour becomes fruitless and unfeasible.

Epistemic sceptics who deny the possibility of knowing anything would not make good scientists. Lack of doubt – maybe termed credulity or gullibility, believing things too easily and unquestioned – can damage the quality of scientific output and the scientist's reputation. The middle way of doubting and questioning may be best understood as healthy humbleness of what we are able to learn from scientific inquiry and what not, and where we better keep doing more research. Case 5.3 provides an example of a student who is not doubtful enough, probably due to the pressure to finish a thesis and to publish a research paper with good results. The PI says something very important: we should be our own harshest critics! The argumentative foundation of this claim is the same as that of this virtue: We have to be critical because that is what makes our knowledge acquisition *scientific*!

Disciplined Self-control: This virtue addresses the scientist's self-perception. If he takes himself as *the greatest*, beyond any flaws and insufficiencies, not able to take criticism, his self-adulation makes him a bad scientist. If he shows a lack of confidence in himself, playing himself down, believing in himself being unworthy or flawed, his self-humiliation inhibits his scientific success. A self-controlled scientist will accept criticism, is eager to improve and get better every day, but also knows his strong points and steps in for his convictions. A lack of self-control is often a potent source of biases and, as such, diminishes the scientist's objectivity and mindful truth-seeking ability.

Fairness: In almost all professional realms, from academia to industry to public service, a scientist is a team player, obliged to be devoted to fairness and cooperation. As we have learned that community and intra-community communication are an inevitable part of scientific methodology, we can understand fairness as a result of this methodology, not merely as a call from common morality. Violating fairness undermines the power of the scientific method. Scientific micro- and macro-communities – scientific institutes in universities, legislations of countries, the global science community – have established clear rules to ensure fairness and sanction unfairness. But also, the excess vice has a negative impact: a scientist who is too submissive, leaving the stage (and the fame) to others, stepping back to favour his colleagues, will have difficulties establishing a smooth and fruitful career.

It should be noted that we can, of course, list other virtues, such as curiosity, courage (think of Galileo who had to defend his science in front of the powerful Roman-Catholic church), perseverance, helpfulness, diligence, assiduity, and many more. However, whereas the eight virtues discussed above are directly derived from the guidelines of science conduct, these other virtues are either *too general* (which job would not demand perseverance and diligence?), or *character dispositions* that don't have the status of a virtue. Moreover, scientists certainly need to have a certain mental and cognitive capacity (in other words: they need to be smart), but this would rather count as a *competence* or *skill* than as a *virtue*. Shamoo and Resnik suggest this list of individual "principles" that are individual traits and virtues, partly overlapping with the list above:[3]

1. Honesty: report only what really is the case, don't make things up.
2. Objectivity: avoid bias.
3. Carefulness: avoid errors and negligence wherever you can.
4. Credit: allocate credit fairly.
5. Openness: share and collaborate, be ready to receive criticism and feedback.
6. Confidentiality: respect and protect confidential information.
7. Respect for colleagues: no discrimination, but support, advise and mentorship.
8. Respect for intellectual property: no plagiarism, giving credit.
9. Freedom: don't interfere with freedom of thought and inquiry.
10. Protection of animals used in research.
11. Protection of human research subjects: minimize risks, obtain informed consent.
12. Stewardship: make good use of resources.
13. Respect for the law: compliance with policies.
14. Professional responsibility: maintain professional competence, expertise, and integrity.
15. Social responsibility: promote good social consequences of scientific activity.

5.3 Communal Virtues of Science

There is at least one alternative approach to formulating virtues of scientific activity. In the previous section, virtues are understood as individual attitudes that are manifested by cultivating behavioural patterns that are judged as *right* by the measures of a professional activity. Admittedly, that is a challenging task. Robert Merton suggested a way to circumvent the demanding call for individual character formation and re-shaping of attitudes. His set of communal virtues for scientists derives its justification from the self-motivated desire to respect and promote the rules and guidelines of the community one is a devoted member of. These virtues are, then, not attitudes but *maxims* or imperatives of conduct.[18]

Universalism: Scientific statements should be valid independent of time, space and cultural framework. Forces undermining this virtue are ideology, dogmatism, or particularism (allowing case-by-case solutions of problems without claiming universalizability). This communal virtue represents the individual and collective efforts of scientists to enlighten knowledge of the world. Objectivity and truthfulness are maintained when the universal attitude of scientists outweighs their cultural, geographic, confessional, and ideological matrices. After all, it shouldn't matter who is generating this scientific insight. Science is not about ego, but about the collective endeavour to increase every human mind's understanding of this world.

Communalism: This virtue stresses the character of science as a collective endeavour in which it is crucial that the members of the scientific community share their insights and have equal chance to both participate and benefit from it. The opposite of this norm is secrecy. Here, we find the abovementioned virtues of fairness, self-control, honesty and truthfulness again. Merton went so far to suggest that all scientists should have common ownership of all scientific goods so that every scientist is motivated to promote collective collaboration.

Organised Scepticism: Scientists should always be their own strictest critics and always question their insights and claims. Gullibility and benevolent but biased evaluation of scientific results contradict this virtue. That theories and models may be flawed, as historical evidences suggest, should be accepted, expected, and carefully considered. Then, critical scrutiny as an inevitable part of the scientific method is embraced and welcomed. In the previous section, scepticism as categorical doubt was mentioned as a vice in the context of systematized doubt. Whereas an individual scientist must be *reasonably doubtful*, knowing when to stop doubting and start stating, the scientific community as a whole takes *institutional scepticism* as a *Leitmotiv* (guiding principle) to be clearly distinguishable from doctrine-driven institutions (like religion) and to be societally legitimised as such.

Disinterestedness: Scientists must not follow individual interests but, in each individual action, support and benefit the community of scientists or the social institution *science* as such. Self-interest, unfairness and opportunism, but also self-aggrandisement and exploitation of the ignorance of scientific laymen undermine this attitude. As a maxim, it also suggests to embrace mutual feedback as peer review and being open for critique and debate.

In later works, Merton added a fifth norm, originality.[19] He was convinced that scientists stick to these norms out of a mix of institutional control (including fear of sanctions) and psychological conflict (personal gains *versus* the community character of the scientific enterprise). The suggested imperatives solve the conflict by pointing out the prospects of successful science and *making this world a better place* as long as the players follow the rules.

5.4 Does Science Need an Ethos?

As mentioned above, a set of virtues that forms a behavioural guideline for a profession is called an ethos. The most prominent professional ethos is that of medical doctors. In the famous Hippocratic Oath, at the beginning of their career, they declare that they will always work in accordance with virtues like non-maleficence, harm-avoidance and benevolence, saving lives and reducing pain as best as they can. We understand that this devotion is necessary for a profession that has a lot of power over the life of others, here in a very direct sense. It helps to establish trust and providing the grounds for patients to hand themselves and their vulnerable integrity over into the competence of a medical doctor. Given the impact that scientific activity and its catalysing effect on technological progress have, would it be reasonable or demanded to ask scientists to swear an oath, too, as the two protagonists of Case 5.1 wonder?

> **Box 5.1** Exercise: Write the Baconian Oath!
>
> Take Jenny from Case 5.1 seriously and write the Baconian Oath that she suggests. Imagine an event, perhaps the graduation ceremony, at which the graduates have to read out a short text in which they declare that they will always, to the best of their knowledge, act professionally in accordance with the virtues of science. What must be mentioned in the text? Are there debatable claims that, probably, not everyone would agree with as part of a binding oath? Moreover, do you believe that swearing such an oath would really impact the attitude of the young scientists? Or, put it this way: is there any way to formulate this oath in a way that impresses the young scientists so deeply that they can never forget it and always act accordingly?

Attempts to formulate and promote concise oath-like codes of conduct for chemists have been communicated.[20] Some of the national chemical societies in Europe have suggested codes of conduct, yet none of them are obligatory for all members, not to speak of every chemistry graduate or professional. A collection of such codes of conduct was compiled by the Organisation for the Prohibition of Chemical Weapons (OPCW)[21] and resulted in The Hague Ethics Guidelines.[22] In chemical education, curricular courses in research ethics have been implemented at some but, by far, not all universities. Study material that attempts to facilitate the cultivation of scientific integrity is available.[23-25] At those institutions that ask their students to complete a course on research ethics, good scientific practice or scientific integrity, passing an exam is required for being awarded the credits. Yet, swearing an oath is a different level, and the conceptualisation and implementation of such a binding ethos is still in its infancy. Whether it is an effective measure to make scientists act more ethically is a question that is difficult to assess. Empirical evidence shows that attending courses on research ethics does not prevent the participants from violating ethical guidelines in their lab practice.[26-28] Yet, maybe, swearing an oath leaves a bigger and more effective impression on the chemical practitioner.

5.5 Extending the Scope

It may seem like the virtues of science apply only to academic science and research at universities and other public or private academic institutes. Certainly, research in the private industrial sector needs to follow the same principles, especially the virtues of objectivity, honesty and fairness. Yet, other guidelines from different fields of applied ethics, for example business ethics or environmental ethics, add to research ethics as a specific branch of profession ethics.[29] (*Note that the term* profession ethics, *here, is used for that field of applied ethics that concerns codes of conduct in professional contexts. Some other authors use the term* professional ethics. *The author of this text prefers using* professional ethics *for the philosophical expertise of ethicists who do ethics professionally.*) In principle, guidelines of professional conduct, no matter in which profession, arise from the specific portfolio of professional activities and impacts. As we have seen, good scientific practice guidelines are built on methodological considerations on good science, justified by the chemical profession as an instance that is firmly embedded in social acceptance and justification. Business ethics, for example, would build its rationale on the specific practices in business and economic interaction, stressing virtues of generosity, respect, fairness, and cooperation, perhaps. Chemists employed in the public service sector face very different professional guidelines.[30] Not all organisations, companies or agencies explicate codes of conduct explicitly. This should not bother the professional practitioner. The contribution of ethics for the elaboration of such guidelines is rather small: good professional conduct and professional integrity are the result of common-sense morality rather than professional ethical reasoning. The more urgent question is: why would someone be motivated to violate the imperatives of professionalism, and how can compliance with the guidelines be incentivised?

In some contexts, for example the collaboration between academic and industrial chemists (as discussed in Chapter 8), behavioural imperatives and maxims may clash and conflict. Profit-oriented research and development (R&D), for example, applies different standards to the disinterestedness of the involved researchers or the degree of mindfulness that is necessary for objective truth-seeking. The virtues introduced in this chapter and those that are not mentioned but come into effect in respective contexts serve an important purpose: they can be employed as the argumentative grounds for outlining differences in professional attitudes and creating mutual understanding of each other's attitudes and approaches to maintain professional integrity.

Exercise Questions

1. Which of the following is not one of the communal virtues proposed by Robert Merton?

   A: Scepticism
   B: Communalism
   C: Fairness
   D: Universalism

2. Identify the vice among the scientific virtues in this short list:

   A: Dedicated disinterestedness
   B: Objectivity
   C: Truthfulness
   D: Bias

3. Which of the following terms represents a vice associated with an excess of 'fairness'?

   A: Babbling
   B: Self-abjection
   C: Credulity
   D: Objectivity

4. Universalism means:

   A: Supporting the entire scientific community.
   B: Valid independent of time, space and cultural framework.
   C: Commonly held beliefs.
   D: Meeting the standard expected for a university education.

5. Which of the following is a vice associated with 'intellectual honesty'?

   A: Opportunism
   B: Dogmatism
   C: Gullibility
   D: Ideology

6. Which of the following can be considered a 'father' of virtue ethics?

   A: Spinoza
   B: William James
   C: Plato
   D: Jeremy Bentham

7. Which of the following is not considered to be a scientific virtue?

   A: Truthfulness
   B: Ideology
   C: Objectivity
   D: Fairness

8. Which of the following would be considered a scientific virtue?

   A: Credulity
   B: Categorical doubt
   C: Gullibility
   D: Systematised doubt

9. Which of the following would be considered a communal virtue?

   A: Particularism
   B: Dogmatism
   C: Universalism
   D: Ideology

10. What is an "ethos"?

    A: In Ancient Greek ethics, Ethos is a hypothetical character that personifies a 100% virtuous man.
    B: An ethos is a feeling of guilt after acting immorally.
    C: Ethos is a term that describes the total of moral rules that a society commits itself to (as, for example, expressed in laws and church doctrine).
    D: An ethos is a set of ethical guidelines and codes of conduct for particular professional groups like doctors, psychologists, engineers or scientists.

References

1. J. Kovac, *The Ethical Chemist: Professionalism and Ethics in Science*, Oxford University Press, New York, 2nd edn, 2018.
2. S. Greer, *Elements of Ethics for Physical Scientists*, MIT Press, Cambridge, USA, 2017.
3. E. Shamoo and D. B. Resnik, *Responsible Conduct of Research*, Oxford University Press, Oxford, 3rd edn, 2015.
4. F. L. Macrina, *Scientific Integrity: An Introductory Text with Cases*, American Society for Microbiology Press, Washington, 4th edn, 2014.

5. C. Russell, L. Hogan and M. Junker-Kenny, *Ethics for Graduate Researchers. A Cross-disciplinary Approach*, Elsevier, London, 2013.
6. Committee on Assessing Integrity in Research Environments (CAIRE), *Integrity in Scientific Research*, The National Academies Press, Washington, 2002.
7. S. Loue, *Textbook of Research Ethics – Theory and Practice*, Kluwer, New York, 2002.
8. National Academy of Sciences, *On Being a Scientist: Responsible Conduct in Research*, National Academy Press, Washington DC, 3rd edn, 2009.
9. D. B. Resnik, *The Ethics of Science – An Introduction*, Routledge, London, 1998.
10. K. S. Shrader-Frechette, *Ethics of Scientific Research*, Rowman & Littlefield, London, 1994.
11. *Research Ethics*, ed. A. S. Iltis, Routledge, New York, 2006.
12. *Handbook of Research Ethics and Scientific Integrity*, ed. R. Iphofen, Springer, Cham, 2020.
13. D. Koepsell, *Scientific Integrity and Research Ethics. An Approach from the Ethos of Science*, Springer, Cham, 2017.
14. R. T. Pennock and M. O'Rourke, Developing a Scientific Virtue-Based Approach to Science Ethics Training, *Sci. Eng. Ethics*, 2017, **23**, 243.
15. J.-Y. Chen, Virtue and the Scientist: Using Virtue Ethics to Examine Science's Ethical and Moral Challenges, *Sci. Eng. Ethics*, 2015, **21**, 75.
16. *The Handbook of Virtue Ethics*, ed. S. van Hooft, Routledge, Abingdon, 2014.
17. D. Fenner, *Einführung in die Angewandte Ethik*, Narr Francke Attempto Verlag, Tübingen, 2010.
18. R. K. Merton, The Normative Structure of Science (1942), in *The Sociology of Science: Theoretical and Empirical Investigations*, ed. R. K. Merton, University of Chicago Press, Chicago, 1973.
19. B. Macfarlane and M. Cheng, Communism, Universalism and Disinterestedness: Re-examining Contemporary Support among Academics for Merton's Scientific Norms, *J. Acad. Ethics*, 2008, **6**, 67.
20. M. Sutrop, M.-L. Parder and M. Juurik, Research Ethics Codes and Guidelines, in *Handbook of Research Ethics and Scientific Integrity*, ed. R. Iphofen, Springer, Cham, 2020.
21. OPCW, *Compilation of Codes of Ethics and Conduct*, 2015, https://www.opcw.org/sites/default/files/documents/SAB/en/2015_Compilation_of_Chemistry_Codes.pdf, accessed on August 25th 2020.
22. https://www.opcw.org/hague-ethical-guidelines, accessed on August 25th 2020.
23. X. Hu, *Smoothing a Critical Transition. Nontechnical Knowledge and Techniques for Student Researchers*, Springer, Singapore, 2020.
24. *A Guide to the Scientific Career. Virtues, Communication, Research, and Academic Writing*, ed. M. M. Shoja, A. Arynchyna, M. Loukas, A. V. D'Antoni, S. M. Buerger, M. Karl and R. S. Tubbs, Wiley-Blackwell, Chichester, UK, 2020.
25. B. G. Blundell, *Ethics in Computing, Science, and Engineering. A Student's Guide to Doing Things Right*, Springer, Cham, 2020.
26. M. D. Mumford, Assessing the Effectiveness of Responsible Conduct of Research Training: Key Findings and Viable Procedures, in *Fostering Integrity in Research (Report of the National Academies of Sciences, Engineering, and Medicine)*, The National Academies Press, Washington, 2017.
27. M. S. Anderson, A. S. Horn, K. R. Risbey, E. A. Ronning, R. De Vries and B. C. Martinson, What Do Mentoring and Training in the Responsible Conduct of Research Have to Do with Scientific Misbehavior? Findings from a National Survey of NIH-Funded Scientists, *Acad. Med.*, 2007, **82**, 853.

28. C. L. Funk, K. A. Barrett and F. L. Macrina, Authorship and Publication Practices: Evaluation of the Effect of Responsible Conduct of Research Instruction to Postdoctoral Trainees, *Acc. Res.*, 2007, **14**, 269.
29. J. Oakley and D. Cocking, *Virtue Ethics and Professional Roles*, Cambridge University Press, Cambridge, 2001.
30. A. W. Dutelle and R. S. Taylor, *Ethics for the Public Service Professional*, CRC Press, Boca Raton, 2nd edn, 2018.

6 Scientific Misconduct

Overview

Summary: After defining what may count as good scientific conduct, it should be clear what scientific misconduct is: acting in contrast to or violation of one or more of the scientific virtues. Yet, this can mean a lot, and very often there are grey zones in which the researcher is not sure whether an intended action is permissible or not. Many cases of downright fraud have been reported throughout the history of science, some of which are individual criminal intent, while others are the result of character dispositions amplified by systemic pressure. Some cases are not only a matter of research ethics, but also illegal. Fabrication of data, falsification of data, and plagiarism are forms of cheating, betraying and stealing, and, thus, can be sanctioned in institutional and legal terms. Other cases like the omission of data, the post-processing of images, copying experimental instructions from other sources, and so forth, are debatable, but not clearly illegal or unethical.

This chapter has two purposes: to show that scientific misconduct is a real problem in the chemical community, and to give guidance for the decision as to whether an intended action in a research context is appropriate or not. The former has received a lot of media attention lately. More importantly, empirical studies on the behaviour of scientists have been conducted, so that data on misconduct is available. The more difficult question is the reason for fraud and misconduct. It is worth highlighting at least some of the motivations so that an awareness of them can protect from falling victim to them. The latter purpose is a matter of discourse. We will see how the science virtues can help one make the right decisions for oneself, but also protect others from slipping into the dark side of betrayal and fraud by seeking goal-oriented mature conversations. Empirical studies have shown that training in research ethics doesn't make researchers commit less fraud. But whistleblowing does! Paying attention to one's surrounding and finding proper strategies to address misconduct is, arguably, the most efficient way to ensure the community's scientific integrity.

Good Chemistry: Methodological, Ethical, and Social Dimensions
By Jan Mehlich
© Jan Mehlich 2021
Published by the Royal Society of Chemistry, www.rsc.org

Key Themes: Scientific misconduct (fabrication and falsification of data, plagiarism, sabotage, conflicts of interest); Reasons for fraud (greed for fame, power, money; institutional pressure; pride); Whistleblowing.

Learning Objectives:

The goals of this chapter are that you:

- Are aware of the possible forms of misconduct that are frequently reported and that are constantly around the corner as options for chemical practitioners;
- Won't fall into the trap of believing that fraud and misconduct can bring you any benefit;
- Have the competence to identify misconduct in your direct environment, for example in the lab where you are working or in research papers you are reading;
- Know how to address forms of misconduct and make convincing arguments that explain in which way they are wrong;
- Can bring forward these arguments as a whistle blower without risking your own integrity.

6.1   Introductory Cases

Case 6.1: Bengü Sezen

[Real case] Reports from Columbia University and the Department of Health & Human Services (HHS) show a massive and sustained effort by Bengü Sezen over the course of more than a decade to dope experiments, manipulate and falsify NMR and elemental analysis research data, and create fictitious people and organisations to vouch for the reproducibility of her results. She was found guilty of 21 counts of research misconduct by the federal Office of Research Integrity (ORI) in late 2010. Amongst other things, she logged into NMR spectrometry equipment under the name of colleagues, merged NMR data and used correction fluid to create fake spectra showing her desired reaction products. How is this possible, both in terms of Sezen's personality/psyche and in terms of peer assessment by colleagues, co-workers, supervisors or manuscript reviewers?

Case 6.2: The Baltimore Case

[Real case] In the late 1980s, MIT Prof. David Baltimore's PhD student Thereza Imanishi-Kari and postdoctoral fellow Margot O'Toole had a disagreement on the validity of data that served as the basis of one of Imanishi-Kari's papers. O'Toole believed to have found discrepancies on 17 pages of one of Imanishi-Kari's lab notebooks. In the first instance, Imanishi-Kari was found guilty of 'serious scientific misconduct', leading to the retraction of the paper. Yet, a federal prosecutor decided not to prosecute her for fraud because of the complexity of the case making it impossible for the jury to judge. Not much later, the ORI concluded that she did fabricate data. The HHS barred her from receiving federal funds for 10 years. However, it revoked this decision after an HHS panel found that the ORI didn't prove misconduct. Why is this case so complicated? What role does record-keeping play here?

Case 6.3: Millikan, the Betrayer?

[Real case] In 1978, historian Gerald Holton reported that Millikan omitted data without apparent reason when publishing his famous oil drop experiment in which he determined the charge of an electron. This could be a case of fraud. Indeed, out of the 107 valid drops listed in his lab notebooks, he reports only 58 in the publication. 27 of those not published have been dismissed after calculating e. For at least five of these Millikan's notes don't explain his decision to omit them. Critics claim that the only apparent explanation for the omission is that they didn't fit Millikan's expectation. How can it be decided whether this is a case of misconduct?

6.2 Scientists Behaving Badly

The virtues introduced in Chapter 5 describe an ideal. We may assume that the majority of scientific researchers comply with these virtues and maintain a reasonable degree of scientific integrity and good scientific practice. Deliberate scientific misconduct is the exception rather than the rule. Yet, as in every large community, there are black sheep. Of course, it is the goal of a book like this one to prevent the reader from even considering misconduct and becoming such a black sheep. Yet, more importantly, it is our goal to equip the reader with the competence to identify misconduct in his or her surrounding, and to know strategies to intervene or find the right support to make wrong-doers get back on the track of good scientific practice. Why is that important? Two reasons seem significant.

The first concerns the community-internal responsibility. Chemistry as either scientific research or part of innovative research and development (R&D) is built on methodological and science-theoretical foundations. As seen in Chapter 3, critical scrutiny by peers and colleagues must be considered an essential element of this methodology as no individual is free from erring. This control instance operates in two ways: it helps to evaluate and improve the epistemic output of scientific inquiry on the content level, and it establishes feedback and review on a practical methodological level. Here, the latter is of interest: chemists judge whether fellow chemists *play by the rules* that arise from the science-theoretical foundations of chemistry in the context of science, research and innovation. This is the *profession ethics* dimension of good scientific practice. Undermining the epistemic rigidity of scientific claims by intended misconduct, deliberate sloppiness, or careless inconsiderateness damages the chemical community by requiring extra efforts and additional labour, resources and time to correct the misleading or erroneous results produced by this inappropriate behaviour.

The second reason why the compliance with and maintenance of scientific integrity is important is an *external* one: scientific misconduct undermines the credibility and authority of scientific inquiry as an important social sphere. In recent years, the public awareness of scientific fraud has grown significantly, amplified by media coverage and more frequent reporting of scandals in newspapers, on TV and in the world-wide-web. Science as an institution is under public scrutiny, more than ever before. Generally, speaking for Europe, the trust in science is still high, but it seems to be crumbling. This consideration goes beyond the potentially harmful effect of fallacious and deliberately faked scientific insight like covering up the toxic effects of chemicals or the mistakenly claimed properties of new materials for construction or engineering. It addresses the role of scientific knowledge and the public trust in it as an indispensable remedy in an age of post-factualism and opinionated ideology. Scientific misconduct does not only put the wrongdoer and his or her career at risk. It also damages the scientific community as such by contributing to the public impression that scientific expertise and competence are no longer epistemic authorities. The effect on policy-making, education, society at large, and on the environment can be devastating.

It is clear, now, that scientific misconduct is problematic in two ways: false scientific insights cause harm when others rely on them; and the fact that chemists deliberately produce them causes harm to the chemical community as such. The key word, here, is *deliberately*. Misconduct needs to be distinguished clearly from error or failure. As discussed in Part 1 of this book, error and failure are inevitable elements of scientific methodology and, thus, make scientific research a community endeavour involving discourse, scrutiny, peer review and cooperation. We usually grant researchers and scientists the possibility of failure because doing research is always also a learning process. In Chapter 5, we have elaborated a virtuous attitude towards scientific conduct that addresses this possibility: with truthfulness, objectivity, dedicated disinterestedness and systematised doubt, the researcher makes sure that scrutiny and discourse play out as the efficient epistemic instruments that they should be. False insights produced deliberately, on the contrary, aim at the opposite. They follow an interest other than truth, often try to deceive, and lack a healthy degree of (self-)scrutiny. It is the intention that makes the crucial difference.

This intention comes in degrees, ranging from downright criminal intent to rather sloppy yet fully aware inconsiderateness. In an influential article published in *Nature* in 2005, the authors provided a list of 16 forms of scientific misconduct that serves well as an overview of common violations of the codes of good scientific practice.[1] In the authors' view, the items are ranked by severity of the fraud, with No. 1 being the most serious and No. 16 a comparably light form of misconduct.

1. Falsifying or 'cooking' research data.
2. Ignoring major aspects of human-subject requirements.
3. Not properly disclosing involvement in firms whose products are based on one's own research.
4. Relationships with students, research subjects or clients that may be interpreted as questionable.
5. Using another's ideas without obtaining permission or giving due credit.
6. Unauthorized use of confidential information in connection with one's own research.
7. Failing to present data that contradict one's own previous research.
8. Circumventing certain minor aspects of human-subject requirements.
9. Overlooking others' use of flawed data or questionable interpretation of data.
10. Changing the design, methodology or results of a study in response to pressure from a funding source.
11. Publishing the same data or results in two or more publications.
12. Inappropriately assigning authorship credit.
13. Withholding details of methodology or results in papers or proposals.
14. Using inadequate or inappropriate research designs.
15. Dropping observations or data points from analyses based on a gut feeling that they were inaccurate.
16. Inadequate record keeping related to research projects.

A form of misconduct that is not in this list, but has received increasing attention in recent years is the sabotage of colleagues' work (experimental setups, lab equipment, data, IT, *etc.*).[2,3] Not all these behaviours are of significance for chemists (for example, Numbers 2 and 8), but most play a role in chemical research. Whereas the Top 8 forms of misconduct are unmistakably identifiable and, once caught, the subject of unambiguous judgment, the misconduct types further down on the list are often difficult to evaluate. For the sake of clarity, and as a structuring element for this chapter, at least three levels of misconduct need to be distinguished in order to address each of them properly and with goal-oriented remedies:

- *Intended Fraud*: Betrayal, cheating, manipulation, deception, stealing, deliberate wrong-doing.
- *Biased Motivation*: Interest-driven, benevolent, opportunistic, non- or pseudo-scientific acquisition, processing and interpretation of data.
- *Deliberate Sloppiness*: An inconsiderate attitude that does not take professional responsibility serious enough.

Fraud committed with criminal intent, fully aware of the violation of the codes of good scientific practice, can be sanctioned or punished. Identifiers of such cases of misconduct – lab mates, colleagues, fellow researchers, readers of manuscripts and published articles – are supported by institutional ethics boards and legal instances. We will address whistleblowing and other preventive measures in Section 6.6. Biased motivation and deliberate sloppiness are misconduct in the sense that they ignore the researcher's responsibility to apply scientific rigour to experimenting, data handling, interpreting, and communicating results. As it must be assumed that a researcher knows the standards of good practice in his or her profession, acting against them cannot be excused by ignorance or accident. The researcher may be held fully accountable for his/her bias and lack of methodological robustness. In case of interest-driven and benevolent scientific judgment, possibly based on unsound methodological approaches and flawed or insufficient data, the judgment of misconduct can be easy. Yet, there is a large grey zone in which it is not so clear whether a scientist may be accused of misconduct. Are inconsiderateness and epistemic carelessness really misconduct or rather just poor skill?

Before going into detail to find an answer for this question, it is insightful to gain an idea of the quantity of scientific misconduct. In the above-mentioned essay from 2005, Martinson and colleagues published the results of a survey with 3247 scientists on their scientific practices. 33% admitted having committed at least one of the 16 forms of misconduct listed above, with the majority of these naming one of the items 13–16 as their wrong-doing.[1] A 2009 study found 2% of surveyed researchers had committed fraud in the form of falsification or fabrication of data, the more serious forms of violation of professional codes of science conduct.[4] Later studies confirm these findings.[5] These results suggest that severe cases of intended fraud are still an exception, but that many scientists operate in a grey zone of what they believe is allowed or slightly (and acceptably) beyond that. If we agree that both the number of cases of fraud and that of misconduct from bias or deliberate sloppiness are too high, two interesting questions arise:

1. What are the reasons for scientific researchers to violate their own standards of professional conduct? Is something fundamentally wrong with the standards (too demanding, perhaps), or are the incentives for compliance with these standards too little, or is something structurally flawed in the way the scientific community (academic and/or industrial) is constituted, organized and/or administrated? This is explored in Section 6.5.
2. How can scientific misconduct be prevented, and scientific integrity be maintained? Are there remedies (for example, effective sanctions and punishments, or institutionally protected peer assessment and whistleblowing)? What can white sheep do to turn grey sheep white before they go black? Section 6.6 introduces strategies as the skills of a good chemist.

6.3 FFP Cases

Speaking of scientific misconduct, most scientists think of the so called FFP cases. FFP stands for **f**abrication, **f**alsification and unauthorised copying (**p**lagiarism) of data and text.[6]

Some prominent cases like that of Korean geneticist and biochemist Hwang Woo-Seok who fabricated an enormous amount of data

> **Box 6.1** Definition: FFP
>
> *Fabrication:* Invention of results. Example: manually creating a dataset of a measurement that has never been performed.
>
> *Falsification:* Data manipulation. Examples: ignoring or deleting undesired data points or entire sets, manipulation of devices and experimental setups, manually editing analytical results (spectra, images, *etc.*).
>
> *Plagiarism:* Copying, using and/or publishing other people's intellectual property (for example, data or text) without allocating credit, pretending it is one's own. Examples: stealing someone's research idea, using someone's data, copying text passages into one's own research paper manuscript.

in order to keep the illusion of the correctness of his revolutionary research alive[7] are obvious and clear in ethical evaluation. Many more examples can be found on internet platforms such as pubpeer.com or on science blogs (for example, forbetterscience.com or scienceintegritydigest.com). Here is a blatantly obvious example of result fabrication by photoshopping from a group around Indian scientist Prashant Sharma. In a 2017 paper,[8] an image was reported to be flawed (Figure 6.1(1A), top left), and Sharma submitted a corrected image (top right). Claiming it to be a different TEM image of gold nanospheres, it is obvious that the distribution pattern of the dots is exactly the same as in the original figure, just a bit zoomed out and each dot slightly less bold compared to the first version. The chance for two different TEM scans of different samples revealing exactly the same distribution pattern is very close to zero. Even more astonishingly, the same pattern can be found in images of a completely different paper in which the dots are claimed to be $MoSe_2$:CdS and WSe_2:CdS composite nanodots.[9] On top of that, the backgrounds show the same impurities as in some figures of that first paper (Figure 6.1C, bottom right), for example those gold nanorods that – no surprise – also all look exactly the same (just randomly rotated).[8] There is only one explanation for it: Photoshopped (not even very well done) from a collection of background images and randomly rotated or scaled multiplication of single dots, spheres and rods. Besides the downright criminal character of this type of fraud, we may also wonder how it is possible that obvious fabrications like these can pass a review process.

Case 6.1 is similar.[11] Using correction fluid to fake NMR spectra appears even more crude than poorly skilled digital manipulation of images. There is no doubt that Sezen's behaviour is ethically despicable. Her fraud translates into stealing (using colleagues' identities to get access to their equipment and data), deception (faking spectra) and lying (creating fictitious people and organisations for the illusion of her results being reproducible). Here are a few more prominent cases:

- *Falsification/manipulation*: Guido Zadel claimed in his doctoral thesis and an article published in Angewandte Chemie[12] to have found a way to influence the enantioselectivity of addition reactions (alkylation of aldehydes, reduction of ketones) by subjecting the reaction to a strong magnetic field. After none of his results could be reproduced, it was found that his experiments were manipulated. His article was retracted (the German version is still available with no retraction note, however) and he lost his PhD degree.

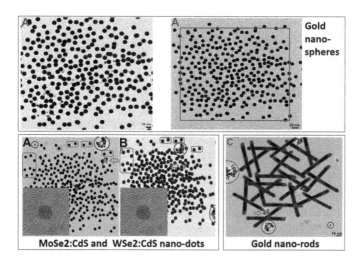

Figure 6.1 Example of fraud. Images from three different research papers and contexts showing the same elements. Top: Figure 1A (left) from ref. 8 and a 'correction' of it submitted later (right), claimed to show gold nano-spheres. Bottom left: Figures 3A and B, claimed to show different composite nano-dots, the same distribution as in the images from ref. 8, the same background dirt, but in different places. Bottom right: Figure 1C from ref. 8, claimed to show gold nano-rods, identical rods in different orientations, with the same background dirt as in the images from ref. 9. Compiled and marked by user '*Hoya Camphorifolia*' on pubpeer.com.[10] Figures 1A (top left, right) and Figure 1C (bottom right). Reproduced from ref. 8 with permission from American Chemical Society, Copyright 2017. Figure 3A and B (bottom left, middle) reproduced from ref. 9 with permission from Royal Society of Chemistry.

- *Sabotage*: As a post-doc, Vipul Bhrigu sabotaged the research of PhD student Heather Ames by secretly pouring alcohol into her highly sensitive samples and switching the labels of probes. He was caught on camera and convicted of a misdemeanour. The judge sentenced him to a psychological evaluation, to probation and to pay 30 000 US$ in restitution to the University of Michigan. He told the group leader Prof. Theodora Ross: "*I thought Heather was so smart and I did it to slow her down. It was because of my internal pressure.*"[13]
- *Fabrication:* H. Zhong, T. Liu and several of their colleagues at Jinggangshan University (China) have retracted at least 70 papers published in Acta Crystallographica after it was revealed that the data had been taken from other valid structures that had been altered by substituting atoms, resulting in impossible or implausible structures.[14]

- *Plagiarism*: Indrajit Chakraborty, Assistant Professor in Organic Chemistry at Malda College, India, copied a research paper including its images and figures from the lab of Alexis Ostrowski and published it in an Indian journal. The original paper is not at all mentioned in the plagiarised one.[15]

All these cases have in common that their ethical evaluation is unambiguous and clear: these behaviours violate even the most basic codes of conduct (not just professional standards). Moreover, in one way or the other, they are all motivated by expected personal benefits, mostly in terms of the wrong-doer's career. The sabotage case might be an exception, here, as it is not very clear how the saboteur has any personal gain from his actions. Plagiarism cases overlap to some extent with publication issues that we will discuss in Chapter 7. Yet, as such, they are clearly fraud in the form of stealing and, thus, listed in the context of the most severe forms of scientific misconduct. FFP cases are always intended fraud (the first of the three degrees of scientific misconduct outlined in Section 6.2) because they must be committed fully aware of their inappropriateness and the possible consequences.

6.4   Grey Zone Cases

As explained above, scientific misconduct goes beyond clearly ruthless fraud. When scientists evaluate and analyse their data, they find data points, spectra or images that do not fit into the measurement series, that do not look presentable, or that contradict plausible expectations. There is a very large *grey zone* of debatable omission of data in the context of scientific reasoning and interpretation! The reflections in Part 1 of this book have shown that all scientific thinking is embedded in presuppositions and cognitive frameworks that either constitute plausible grounds for operation or, potentially, a threat for epistemic integrity from biases and fallacies. When does *manipulation of data* start? It must not always be the intended manipulation of a device or the direct fabrication of results. There are countless ways for researchers to process data benevolently within the boundaries of scientific standards. The sophistication of spectrometers and other imaging devices as technical extensions of our limited senses turns them more and more into black boxes. This convenience bears the danger of a temptation to interpret pictures benevolently and in accordance with the expectation (or the desired finding) rather than with the necessary critical analysis – a perfect example of the impact of philosophy of science on the ethical conduct of science. *Trust* – in oneself, one's colleagues, the applied methodologies, experimental setups, the equipment and the technical devices – and *good will* are non-scientific categories that are subtly pervading all research activity, yet extremely difficult to evaluate in scientific terms.

The logic of the argument is this: scientists should be aware that epistemic agency (here: knowledge production) bears the risk of biases and fallacies while, at the same time, it is their professional responsibility to avoid these biases and fallacies. When a scientist ignores this and, instead, undermines scientific rigour by applying interest-driven, opportunistic, goal-oriented reason, where he or she could do otherwise, it deserves the label *misconduct*. If, on the other hand, there are scientific reasons for decisions on data processing (acquisition, calculations, illustrations, interpretation) and these reasons are sound (which means they withstand critical scrutiny), scientific integrity is maintained.

A textbook example for this grey zone of acceptable or unacceptable data omission is Millikan's oil drop experiment with which he determined the charge of an electron (see Case 6.3). In 1978, historian Gerald Holton reported that Millikan omitted data without apparent reason, which could be a case of fraud.[16] A deeper analysis of his lab books revealed the following (according to Franklin, 1981):[17]

- Total number of drop experiments in the notebooks: 175
- Number of valid drops: 107
- Of valid drops, number published by Millikan: 58
- Of valid drops, number excluded by Millikan: 49
- Of valid drops, number excluded with no calculation: 22
- Of valid drops, number excluded after calculating *e*: 27

It is the last 27 drops that are of interest here: are they omitted because their calculated values were in conflict with Millikan's expected findings? Franklin found:

- 12 drops: inappropriate pressure and radius (would require 2nd order correction).
- 2 drops: on experimental grounds.
- 5 drops: too few reliable measured values of t_f (rising time).
- 2 drops: no apparent reason (probably not needed for calculating *e*).
- 1 drop: "anomalous" (Franklin provides no further information).
- 5 drops: Franklin: *"His only evident reason for rejecting these five events is that their values did not agree with his expectations."* (ref. 17, p. 195).

There is an obvious difference between experienced and competent scientific judgment on whether an experimental event may count as valid or – for experimental, technical, circumstantial, or alike reasons – should be ignored in the analysis, and result-oriented selection of favourable data. Defenders of Millikan claim that he did the former: his profound expertise as an experimenter provided him with the necessary scientific intuition and the ability to make reliable and justified judgments. Critics focus on those last five points: if it is true that he omitted them because they would have made his calculations more susceptible for attacks and criticism, then it would be a case of scientific fraud. Yet, it is only five values, not the 49 claimed by Holton. Others have shown that Millikan's final value for *e* would not have been different had he included those five measurements, but that only the standard mean value would have been bigger, making his value an easier target of criticism.

The basis for the dispute about the Millikan case are Millikan's notebooks in which he documented his experiments. As mentioned in Chapter 3, keeping a complete and clear record of one's lab activities and experimentation is a crucial element of scientific methodology. Here, we see the ethical dimension of that claim. Researchers held accountable for what they were doing (we will see in Chapter 12 that accountability is the past-oriented ascription of responsibility) need to be able to provide evidence for the scientific plausibility of their data operation decisions. In extreme cases, this can decide research integrity judgments in court. A prominent historical example is the incident in the lab of Prof. David Baltimore at MIT (Case 6.2 above).[18,19] Over almost a decade, this case has been debated, first in Prof. Baltimore's group, then at the institute's ethics board, at the ORI, the HHS, and in the court. This case is so complicated because it could not be decided whether Imanishi-Kari committed fraud or was just sloppy with her lab books and hasty with her conclusions. This uncovers some very basic questions of scientific agency and how we understand our profession:

- *Can and Ought Scientists Police Themselves?* The whistle blower in this case, Dr O'Toole, was involved to a much larger degree than she could imagine when she raised her concerns in front of principle investigator (PI) Baltimore, the institute, and the legal instances. All she wanted was an honest discussion and clarification of what she found to be irregularities. This case may discourage peers to scrutinize colleagues' works even though this is a crucial element of scientific practice. Scientific discourse does not mean *policing each other.* Yet, it is clear that misconduct prevention starts with internal control and attention. These mechanisms work well when teams collaborate well and with trust. Here, personal conflicts and, perhaps, simply antipathy to each other interfere with finding reasonable and efficient solutions to cases of 'irregularities'.

- *Who, if Not Scientists, Should Police Scientific Practice?* Legal organizations and even the secret service have been involved in this case in quite sophisticated detail, for example the examination of Imanishi-Kari's thin layer chromatography plates, only to find after almost 10 years that the case is too complex for them to judge. The inefficiency of legal investigations into practical matters of doing experiments suggests that internal peer review of scientific methodology and reasoning is still the more appropriate choice. Only when a criminal intent is reasonably suspected a case should be subjected to legal investigation.

- *What are the Legitimate Rights and Responsibilities of Researchers in the Natural Sciences?* The right of fair unbiased treatment and judgment does not free the researcher from the duty to respond aptly to scrutiny and accusations of improper conduct. In this respect, it is certainly within the researcher's responsibility to be ready at any time to provide evidence for one's scientific claims and the integrity of one's methods of acquiring that evidence. A special responsibility lies on team leaders, here represented by Prof. Baltimore. Mentorship and leadership in the context of a collaborative endeavour like chemical research will be discussed in greater detail in Chapter 8.

6.5 Reasons for Misconduct

Why would scientists be susceptible to improper conduct of research or even fraud? Apparently, there is a lure or an incentive that makes the originally innocent scientist believe that a certain goal can be reached by choosing misconduct over scientific integrity. From another view, there might be a lack of control or force that ensures that researchers follow the codes of good scientific practice under all circumstances. These two perspectives represent the major two approaches to the role of governance and the nature of humankind: basically, one view holds that humans are intrinsically good and virtuous, and if they commit wrong or unethical deeds there must be something that interferes with human goodness (remedy: eliminate the lure towards vice and crime by offering stronger incentives for goodness); and the other view regards human nature as intrinsically evil (for example, self-interested, non-cooperative, greedy, protective) and people must be kept on the track of goodness by force of law and order. We need both views to understand the motivation to ignore guidelines of research integrity.

Many of the scientists convicted of fraud claim that they were driven by an immense pressure. This pressure can be caused by various influences. Basically, scientists in both academia and industry feel the need to succeed on the content level of their research: They need to produce good results. Sometimes, this pressure is real in the sense that milestones of a collaborative research consortium have to be achieved in certain time frames, determined and written down in contracts and agreements. As will be outlined in Chapter 8 in greater detail, the competitive environment in private sector R&D creates time and resource constraints for research groups. But also in academia, the struggle for research grants and the fast-paced progress in the international scientific community put individual scientists under stress, especially in their early career stages. In the case of bias-driven ill-motivation, of course, the bias is the superficial cause of the misconduct: being interested in personal fame, financial reward, a smooth career, impressive CV, or confirmation of (scientific) worldview – interests other than a purely epistemic-scientific one – may make researchers lose their professional commitment to good scientific practice. Yet, on a deeper level, we may ask what causes these biases. Here, the character or personality of the researcher plays a role, and it is often ego, pride and/or too high self-expectations that make scientists engage in misconduct. Students are a special case: Diploma, Masters or PhD students feel pressure to achieve good grades with their thesis, so they feel like they have to obtain good results in their research project, which is often limited in time.

The suggested image, here, is that scientific researchers would, actually, like to maintain a high professional integrity but fail because the way science is organised is fundamentally flawed. Funding practices, institutional customs, limited resources (including money and workforce), and the education system (in the case of university research) as elements of the system *science* make it impossible to (always) reasonably stick to the virtues of scientific practice. Endorsers of this view claim that something has to change structurally in the scientific community and its governance, either on the small scale (for example, the peer review system of publishers, or the custom of measuring research output quality by impact factors of academic journals) or on the large scale (for example, allocating more public funding to academic science, or reforming the higher education system). While many possible plausible scenarios that incentivise good scientific practice by reducing pressure can be constructed, this view bears one danger: understanding misconduct solely as the effect of internal and external pressure might degenerate this cause into an excuse. Understanding a behaviour's root causes that lie outside of the control of the subject often lead to sympathy with the culprit and an easier forgiveness of the misconduct. In fact, it leads peers to remain silent over observed misconduct because "*he is, actually, a good person, but the pressure is just too high! In his position, who wouldn't...?!*". Fraud and other forms of misconduct are still wrong, no matter how much we sympathise with the afflictions of the wrong-doer.

Moving beyond pressure as the sole reason for misconduct, we turn to the second possibility: improper behaviour as the result of a lack of control, regulation, sanctioning, or punishment. If we accept it as a fact that scientists under pressure occasionally feel urged to commit fraud, we may feel the need to prevent that by implementing mechanisms that stop them from doing so. These can range from group-internal guidelines to institutional organisations (ethics boards, offices of research integrity) to legally binding publishing policies (like the duty to provide disclosure statements) to laws and their rigorous execution (for example, loss of a PhD title when being convicted of plagiarism) to obligatory research ethics courses in the curricula of science majors (such as a course on '*Good Chemistry*').

Not all of these measures are effective.[20] Empirical studies indicated that formal ethics training is ineffective in the prevention of misconduct.[21,22] Some overshoot the mark by undermining trust and stigmatising all scientific researchers as intrinsically ill-motivated. Moreover, the simplistic *keep-them-on-track* rhetoric of the law-and-order-approach underestimates the real complexity of the misconduct causes. Many reported cases suggest that – in view of the expected prominent visibility and application of (fabricated or falsified) findings and the subsequent fame, such as Hwang's case – the researchers committing fraud must have been fully aware that their misconduct would be uncovered, hinting at pathological behaviour that requires therapy and treatment rather than punishment and dismissal from their academic positions. This attitude brings to mind a phenomenon called *paraphrenia*: even though there is clear evidence, the culprit is downright convinced that he/she didn't commit fraud, as if the act is entirely deleted from the memory by explaining it away with rational reason.

6.6 Prevention of Misconduct, Whistleblowing

In the previous section we reflected on causes of misconduct, and, depending on that, countermeasures have been suggested, for example, reduction of pressure or implementation of institutional policies and research integrity boards. Some of the suggested responses apply *after* a case of misconduct has been uncovered but have little preventive power. In this section, we want to learn what an individual is empowered to do to prevent oneself and, perhaps more importantly, others from entering the realms of scientific misconduct.

It was stated that ethics courses are of limited success. The effect on the prevention of misconduct could simply not be observed. Yet, even if (or, perhaps, *especially* if) it does not have the power to outweigh a fraudulent intention, it may facilitate the competence to identify misconduct and speak up against it. A practice-oriented course on scientific integrity sharpens the awareness of scientific misconduct and the sometimes rather hidden and subtle forms of it. This has at least three effects, all of which have to do with discourse skills:

1. The mindful chemical scientist maintains a professional attitude in which decisions are made and operations are executed under consideration of the virtues that constitute and secure the very foundation of the profession. This should be understood as an **internal discourse** and continuous scrutiny of personal goals, motivations and mindsets. *Why am I inclined to discard this measurement series? What effect does the way I describe my research project have on the potential collaborator at the conference? Does my data omission have scientific reasons or am I opportunistic in some way? Do I want to add that author to my paper because I hope to benefit from this favour or because it truthfully represents an important contribution?* Keeping a neutral distance from one's own motivations increases the chance of identifying biases and unscientific motives.

2. As explicated in detail in Chapter 8, chemistry is almost always a community endeavour. Within teams, the customs and practices should be discussed on both the content level (the research topic itself, the meaning of observations, the interpretation of data) and the methodological level (what methods are applied and why, what additional tests might be necessary, where do pitfalls and errors lurk). Facilitated by good leadership, groups build trust that allows the members to distinguish critical issue-related and

goal-oriented feedback from personal attack and resentment. Tendencies towards improper research conduct by a few can be corrected easily and quickly by the majority that endorse the virtues of science and promote them *via* **group-internal communication.** Here, these virtues serve as a tool for argumentation. Someone who is not sure how to address the issue of apparent misconduct may gain logic clarity and confidence by assessing the criticised practice on the basis of good scientific practice guidelines. This prevents the whistle blower from appearing like a troublemaker or as having a personal aversion against a teammate. Moreover, hierarchic obstacles in bottom-up communication (from student to professor, or from employee to boss) are easier to deal with when the inferior prepares a statement based on scientific reason and pragmatic plausibility.

3. A third competence is the stirring of a fruitful **discourse across the community** (here: the community of chemists, or chemical researchers, or academic chemists, *etc.*) with the goal to induce changes that effectively reduce the incentives for scientific misconduct and substitute them with more acceptable customs and approaches. A responsible chemist with convincing and plausible ideas concerning peer review, merit allocation, harassment protection, fair funding that endorses academic freedom, or the administration of large projects in science-industry collaboration certainly has some influence on the customs and self-set guidelines within the chemical community and beyond.

Taking the role of a whistle blower deserves a few more detailed considerations.[23] Many scientists, especially science students and early-career researchers, worry about retaliation in the form of civil lawsuits (for example, for slander), being fired or being black-listed.[24] After stepping up and reporting misconduct, the accused may try to defend him- or herself by instituting a lawsuit for slander. Yet, whistleblowing, if done well, is one of the best protections against misconduct. *Done well*, here, means in the interest of science, focused on finding a pragmatic solution, free of personal attack, exaggeration, evidence-lacking accusations and reproaches. A good whistle blower is well prepared, informed about legal and organisational options for further action, emotionally calm, and quick-witted enough to respond to (occasionally unfair) counterattacks. A whistle blower should seek support and avoid acting alone. If, for example, a lab mate is suspected to have committed some kind of misconduct, the first observer should first address the suspect directly and, if that hardens the suspicion but leads to no solution, inform other trusted co-workers. With sufficient evidence, the group as a whole may then approach the PI or head of the lab. If something is at stake, for example a grade, a research grant, a position, a career, or a research collaboration, it is advised to address these justified worries directly when filing the misconduct report. Point out that the goal of the whistleblowing is scientific integrity and concern about the research group's reputation and professionalism, and that it is hoped that the action taken has no repercussions on the research progress and integrity of the whistle blower(s). Usually, the authority (professor, boss) will understand the situation. In any case, being firm and factual rather than emotional and personal will definitely help the whistle blower to be taken seriously and be institutionally protected. This kind of conversation can and should be practiced ahead.

It is worthwhile at this point to reiterate the line of reasoning again: starting from the philosophical foundations of science and its power

---

Box 6.2 Exercise: Whistleblowing

Imagine a case of scientific misconduct in your laboratory or in your research consortium, or, if that is the case, unfortunately, take a real case of misconduct happening in your research unit. You are the first person to suspect a colleague or lab mate of misconduct, and you have evidence for it (for example, a suspicious spectrometric result for this week even though the department's spectrometer has been out of service since last week).

Step 1: Face the suspect and inquire about your observation. Find out if the whole story is, perhaps, a misunderstanding. Prepare questions and statements beforehand. Try to start every sentence with "I" (*I have observed that... I got the impression that you... I think we need to talk about...*) and avoid "You"-claims (*You did...! Your results are...! You must explain why and how...!*). If the suspect cannot respond convincingly to your accusation, proceed to the next step.

Step 2: Talk to peers whom you trust. Explain the situation. Try to be diplomatic, still assuming the possibility that you are erring. Convince them to keep an eye on the case together with you and collect more evidence. Extra challenge: Consult someone in the team who is rather close with the suspect and will probably side with the suspect. Reconfirm that your goal is scientific integrity, not personal attack.

Step 3: Inform the PI or head of the lab/project. Prepare well for this conversation. If helpful, write down the statement, either as a practice or as a document for the records. Play through different scenarios (boss is open and requests more input, boss gets upset and questions your integrity, boss turns out to be part of the misconduct case, boss wants to make you a mole in the team, *etc.*) and formulate appropriate reactions. Remember: the goal is to keep your own face while at the same time stand firm in your defence of scientific integrity. Extra challenge: Insist on having this meeting including the suspect.

and limits we agree on a methodology that deserves the label *scientific*, and we ascribe the responsibility to its practitioners to manifest and embody the purposes and goals of science by committing themselves to the virtues that support this methodology with the unbiased attitude of an 'ambassador of science' in each and every professional decision or act. It is hoped, of course, that good scientific practice can be maintained out of an ethical motivation of scientists, and not out of fear of sanctions and penalties. The scientific community should set incentives for ethically sound conduct of research (from grant application to publishing), so that reasons and motivations for intended fraud are reduced to a minimum.

Exercise Questions

1. Which of the following terms is not included in the FFP-definition of scientific fraud?

   A: Fantasy
   B: Falsification
   C: Plagiarism
   D: Fabrication

2. Which of the following is unlikely to provide a reason for scientific misconduct?

   A: Greed
   B: Pressure from funding sources
   C: Ambition
   D: Professionalism

3. Paraphrenia is associated with:

   A: Delusions of grandeur.
   B: Manipulating results by photoshopping.
   C: Self-denial in the face of convincing evidence to the contrary.
   D: Multiple publications of the same basic data.

4. Which of the following factors are likely to promote scientific misconduct by students?

   A: Time pressure.
   B: Laziness.
   C: The need to succeed.
   D: All of A–C.

5. According to empirical studies, which of the following would be unlikely to decrease the incidence of scientific misconduct?

   A: Institutional controls.
   B: Effective peer-review procedures.
   C: Courses on ethical behaviour.
   D: Rewarding ethical behaviour.

6. According to surveys (*e.g.*, Martinson *et al.*, 2005), what percentage of scientists, within the established scientific community, have committed scientific misconduct in one or the other form?

    A: 100%
    B: ≥30%
    C: 10%
    D: ≤2%

7. Fabrication means:

    A: Ignoring inconvenient results.
    B: Manipulation of instruments.
    C: Invention of results.
    D: Data manipulation.

8. Among the following reasons for discounting experimental data, which one would be considered questionable or "bad scientific practice"?

    A: Instrument malfunction.
    B: Without the data, a curve looks smoother and more convincing.
    C: A controlled variable was accidentally not kept constant during the experiment.
    D: Data were obtained using a deviation from standard procedure.

9. Which of the following practices might be considered unethical?

    A: Withholding details of industrial sponsorship.
    B: Using data provided by a sponsor without acknowledgement.
    C: Suppressing experimental data to please a sponsor.
    D: All of A–C.

10. Which of the following constitutes the most serious type of scientific misconduct?

    A: Averaging results.
    B: Inappropriately assigning authorship credit.
    C: Invention of results.
    D: Inadequate record keeping.

References

1. B. C. Martinson, M. S. Anderson and R. de Vries, Scientists Behaving Badly, *Nature*, 2005, **435**, 737.
2. B. Maher, Research Integrity: Sabotage!, *Nature*, 2010, **467**, 516.
3. E. Wallace, M. Hogan, C. Noone and J. Groarke, Investigating components and causes of sabotage by academics using collective intelligence analysis, *Stud. High. Educ.*, 2019, **44**, 2113.

4. D. Fanelli, How many scientists fabricate and falsify research? A systematic review and meta-analysis of survey data, *PLoS One*, 2009, **4**, e5738.
5. B. Hofmann and S. Holm, Research integrity: environment, experience, or ethos?, *Res. Ethics*, 2019, **15**, 1.
6. J. D'Angelo, *Ethics in Science. Ethical Misconduct in Scientific Research*, CRC Press, Boca Raton, 2012.
7. R. Saunders and J. Savulescu, Research ethics and lessons from Hwanggate: what can we learn from the Korean cloning fraud?, *J. Med. Ethics*, 2008, **34**, 214.
8. E. Roy, S. Patra, R. Madhuri and P. K. Sharma, Anisotropic Gold Nanoparticle Decorated Magnetopolymersome: An Advanced Nanocarrier for Targeted Photothermal Therapy and Dual-Mode Responsive T MRI Imaging, *ACS Biomater.-Sci. Eng.*, 2017, 2120–2135.
9. P. Karfa, R. Madhuri and P. K. Sharma, Multifunctional fluorescent chalcogenide hybrid nanodots ($MoSe_2$:CdS and WSe_2:CdS) as electro catalyst (for oxygen reduction/oxygen evolution reactions) and sensing probe for lead, *J. Mater. Chem. A*, 2017, **5**, 1495.
10. https://pubpeer.com/publications/33CB940D42723BAB8B4F737B434E86, accessed on August 10 2020.
11. W. G. Schulz, A Puzzle Named Bengü Sezen, *Chem. Eng. News*, 2011, **89**, 40–43.
12. G. Zadel, C. Eisenbraun, G.-J. Wolff and E. Breitmaier, Enantioselective Reactions in a Static Magnetic Field, *Angew. Chem. Int. Ed.*, 1994, **33**, 454.
13. T. Ross, *A Crime in the Cancer Lab*, The New York Times, published on January 28 2017, https://www.nytimes.com/2017/01/28/opinion/sunday/a-crime-in-the-cancer-lab.html, accessed on August 25 2020.
14. W. T. A. Harrison, J. Simpson and M. Weil, Editorial, *Acta Crystallogr. E*, 2010, **66**, e1.
15. E. Bik, *Plagiarism in Chemistry: A Case Report*, Science Integrity Digest (Blog), published on July 28 2020, https://scienceintegritydigest.com/2020/07/28/plagiarism-in-chemistry-a-case-report/, accessed on August 10 2020.
16. G. Holton, Subelectrons, Presuppositions, and the Millikan-Ehrenhaft Dispute, *Hist. Stud. Nat. Sci.*, 1978, **9**, 166.
17. A. Franklin, Millikan's Published and Unpublished Data on Oil Drops, *Hist. Stud. Nat. Sci.*, 1981, **11**, 185.
18. S. Lang, Questions of Scientific Responsibility: The Baltimore Case, *Ethics Behav.*, 1993, **3**, 3.
19. K. Goodman, *The Baltimore Affair*, University of Miami, Institute for Bioethics and Health Policy, https://bioethics.miami.edu/education/timelines-project/the-baltimore-case/index.html, accessed on August 11 2020.
20. National Academies of Sciences, Engineering, and Medicine, Addressing Research Misconduct and Detrimental Research Practices: Current Knowledge and Issues, in *National Academies of Sciences, Engineering, and Medicine. Fostering Integrity in Research*, The National Academies Press, Washington, USA, 2017, ch. 7, pp. 105–146.
21. M. S. Anderson, A. S. Horn, K. R. Risbey, E. A. Ronning, R. De Vries and B. C. Martinson, What Do Mentoring and Training in the Responsible Conduct of Research Have to Do with Scientific Misbehavior? Findings from a National Survey of NIH-Funded Scientists, *Acad. Med.*, 2007, **82**, 853.
22. C. L. Funk, K. A. Barrett and F. L. Macrina, Authorship and Publication Practices: Evaluation of the Effect of Responsible Conduct of Research Instruction to Postdoctoral Trainees, *Account. Res.*, 2007, **14**, 269.

23. M. V. Dougherty, Academic Whistleblowing, in *Correcting the Scholarly Record for Research Integrity*, Research Ethics Forum, Springer, Cham, Switzerland, 2018, vol. 6, ch. 5, pp. 117–151.
24. S. P. J. M. Horbach, E. Breit, W. Halffman and S.-E. Mamelund, On the Willingness to Report and the Consequences of Reporting Research Misconduct: The Role of Power Relations, *Sci. Eng. Ethics*, 2020, **26**, 1595.

7   Scientific Publishing

Overview

Summary: We have learned that communicating scientific findings is a crucial step of scientific methodology. Scientific claims gain their universal validity only through passing critical review by fellow experts. In addition to doing science, *writing science* is one of the main activities of researchers and scientists. Thus, it is not surprising that many cases of scientific misconduct are committed in the context of publishing. Authorship decisions, citation practices and plagiarism, but also peer review and the benefits and dangers of impact factors are frequently discussed among chemists. This chapter focuses on the intra-community aspects of scientific publishing, whereas Chapter 15 addresses issues regarding the public communication of chemistry. The issues in that field are very different.

We will see how the scientific virtues introduced in Chapter 5 can inform decision-making and discourse on publishing issues. Fairness, disciplined self-control, and communalism play the most important role in this context. Yet, self-interests can cause biases that impact the choice of authors for a paper, the choice of references given in an essay, the review process of competitive papers, or publishing practices that increase a researcher's visibility in the form of impact factors. A special topic, here, is the publication of research that has obvious dual-use potential and is, thus, controversially discussed.

Key Themes: Doing science *versus* writing science; authorship; citation practices; peer review; impact factors; predatory publishing; publishing controversial research.

Good Chemistry: Methodological, Ethical, and Social Dimensions
By Jan Mehlich
© Jan Mehlich 2021
Published by the Royal Society of Chemistry, www.rsc.org

> **Learning Objectives:**
>
> This chapter supports you in:
>
> - Becoming aware of publishing-related ethical issues.
> - Learning possible solutions for arising conflicts such as authorship discussions or peer review problems.
> - Applying the virtue approach to publishing-related professional conduct.
> - Becoming a responsible member of the scientific community by engaging in improving the fairness and ethical integrity of practices such as peer reviewing and impact factors.

7.1 Introductory Cases

Case 7.1: The Unknown Collaborator

Belinda just published her first research paper. She wrote the first draft, including the introduction, the experimental section, and the discussion of the results. Her principal investigator (PI) 'polished' the paper and submitted it. After being accepted, she receives the 'proofs', and to her surprise there is an author X on the list of whom she had never heard before. Immediately she consults her PI to ask about X who, according to the information on the paper, works at another university. Her PI says that X is a friend and collaborator who contributed to this paper by inspiring him to do this research and also *"reviewed the introduction and suggested some changes"*. Yet, Belinda finds, the introduction as printed in the proofs is almost 99% identical with her first draft. She assumes that her PI adds X as a personal favour. Yet, as X may increase the visibility of the paper with his name, she doesn't object it. However, when she presents a poster at a conference a few weeks later, another researcher approaches her and inquires about her collaboration with X: *"It is really interesting that you worked with X! What is your experience with him? Have you visited his labs? Pretty amazing equipment, right?"*. She feels embarrassed and has a tough time finding the right replies.

Case 7.2: Temptation

While writing a research paper on the kinetics and mechanism of a novel organic catalysis reaction – a project that required hard work and included a good postdoc and the best PhD candidate – Prof. Smith receives an article to review for the *Angewandte Chemie* journal. To Smith's horror, the paper, written by a senior colleague at another university, reports the same reaction. Apparently, she has been faster! Smith knows that he could simply raise objections to the paper and delay its publication for months, giving time to submit his own article to another journal. On the other hand, working on a very similar project, he knows that the colleague did very excellent work, actually.

Case 7.3: Careful with that Pox, Eugene!

[Real case] Ryan Noyce, Seth Lederman and David Evans published the complete chemical synthesis and fabrication of the horsepox virus in January 2018. Immediately, concerns were raised that this information could be used to construct variola, the virus that causes the eradicated disease smallpox. In a response to the objections, communicated in October 2018, they argue that: (1) available knowledge is not sufficient, but the construction needs skills that require advanced scientific training and insider knowledge that is not widely available; (2) that the benefits of the availability of this synthetic virus (developing vaccines) outweigh the risks; and that (3) there are no regulation issues as the research team has thoughtfully obtained all necessary legal reviews of relevant legislation. Is this convincing enough to dispel the worries concerning bioterrorism and biosafety?

7.2 Doing Science *Versus* Writing Science

As we have pointed out on several occasions in the previous chapters, communication of research, discourse and subjection of knowledge into critical scrutiny are inevitable and integral elements of scientific inquiry. Research is published in order to make it accessible for others for various purposes. The general public as such has an interest in scientific knowledge. It is even regarded as *public property*. Other scientists and researchers are interested in our contribution to knowledge creation. They read and evaluate it and use it for other research processes (remember the web of scientific knowledge acquisition, Chapter 3). As such, publishing one's research is part of an internal control and review system: only when research insights are communicated can they be refined, discussed, and exploited. Last but not least, publications are the merit of a scientist or researcher. It is the final product of our professional activities. Whereas engineers can sometimes see their work manifested in the form of technical artefacts or constructions, scientific researchers see their work manifested in journal articles, books, and their citations in other scientists' work.

Before going into detail on elements of common publishing practices that are relevant in the context of scientific integrity and good professional practice, it is worthwhile reflecting on how research reports, academic journal articles or books are written. In an empirical study of such practices in the early 1980s, Karin Knorr-Cetina found and pointed out that a scientific report or article does not resemble the real research process, but describes an idealised course of action through a selective filter.[1] We may say: we tend to write research backwards. We take selected experimental data (that which satisfies our expectations and looks presentable), elaborate on its interpretation, then write the frame narrative (research context and idea), so that in the end an essay delivers an image of a flawless research process. This is almost never the reality of our lab practice. On the one hand, we may say it is good like this for pragmatic reasons: research reports are not meant to be exciting narratives or tales. We don't write novels or short stories to entertain anyone. On the other hand, this practice bears certain dangers. We tend to idealise the course of our research and leave out critical or undesired occurrences and observations in our description, unless they are part of the study design. Then, our interest in successfully publishing a journal article may create a bias in the way we write the introduction (exaggerations, hypes and trends, overestimation of the significance of one's findings) and the result discussion (meaningfulness, explanatory power, *etc.*).[2] In the worst case, after yielding experimental data and at the end of writing the report, we change the initial hypotheses into new ones that better fit the obtained data. Another consequence of this way of writing science is that negative results – that did not work, or that did not lead to useful results – are almost never published unless they are part of a study that aimed at identifying the best alternative among several strategies. Attempts to establish publishing platforms for negative results have always failed, so far.

We learned before, that scientific knowledge may be regarded as constructed rather than representing truth or reality. The fact that all reported scientific knowledge comes into existence in a writing process that constructs a narrative that is meaningful and convincing, rather than resembling the actual course of research and lab experience, supports this view. *Constructed*, here, does not mean that it is unsubstantiated or invalid, wrong, or unreal. Yet, an awareness of the constructive character of writing down scientific research may help us to be better and more careful – thus, more scientifically credible – scientists.

There is an abundance of literature on *how to write science* (reports, journal articles, presentations, grant proposals, *etc.*). Some instruction manuals focus on linguistic aspects like correct academic English, style, clarity, and convincingness.[3–5] In an international community in which the majority of the contributors are using a foreign language to communicate, training in the craft of writing is certainly of great importance. We can even imagine ethical issues arising from intercultural scientific communication or accusations of plagiarism when students whose English is not very good use formulations they find in published works. Some authors provide strategies for structured writing that make the process of writing more efficient and, at the same time, ensure that the final result is well-structured, logically sound, and most importantly, convincing for publishers, editors, peers and other readers.[6,7] Other books on scientific writing make a clearer connection between the scientific rationale of the research project, the experimental practice, and the writing process.[8-10] In recent years, the academic field of *science studies*, a subdiscipline of sociology, has elaborated useful empirical insights into the processes going on in and around research laboratories that can be exploited to make the scientific enterprise more efficient. This includes an understanding of the scientific writing process and its optimisation.[11,12] Note that all the referenced books in this paragraph focus on writing for peers (academic publishing). Writing for other audiences will be addressed in Chapter 15.

There is a direct implication in the way chemical and other scientists write their research reports and articles: the transition from the practical experience of *what is really happening* to the literary narrative of *what has happened* bears the risk of deviation from scientific facts. That is why the compliance with the scientific virtues is of crucial importance, not only in the design and experimentation phase of a research project, but also in its communication and publication. Honesty, truthfulness, objectivity, disinterestedness and systematised doubt prevent the scientific writer from idealisation, exaggeration, falsification, hasty generalisation, and other fallacies and biases. Fairness stands against plagiarism of any kind. Disciplined self-control keeps the scientist's focus on the scientific endeavour and away from the prospect of fame, power, money, or other non-scientific lures.

Unfortunately, the reality of those who pursue a career in academia, and in different ways also of those chemists in the private sector, looks different. Known as the publish-or-perish-syndrome, the publication record of a scientist more often than not decides over the success of job and grant applications or of collaboration agreements and the acquisition of third-party-funding. Research papers are the currency in science, while patents are often the goal of research and development (R&D) in private sector innovation. This pressure leads some scientists into improper behaviour in the context of publishing. Copying text, data, and even images and figures (plagiarism), is much faster than writing or editing them by oneself. Publishing the results of one research project in four or more 'smaller' essays while it could be plausibly and comprehensively described in one research article artificially makes the publication record of a scientist look better. Reviewing manuscripts of colleagues who work on similar research topics might tempt the reviewer to delay the publication, place a similar paper in the meantime, and look like the faster one. In the following, cases like these are discussed in detail. Besides raising awareness of the ethical pitfalls within those regimes, the practical goal of this chapter is to provide a strategy on how to behave or how to argue when it comes to dilemmas or conflicts.

7.3 Ethical Issues in Publishing Science

In this section, ethical issues are discussed that arise in the process of writing and publishing itself. These include authorship,[13] citation practices,[14] and peer review.[15] These topics have in common that the responsibility to act in accordance with the codes of professional conduct lies entirely on individuals, thus requiring orientation from the individual virtues introduced in Chapter 5. In contrast, the next section will survey publishing-related ethical issues that require a community-wide discourse to be solved.

7.3.1 Authorship

What qualifies for authorship, or: who should be legitimately listed as an author on a publication? Some say it is a matter of workload: those who do significant work for the study or project that is described in the text should be listed as an author. However, a lot of work is done by technicians and non-academic staff. These workers usually do not appear as authors, with the argument that only *intellectual contributions* count. Sometimes, significant parts of the study (experimental work or literature research) are carried out by students (bachelor, masters, PhD level). Some senior researchers and principle investigators (PIs) might exploit their work without giving credit to them. Departments and faculties may give themselves binding guidelines on how to handle these problems. For example, bachelor (undergraduate) students' theses that often include literature research may be used by a professor in the introduction section of research papers without crediting the student with authorship, while Masters and PhD students have a right to appear as an author as soon as they produce data or contribute otherwise to the publication, even when it is not particularly part of their thesis work.[16] Proofreading and commenting on the text do not qualify for authorship, but may be credited in the "Acknowledgements", a section in the article usually accepted by the publisher. It is also important to note that all authors are usually expected to be able to explain and defend the entire work, even when they wrote only parts of it. All authors are held accountable for what is published under their name.

Intellectual contribution is often equivalent to *producing new knowledge*. A form of knowledge that is clearly distinguished from that is so called *driver's knowledge*. The term originates from a true story:

Box 7.1   Excursion: Driver's Knowledge

After receiving the Nobel Prize for Physics in 1918, Max Planck went on tour across Germany. He was provided a Chauffeur (driver) who drove him around the country. Wherever he was invited, he delivered the same lecture on new quantum mechanics. Over time, his chauffeur who always listened to it grew to know it by heart: "It has to be boring giving the same speech each time, Professor Planck. How about I do it for you in Munich? You can sit in the front row and wear my chauffeur's cap. That would give us both a bit of variety!" Planck liked the idea, so that evening the driver held a long lecture on quantum mechanics in front of a distinguished audience. As there was no internet at that time, people did not know what Max Planck looked like, so they couldn't identify the prank. Later, a physics professor stood up with a complex and smart question. The driver – no idea of science at all – recoiled: "Never would I have thought that someone from such an advanced University as Munich would ask such a simple question!", and pointing at Max Planck with the chauffeur cap: "Even my driver can answer it! I'll pass the question to him!"

The difference between the driver and Max Planck is that the former just articulates knowledge without having elaborated the insights. A similar example is the News reader (on TV or the radio) who can do his job without being an expert on world politics. Scientific essays and research articles usually require (as demanded by the publisher) originality, that means they must present new knowledge. Driver's knowledge does not qualify for publication (one exception: review articles) or for authorship; for example, someone who performed a literature search to substantiate the introduction of a research paper.

When the authors of a paper are determined, the next question that can cause trouble is the order of authors. It can cause massive disputes among co-authors! Usually, the first author in the list is the one with the most significant contribution, often also the *corresponding author*, the one that should be contacted in any case of inquiry. The first position is important because in citation practice, it is common style to abbreviate a paper as "[Author 1] *et al.* [Year]", so that the first author is usually the most credited one. For PhD students, it is often important to be first author of their paper because professors or departments have rules concerning the preconditions for finishing a thesis, for example *"The candidate needs to have at least 3 published works as first author."*. The last author is often a senior researcher or PI. As most fellow scientists will usually notice the first and last author of a paper upon scanning the literature, this position is almost as popular as the first. To avoid conflict, in many cases, collaborators agree upon authorship and authors' order before they start a project.

There are two forms of authorship that are regarded as highly unethical because they violate the virtues of fairness and communalism: honorary authorships and ghost authorship.

- *Honorary Authorship* (sometimes called *guest authorship* or *gift authorship*): A person is listed as an author on a paper even though he or she has not made any significant contribution to the research. There are several reasons why people may be authors in name only. Sometimes a person is listed as an author as a sign of respect or gratitude. Some laboratory directors have insisted that they be listed as an author on every publication that their lab produces. Sometimes individuals are named as authors as a personal or professional favour. Some researchers have even developed reciprocal arrangements for listing each other as co-authors. Finally, a person with a high degree of prestige or popularity may be listed as an author in order to give the publication more visibility and impact.
- *Ghost Authorship*: A person is not listed as an author, even though he or she has made a significant contribution, in order to hide the paper's financial connections or obscure the participation of researchers with conflicts of interests. In extreme cases, all the involved researchers are ghosts. Ghost authorship is common in industry-sponsored clinical trials. One study found that 75% of industry-initiated clinical trials had ghost authors.[17]

The example in Case 7.1 involves a gift author and illustrates why it is violating scientific integrity. The motivation to add a prominent name is clearly a non-scientific one: fame, visibility, influence. Additionally, the decision of a superior to add an honorary author may cause severe trouble for the other co-authors. As the PI is playing out the power card, there is not much the student can do besides trying to bring forward arguments for good scientific practice that remind the superior of the virtues of science. Indeed, students and other inferiors often accept the authorship decisions of the authority as the positive effect of the added author is, usually, bigger than the inconvenience and the risks. Yet, the PI exhibits a poor example of mentorship, of course, delivering the message to the students that this act of unfairness that undermines the credibility of scientific merits is, after all, a peccadillo.

7.3.2   Citation Practice

There is no research paper that has no references in it. A manuscript without citations would simply not be credible and cannot pass any review process. Why do we cite other works or quote other scientists? Here is an incomplete list of the most important reasons (see ref. 14 for greater detail):

- *Tradition of Scholarship*: All academic work is based on a rich tradition of practices and achievements that constitutes the roots of current and new progress. Referring to those roots has ever since been part of written elaborations. In other words: *"We have always done it like that!"*.
- *History*: Despite the newness of certain insights and knowledge, everything (!) has antecedents, everything has a history. Reference to that history is due!
- *Utility*: Instead of inflating papers to improper length by describing established procedures (*e.g.* experimental protocols) or complex interrelations over and over again, it is much easier and convenient to refer to other works that the interested reader may turn to in case he or she needs or wants further input.
- *Avoidance of Duplication*: In order to illustrate the originality and newness of a contribution, it should be compared to existing work so that the differences become clear. Citing others shows in which way the own work is not simply a duplication of their work.
- *Establishing Credentials*: Citing proves solid background knowledge of one's academic field. Here, it is possible to cite too little (not enough background knowledge), too much (most likely just copying citations from other sources without critical reflection), mostly one's own work (self-adulation), or just enough.
- *Priority*: Our selection of works that are worth citing constitutes an additional way of quality assurance and peer review. Academics tend to cite works of good quality and avoid mentioning poor quality publications in order to give them a lower visibility. It is tempting, of course, to cite the not-so-good papers to make one's own paper appear in better light.

A crucial point in the citation is the completeness of references. Many scientists perform very selective literature searches, ignoring important work unconsciously or even consciously. Nowadays, there is no excuse for that as the technical facilities (internet) enable a quick and easy overview to be gained, for example through the "web of science" search engine or institute-specific (online) library tools. Scientists have the professional responsibility to perform a background knowledge check and to know their academic field well. Whether there is also the obligation to actually read every cited item is a different question. It is downright impossible to read every article and book completely and thoroughly. However, it is recommended to read at least the abstract and the conclusion. Citing works without having a clue what's written in it, for example by trusting the significance of the title or copying it from other sources that cite this work, can be dangerous when it turns out that either it doesn't have any connection to the statement that it is associated with, or it even states the opposite. A common situation in which this could happen is writing the introduction section of a research article or thesis after reading one good review essay and citing the references provided within that review. The case study on carbon allotropes provided by Hoffmann *et al.*,[14] which the author strongly recommends the reader should read and think about, is an insightful example of the damage caused by careless or inconsiderate citation practice.

Most publishers have clear rules for the layout of references, that means the way the authors' names, article title, publisher, place and year of publication, and so forth, is presented. The same is for quotations, presenting entire statements from other sources than oneself. Quotations are a more sensitive topic as improper labelling and clear mark-up as a quotation may lead to an accusation of plagiarism. Every written word that represents an idea, thought, and claim that is not one's own must be marked as a quotation. In the case of images, giving a reference to the original source is not sufficient as published images come under copyright law. Before publishing someone else's image (including figures, sketches, photographs, diagrams, *etc.*), the permission must be obtained from the copyright holder (often the publisher, not the author).

Subjecting chemical research publications to plagiarism detection software often results in an alert: significant overlaps of wordings and phrases are detected in the methods and experimental protocols sections. Chemical scientists are eager to point out that this does not count as plagiarism but as scientific standard: there is no better way to describe experiments concisely. When protocols are used that are taken from other published works, providing the reference is necessary, of course.

7.3.3   Peer Review

After submitting a research article, book draft, grant proposal or the application for tenure or other academic positions, it is reviewed and evaluated by *peers*, fellow scientists who – ideally – are experts in the field of the submitter/applicant. In the case of publications, these reviewers are chosen by the publisher and asked to express their professional opinion on a piece of work. Most publishers explicitly ask them not to make a recommendation on "publish" or "not publish", but rather provide detailed remarks on the originality and quality of the presented work, point out flaws and insufficiencies, and ask for more detailed input for particular parts of the paper. Based on these comments, the publisher may decide to reject the paper or accept it just as it is, but in the majority of cases the draft with the comments is returned to the submitter for him or her to revise the draft and re-submit an improved version. In the past, most review procedures were *single-blind* reviews in which the reviewers know the name of the author(s) while they themselves remain anonymous. Today, most publishers switched to a *double-blind* review, in which the author(s) also remain(s) anonymous.

There is a good argument for this practice: it is the most effective internal control system of science. Who else could evaluate the quality of research better than fellow researchers? However, there have been many contra arguments against this system, some based on negative experiences with it.[18] Some say, it is impossible for reviewers to be free from **bias**. While in the strictly empirical sciences (natural sciences) it is less of a problem, it is very difficult for scholars in the normative sciences (humanities, philosophy, *etc.*) to evaluate a draft solely in terms of logic consistency and quality of the presented arguments without seeing it through the glasses of their own academic, professional and personal viewpoints and preferences. Then, a draft is not evaluated concerning its academic value but concerning its particular content and position, which is unfair. With this bias, the review system becomes **unreliable**: not the *best* science is supported, but the *most conform*. Unfairness might also arise from another factor: personal academic (often career-related) **interests** of the reviewers. Sometimes, they might see themselves as competitors of the evaluated author in a particular research field. Maybe they plan or currently conduct a study that is very similar to a study presented in a paper that they are asked to review, as described in example Case 7.2. They might feel tempted to artificially prolong the review process in order to get their own paper published first. There have even been cases of **theft** of data and research ideas from drafts that have been sent to reviewers. Further objections against peer review are that available reviewers often lack the competence that would be necessary to evaluate a paper properly, and that it often takes a long time. Sometimes, reviewers return their comments more than six months after they received the draft from the publisher.

In recent years, more and more publishers have offered their authors the choice to suggest reviewers for their own submissions.[19] This practice responds to the risk of a manuscript being sent for review to someone who is negatively biased and would certainly judge this author's submission unfairly. On the other hand, critics say that it shifts the bias to the other end: it opens the door to fraud as authors can form networks of positive reviews, unknown to the editor/publisher. In analogy to gift authorship, this may lead to 'gift reviewerships', as in *"You give me a good review, then I write a good one for you!"*.[20]

Hence, peer review only works when certain ethical duties are enforced. Publishers must ensure a high degree of confidentiality, making it impossible for reviewers to misuse the yet unpublished drafts they receive. Reviewers must respect the intellectual property rights of authors. Publishers should try to ensure punctuality in the review process, for example by setting strict deadlines for communications and submissions. All in all, the review system requires a high level of professionalism among all involved parties (authors, publishers, reviewers), setting aside personal preferences and selfish interests. Prof. Smith in Case 7.2 is obliged to evaluate the colleagues draft as if he wouldn't know who he himself is and what he himself is doing research on. Instead, the only parameters that are applied to the judgment are expertise and knowledge. This attitude resembles an important principle of justice coined by John Rawls: just is what people agree upon and decide behind a veil of ignorance, not knowing which position they are in until the decision is implemented and binding for all involved parties. Then, the veil is lifted, and everybody is back in his or her real identity, accepting the consequences of the decision that has been made. That means, the reviewer of a scientific article should eliminate personal biases by pretending not to know anything about personal circumstances but focus entirely on the scientific value of the evaluated contribution instead. Thus, it is clear what Prof. Smith should do: acknowledge the good work, suggest improvement where it is due, and use his professional scientific expertise and creativity to plan his own next step, for example, a new research project, a collaboration with that senior colleague, a modification of the affected project that justifies a slightly different publication, and so forth.

7.4 Publishing and the Scientific Community

The previous section discussed ethical issues that play a role in individual decision-making in daily professional practice. The responsibility in those cases is on the individual (author of a paper, reviewer) and conflicts arise between two or a few people. This section will highlight publishing-related problems that are rather of a community character: the role of impact factors,[21] the recent phenomenon of predatory publishing,[22] and research that is too controversial to be published.[23] Responsibilities are still on the individual chemist but in a different way: the only way to tackle the ethical conflicts is to induce changes in community-wide practices and attitudes as individual decisions alone wouldn't change the overall adverse custom.

7.4.1 Impact Factors

Impact factors are used to get a measure of the quality of a researcher's work. It is based on the assumption that good articles find their way into good journals while not so good articles can only be published in not so well-established journals. Therefore, journals calculate their impact factor as the average number of citations that articles in this journal get. The most renowned journals are *Nature* and *Science* with impact factors higher than 30. An impact factor lower than 1 is considered bad. A scientist will be happy to place an article in a journal with an impact factor higher than 5. When researchers present their achievements (for example, when they apply for a research grant or a better position), they add up all their articles' impact factors (that means, those of the journals they are published in) as a measure of the quality of their research.

This practice has several disadvantages. First, it motivates scientists to split projects or studies into several sub-projects and publish them in many *small* essays rather than in one. The idea is that many low impact essays might add up to a higher impact than one medium impact paper (*e.g.*, four times "1" is still better than one "3"). Moreover, authors are tempted to submit drafts to several journals at the time, thus causing a lot of editorial and reviewing work, just to see which journals accept their paper, then choose the one with the highest impact factor and withdraw their submission from the other journals. As mentioned in the section on authorship, the practice of adding honorary authors is motivated by increasing the chance of an article being accepted by a high-impact journal. Also, many researchers tend to over-interpret and enthusiastically over-emphasise the importance of their studies, inflating the dramatic narrative by using vocabulary like "revolutionary" and "major break-through", in order to give the illusion of extraordinary scientific, economic, societal or global significance, thus creating a bias. Another problem associated with the power of impact factors and the pressure created by their existence is the phenomenon of self-plagiarism.[24,25] The intended effect of this practice is to increase the apparent impact of one's scientific output by increasing the citation number and, thus, the impact factor of journals. The side effect that makes this practice unethical is the distortion of the significance of research findings, for example the exaggeration of the efficacy of clinical trials.

As an alternative, it has been suggested to substitute impact factors by a productivity index.[26] This *h index* would be calculated as the average number of citations per article of an author. First, this would shift the factor of significance from the journals to the individual authors. Second, this might prevent the practice of excessive publishing of small publications since those kinds of articles potentially reduce a researcher's *h index*. Yet, it does not tackle the problem of self-plagiarism, but might even provide more incentives for this practice.

In any case, changing disciplinary customs takes a lot of effort and time. Today (2020), productivity indices gain more acceptance and significance – a long time since the first proposal in 2005.[26] What it needs is practitioners with voices that have influence in the scientific, academic, chemical, innovation, and so on, community. The discourse skill required for this positive impact consists of a firm endorsement of the scientific virtues that are at play, plus the mindful awareness of the mechanisms, motivations, incentives, and pitfalls, of scientific practice. Understanding both empowers a chemist to make fruitful suggestions and induces sustainable change towards a more efficient research environment with high professional integrity.

7.4.2   Predatory Journals

Some journals aim at exploiting the desperation of scientists to build a long publication record. With a clear commercial interest, they provide a platform for scientists to publish articles quickly and easily, often lacking adequate trustworthy peer review procedures, and often offering open access publishing for a fee. In some cases, predatory journals fake their impact factors to attract more careless researchers. Without doubt, these kinds of journals damage the scientific community in many ways: they undermine the scientific method in which rigorous scrutiny and discourse is crucial; they undermine intra-community and public trust in published scientific knowledge; they exploit the pressure of mostly young academics for non-scientific interests (especially financial); and they make quality control in scientific research increasingly difficult. The website predatoryjournals.com that is devoted to identifying and listing predatory publishers and journals provides this list of behaviours as criteria that indicate predatory intentions:

- Charging exorbitant rates for publication of articles in conjunction with a lack of peer-review or editorial oversight.
- Notifying authors of fees only after acceptance.
- Targeting scholars through mass-email spamming in attempts to get them to publish or serve on editorial boards.
- Quick acceptance of low-quality papers, including hoax papers.
- Listing scholars as members of editorial boards without their permission or not allowing them to resign.
- Listing fake scholars as members of editorial boards or authors.
- Copying the visual design and language of the marketing materials and websites of legitimate, established journals.
- Fraudulent or improper use of ISSNs.
- Giving false information about the location of the publishing operation.
- Fake, non-existent, or mis-represented impact factors.

Authors of chemical research articles, usually, know from their academic experience what good, trustworthy, established, credible journals are. Yet, identifying predatory journals is increasingly difficult as the publishers become better at deception. The *European Journal of Chemistry*, for example, sounds like a big name, has an ISSN number, and a solid web presence. Yet, a closer look reveals that it is a predatory journal, not to be mixed up with *Chemistry – A European Journal*, entirely *open access*, and registered at an address in the residential area of a small town near Atlanta, USA.

A comment on the trend towards open access publishing of scientific research is apt, here. Generally, open access to scientific data and knowledge has many desirable advantages, including transparency and response to a growing public interest in scientific insight and progress.[27] Scientific integrity is only violated in those cases in which the opportunity to buy the space in a journal is used to circumvent critical scrutiny in the form of rigorous peer review. Of course, there can be other motivations to pay for one's research paper being publicly accessible that are in accordance with good scientific practice: important insights that every member of a target group should read (for example, an empirical study that shows how damaging predatory publishing strategies are), specifically targeting potential readers with a low income or from poor institutions that can't afford journal subscriptions and licenses, or publications that serve at the same time as study material or conference proceedings. Here, again, reflections on the scientific virtues reveal whether a motivation is within or beyond the codes of good scientific practice.

7.4.3   Controversial Research

Is there anything that cannot or should not be published? Can research findings be so controversial that it is advised to refrain from making them accessible to the public? Certainly, there are only very few examples at the edge of acceptability. Potential misuse of scientific insight in general cannot be a reason for holding back publication as with this argument almost nothing can be published. However, extremely dangerous or risky matters that could potentially facilitate severe cases of bioterrorism, for example the instructions for the production of the bird flu virus H5N1, have been critically discussed among scientists, peers, publishers and governmental agencies. In the context of chemical research, similar discussions may occur for the synthesis of pathogens or potential chemical warfare agents. Case 7.3 describes the controversy that resulted from the publication of the chemical synthesis of the horsepox virus.[28] A way to deal with the conflict between the researchers' interest in getting their findings published and the publisher's or the public's concern on safety or health may be to publish the report without all details of the synthesis in question, so that others can't repeat or reproduce the fabrication of the harmful substance. This is the direction that argument (1) from the authors of the study takes: the communicated description is not equivalent to knowing how to synthesise the virus as it requires insider know-how rather than know-that. Readers of the article won't get the know-how, not to speak of the practical experience and the required lab equipment.

Yet, arguments (2) and (3) are less plausible. Whether the potential application that the availability of the synthetic virus enables – vaccine development – outweighs the potential risks (bioterrorism, biosafety) requires a much more careful analysis under consideration of many factors and stakeholders and exceeds the judgment capacity of the scientists. We will learn in Chapters 13 and 14 how chemists contribute with their knowledge and expertise to resolving normative questions like these. In any case, it is much more complex than the authors of the horsepox study seem to take it. The third argument that there are no regulation issues *because* the team thoughtfully obtained all legal reviews is logically unsound. The idea that something is allowed because there is no law prohibiting it ignores that in the course of scientific and technological progress we face decision-making dilemmas that are unprecedented and, thus, require normative-ethical discourse ahead of legal advice.

Other reasons for not publishing research are possible conflicts and disagreements among collaborators on fashion, interpretation or presentation of the study and its findings, or aimed at other forms of merit, for example patents. These are aspects we will address in the next chapter on various forms of collaborations in chemistry as a network activity, with arising issues concerning conflicts of interest and intellectual property rights.

Exercise Questions

1. Which of the following criteria would be unlikely to be associated with ethical problems for scientific publishing?

   A: Authorship
   B: Plagiarism
   C: Nationality
   D: Sponsorship

2. Which of the following may be considered an ethical problem in the context of scientific publishing?

   A: Length of the manuscript.
   B: Multiple submissions.
   C: Age of the author(s).
   D: Multidisciplinarity.

3. Which of the following would not normally be considered as a suitable author for a research paper?

   A: The research director.
   B: A research student.
   C: A proof-reader.
   D: A collaborator.

4. What term is used to describe the author of a paper who has been added to the list by the PI without having contributed anything to the work being described?

   A: Honorary author.
   B: Guest author.
   C: Ghost author.
   D: Virtual author.

5. What constitutes the best order for the authors of a scientific paper?

   A: The senior author first.
   B: Any order that is agreeable to all the authors.
   C: Alphabetical order.
   D: The person who has done the most work first.

6. Where might providing reference to the original source not be sufficient in a scientific publication?

   A: For a diagram.
   B: For a book.
   C: For a quotation.
   D: For a theory.

7. Which of the following would make someone unsuitable as a peer reviewer?

   A: Expertise in the appropriate field.
   B: Freedom from bias.
   C: A friend of one of the authors.
   D: Good language skills.

8. Which of the following represents the 'Impact Factor' for the science journal *Nature*?

   A: ≥ 30
   B: ≤ 1
   C: 5
   D: ≥ 50

9. Which of the following is not necessary for the reviewer of a scientific journal?

   A: Forgiveness
   B: Confidentiality
   C: Professionalism
   D: Punctuality

10. For which of the following would *Drivers' knowledge* be needed?

    A: To carry out complex experimental techniques.
    B: To be an author of a scientific paper.
    C: To provide an overview of a scientific project.
    D: To advise on the significance of surprising data.

11. Which of the following would not be considered as grounds for refusing publication of a manuscript?

    A: Evidence of plagiarism.
    B: Lack of originality.
    C: Erroneous conclusions.
    D: Potential misuse of scientific insight by readers.

12. Which of the following is likely to be of ethical concern when submitting a manuscript for publication?

    A: The referencing system being used.
    B: Exceeding the word count limit.
    C: Withholding key experimental data.
    D: The journal the manuscript is being submitted to.

13. An individual who has contributed significantly to the work but has been excluded as an author on a submitted manuscript is known as:

    A: A guest author.
    B: An honorary author.
    C: A ghost author.
    D: A supervisor.

14. Which of the following is not an appropriate reason for including a citation in a publication?

    A: To establish the historical development of a field.
    B: To make clear one's own contribution to the development of the field.
    C: To establish the academic credibility of the author.
    D: To highlight earlier work published by the author.

15. Which of the following is NOT an issue that should be considered by the peer reviewer of a manuscript?

    A: Appropriateness of the citations.
    B: Suitability for publication in the journal in question.
    C: Suitability of the authors.
    D: Quality of the research described.

16. The average number of citations per published article for an author is known as...

    A: ...the productivity index.
    B: ...the author index.
    C: ...the impact factor.
    D: ...the output factor.

17. A *single-blind* manuscript review process means...

    A: ...the reviewer doesn't know the identity of the authors.
    B: ...neither the reviewer nor the authors know the identities of each other.

> C: ...the editor of the journal doesn't know the identity of the authors of a submitted manuscript.
>
> D: ...the authors of a manuscript don't know the identities of the reviewers.
>
> 18. Who or what has an *impact factor*?
>
>   A: A university department.
>   B: An academic journal.
>   C: An individual researcher.
>   D: A single publication.
>
> 19. Which of the following is unlikely to give rise to ethical issues when submitting a manuscript?
>
>   A: Omission of unhelpful results.
>   B: Inclusion of appendices.
>   C: Overestimation of the significance of results.
>   D: Selective inclusion of references.
>
> 20. Which of the following topics has the potential to raise ethical concerns on the appropriateness of publication?
>
>   A: Synthesis of a pathogen.
>   B: The mechanism of a catalyst for polymerisation reactions.
>   C: The crystal structure of a gemstone.
>   D: Toxicity of protein-coated inorganic nanoparticles.

References

1. K. D. Knorr-Cetina, *The Manufacture of Knowledge*, Pergamon Press, Oxford, UK, 1981.
2. J. Schickore, Doing science, writing science, *Philos. Sci.*, 2008, **75**, 323.
3. M. Alley, *The Craft of Scientific Writing*, Springer Science+Business Media, New York, USA, 4th edn, 2018.
4. L. Olson, *Guide to Academic and Scientific Publication*, e-Academia (Oriental Press), Letchworth Garden City, UK, 2014.
5. M. J. Katz, *From Research to Manuscript. A Guide to Scientific Writing*, Springer Science+Business Media, New York, USA, 2009.
6. A. J. Friedland, C. L. Folt and J. L. Mercer, *Writing Successful Science Proposals*, Yale University Press, 3rd edn, 2018.
7. H. Silyn-Roberts, *Writing for Science and Engineering. Papers, Presentations and Reports*, Elsevier, London, UK, 2nd edn, 2013.
8. M. Cargill and P. O'Connor, *Writing Scientific Research Articles. Strategy and Steps*, Wiley-Blackwell, Chichester, UK, 2nd edn, 2013.
9. E. Lichtfouse, *Scientific Writing for Impact Factor Journals*, Nova Science Publishers, New York, USA, 2013.
10. J. Schimel, *Writing Science*, Oxford University Press, New York, USA, 2012.
11. J. Blackwell and J. Martin, *A Scientific Approach to Scientific Writing*, Springer Science+Business Media, New York, USA, 2011.
12. *A Guide to the Scientific Career. Virtues, Communication, Research, and Academic Writing*, ed. M. M. Shoja, A. Arynchyna, M. Loukas, A. V. D'Antoni, S. M. Buerger, M. Karl and R. S. Tubbs, Wiley-Blackwell, Chichester, UK, 2020.

13. M. Hosseini and B. Gordijn, A review of the literature on ethical issues related to scientific authorship, *Acc. Res.*, 2020, **27**, 284.
14. R. Hoffmann, A. A. Kabanov, A. A. Golov and D. M. Proserpio, Homo citans and carbon allotropes: for an ethics of citation, *Angew. Chem., Int. Ed.*, 2016, **55**, 10962.
15. J. M. Starck, *Scientific Peer Review*, Springer Fachmedien, Wiesbaden, Germany, 2017.
16. S. Baykaldi and S. Miller, Navigating the decisions and ethics of authorship: an examination of graduate student journal article authorship, *J. Mass Commun. Educ.*, 2020, 1–17.
17. P. C. Gøtzsche, A. Hróbjartsson, H. K. Johansen, M. T. Haahr, D. G. Altman and A.-W. Chan, Ghost authorship in industry-initiated randomised trials, *PLoS Med.*, 2007, **4**, e19.
18. D. B. Resnik, C. Gutierrez-Ford and S. Peddada, Perceptions of ethical problems with Scientific journal peer review: an exploratory study, *Sci. Eng. Ethics*, 2008, **14**, 305.
19. J. Shopovski, C. Bolek and M. Bolek, Characteristics of peer review reports: editor-suggested *versus* author-suggested reviewers, *Sci. Eng. Ethics*, 2020, **26**, 709.
20. J. A. Teixeira da Silva and A. Al-Khatib, Should authors be requested to suggest peer reviewers?, *Sci. Eng. Ethics*, 2018, **24**, 275.
21. J. F. X. Jones, The impact of impact factors and the ethics of publication, *Ir. J. Med. Sci.*, 2013, **182**, 541.
22. L. E. Ferris and M. A. Winker, Ethical issues in publishing in predatory journals, *Biochem. Med.*, 2017, **27**, 279.
23. L. J. Miller and F. E. Bloom, Publishing controversial research, *Science*, 1998, **282**, 1045.
24. S. R. Robinson, Self-plagiarism and unfortunate publication: an essay on academic values, *Stud. High. Educ.*, 2014, **39**, 265.
25. W. C. Lin, Self-plagiarism in academic journal articles: from the perspectives of international editors-in-chief in editorial and COPE case, *Scientometrics*, 2020, **123**, 299.
26. J. E. Hirsch, An index to quantify an individual's scientific research output, *Proc. Natl. Acad. Sci. U. S. A.*, 2005, **102**, 16569.
27. M. Parker, The ethics of open access publishing, *BMC Med. Ethics*, 2013, **14**, 16.
28. R. S. Noyce and D. H. Evans, Synthetic horsepox viruses and the continuing debate about dual use research, *PLoS Pathog.*, 2018, **14**, e1007025.

8   Chemistry as a Network Activity

Overview

Summary: Chemistry is – on several levels – teamwork, and as such embedded into a wide network of actors and stakeholders. This chapter will focus on issues that arise in the context of collaborations and co-operations across these levels. We will see what kind of conflicts can arise when chemists work with fellow chemists (including professor-student interaction and mentorship), with other natural scientists, or with completely different scientists (social sciences, humanities). Then, we turn to two instances in the network that are outside the academic community: politics (and its role for chemistry), and industry. Typical ethical issues arising in these contexts are conflicts of interests, academic freedom in the light of contemporary science funding practices, and intellectual property right protection.

Every network of people is, almost necessarily, also a network of interests. Sometimes, these interests overlap, and people pull in the same directions. Yet, at other times, interests clash, and collaborative work becomes inefficient, exhausting, or unfair. Throughout one's career as a chemist, whatever that career looks like, every chemist faces various situations that bear risks of conflicts and dilemmas. For most of us, the first time is the research work in a professor's group as graduate or PhD students. Besides conflicts arising from personality dispositions and competition, an important aspect is the power imbalance between mentor and student. Both mentor and student need skills in professional communication and conflict solving to reach their goals to the satisfaction of both. At all stages of the chemical career, multi-, trans- and inter-disciplinary collaborations, nowadays, are rather the rule than the exception. These span a wide variety of experts, non-experts, interest groups and stakeholders, posing different challenges on the conduct of the chemical practitioner. This chapter attempts to apply the scientific virtues of Chapter 5 to this realm of professional integrity. Necessarily, we will extend the scope of research ethics to issues of business ethics and professional ethics.

Key Themes: Mentorship and PI-student relationship; multi-, trans- and inter-disciplinary collaborations; public *versus* private funding; academic freedom; academic science *versus* industrial research and design (R&D); conflicts of interest; intellectual property.

Good Chemistry: Methodological, Ethical, and Social Dimensions
By Jan Mehlich
© Jan Mehlich 2021
Published by the Royal Society of Chemistry, www.rsc.org

> **Learning Objectives**
>
> After this chapter you will be:
>
> - A better mentor/superior, or a student/inferior with the ability to solve conflicts with convincing discourse skills and good arguments.
> - A better collaborator with high scientific integrity, communicative skills and positive influence.
> - An open-minded interdisciplinary bridge builder that can see beyond the narrow margin of your own professional expertise and competence.
> - Able to identify potential conflicts of interests that underlie your motivations for and decision-making in your research activities and collaborations.
> - Able to distinguish between interest- or purpose-driven science and academic curiosity-driven basic science, and choose your job according to your preferences.

8.1 Introductory Cases

Case 8.1: No Love!

Prof. Miller has a very strict rule for students joining his research group: he doesn't allow love-relationships between two members of the group. If two form a couple, one must leave the group. For some students, this is reason enough not to join Prof. Miller's group. Some think that it intrudes the students' privacy and that it is not a professor's business how close his students are with each other. Yet, after having bad experiences with couples in his group, leading to trouble and low work efficiency whenever the couple had private problems, he claims that he has good reasons to establish this rule. Is this appropriate leadership and mentorship?

Case 8.2: Money Matters!

Graduate student Alan is working with Dr Lee on a DFG-funded (German Research Foundation, public) research project on mass spectral identification of biomolecules. Yet, after starting the work, Dr Lee is asking Alan to study a series of organometallic cluster compounds which are important in industrial catalysis but have no direct biological significance. Upon Alan's inquiry, Dr Lee explains that, if the project works, she would be able to obtain contract funding from major oil companies which is important in times of tight government funds. Alan wonders whether it is acceptable to be paid from the DFG grant but work on something very different. Dr Lee says that she would be able to make up a good biochemical reason for Alan's work, and that the project manager would never know the difference. *"Everybody does it like this!"*.

Case 8.3: Not my Ligand!

[Real case] Two students in the same group were funded by different companies with individual confidentiality agreements. Student A was trying to find new homogeneous catalysts for a particular reaction; student B was trying to find new uses for a particular ligand. Student A asked the supervisor if it would be OK to use student B's ligand (provided by the company funding B) in the reaction being studied. The supervisor agreed to allow the use of student B's ligand by student A and it gave better results than any other. Without naming the other company, both companies were informed. They were both extremely upset and threatened legal action. In the end, they agreed that the intellectual property right (IPR) should be taken out by the company funding student B. The company who funded student A never funded more research in the group (after about 15 years of continuous funding) whilst the other company continued to fund the group. If student A had used a commercial sample of the ligand rather than the sample that had been supplied by the company funding student B, there would have been no problem.

8.2   Chemistry and its Actors

Chemistry is a community endeavour that is performed in a network of people, institutions, social spheres, and interests. Almost no chemist is working on his or her own, isolated, without any connections to the world outside of the academic ivory tower or the walls of a corporation's R&D labs. Figure 8.1 illustrates the main stakeholders in this network and how they are connected.

In the public sector, chemical research is done at universities and public research institutes. Here, we find what is usually coined academic science (the bottom of Figure 8.1). Academic chemists usually work in teams within their institute. Even though different groups, headed by senior scientists (professors, principle investigators (PIs)), within one faculty or institute work on different topics and projects in slightly different fields of expertise, they still share facilities, devices, analytic service competences and – for example, in institutional meetings, symposia, colloquia, and so forth – ideas and visions. A special feature of universities is the presence of students who join the research groups, receive their practical and theoretical education, but also contribute with their experimental work to the research progress of the group.

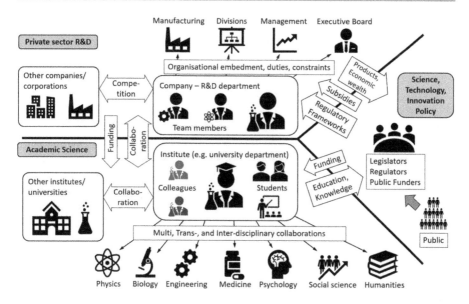

Figure 8.1 Network of different chemical professions and their organisational and institutional environments.

Then, there are colleagues and peers outside of the 'home institute': those in neighbouring institutes of the same university or research facility, those in similar institutes in other cities or countries. This includes all fellow scientists of the same orientation (for example, natural sciences), but not necessarily the same discipline (for example, physics, chemistry, biology). These collaborating scientists team up because their combined competences can result in something that they would not be able to achieve individually. An example could be the *Human Genome Project* in which scientists (biochemists, molecular biologists, geneticists, *etc.*) all over the planet worked together in order to decipher the human genetic code.

Another form of collaboration happens when *very different scientists* work together, for example, natural scientists with social scientists or philosophers. As we will see, the demands on this high degree of interdisciplinarity are very different from the kind of collaboration mentioned above. An example for this field of networking is technology assessment in which the social and ethical implications of scientific and technological progress are analysed and controlled by experts with very different knowledge backgrounds.

Chemical research in the private sector is embedded in quite different environmental conditions. Whereas the faculty at a university department is compiled along epistemic rationales (basically: what are interesting research areas that add up usefully to our local portfolio?) and a plausible academic profile (for example, a focus on organic synthesis), R&D departments of corporations hire researchers for functional reasons and in accordance with the requirements and demands of the company's industrial sector. As a consequence, teams are more diverse in the compilation of member competences than academic research teams with their disciplinary boundaries. Research that is part of the innovation efforts of companies has goals that are very different from academic research. Other companies and their R&D activities are rather competitors than collaborators, unless venture agreements, for example to cooperate on benchmarking studies, have been made. Within the organisational structure of the corporation, the R&D department is more or less closely collaborating with manufacturing/production units, other operational units like sales, logistics, marketing, customer relations, and so forth, the management (financial, supply, human resources, *etc.*), and the executive board that has the decision-making power. Depending on the size of the enterprise, this organisational embedment implies distinct duties and constraints, but most of all a company- and sector-specific innovation culture that determines how research is done and exploited in and by the company.

There is a tight relationship between academia and industry. It is estimated that 80% of the financial sources for performing academic research comes from companies and corporations.[1] It is obvious that industrial partners have very clear and particular interests in product innovation that creates functional (utility) and economic (profit) value – goals that may stand in sharp contrast to the motivations of academic science. Some worry that the industrial involvement in academic science, especially at universities, shifts the focus of science too much away from *basic research* and too much towards *applied research*. A good example for a field in which conflicts naturally arise is pharmacology. Huge pharma corporations dominate not only the drug market, but also the related research field and what kind of pharmacological research and development is performed (mostly for profitable drugs that have a big market potential rather than cures for diseases prevailing in developing countries).

Both of these two realms of chemical research are impacted by a third player. Policymaking and governance related to science, technology and innovation are the tasks of political organs that are legitimised by the public in democratic societies. Depending on the political structures that countries have established in particular, scientific activities in academia and industry are given regulatory frameworks by parliaments, governments, or agencies, ranging from safety policies, environmental protection policies, disposal and recycling regulations, labour policies, to funding and subsidy programs. Chemists working in academia or in industry in the European Union will be affected by both national and EU-level policies and regulations. This mix of control and support is, of course, motivated by balancing the various public interests like legal security, societal integrity, health and safety on the one hand, and education, scientific and technological progress, and economic wealth on the other hand. The former is achieved by regulations and laws that are sometimes perceived as constraints or limitations, while the latter is facilitated by financial support in the form of programmatic funding initiatives (for example, the nanotechnology program) or subsidies in times of crisis (for example, supporting the economy during the Covid-19 pandemic).

In this overview, we can see two basically different motivations for collaborating and forming networks. Some collaborations, mostly those with colleagues and other fellow scientists, are formed for procedural or epistemic reasons. Scientists share knowledge, know-how, resources, facilities, and competences in order to create more than they would individually be able to. In other cases, links are formed mostly for structural and financial reasons. Scientists and their institutions simply depend on money sources and on support from governance and policymaking.

In this chapter, we want to zoom from the inside out. We will start with teamwork at the microlevel and the ethical conflicts that may arise (Section 8.3): relationship management, the impact of power, clash of interests, responsibilities arising from duties and rights (including intellectual property). Then, two topics that play a bigger role in academic (university) settings, but not exclusively there, are discussed: mentorship, and interdisciplinarity (Section 8.4). Section 8.5 covers issues that are of significance predominantly in private sector chemical research and innovation. Last but not least, the impact of policy and governance, like the implications of funding practices on scientific integrity and academic freedom is surveyed in Section 8.6.

8.3 Teamwork

The success of teamwork depends on many factors. It would be a platitude to state that members of teams in professional realms should have traits like kindness, respect, empathy, or communicative competence. The more interesting question in terms of professionalism is how to deal with people who do not exhibit these attitudes in interaction with colleagues and collaborators. This is always then a matter for scientific integrity when the clash of personalities, attitudes, behaviours and worldviews impacts the efficiency and productivity of the team and interferes with the quality of scientific knowledge production. In the worst case, research projects fail because the involved people cannot get along with each other and inhibit the work process rather than fruitfully enhance it. Here, we take for granted that people are different, especially in a professional field with an extraordinarily diverse workforce, often international, involving all genders, a wide range of age groups, and all kinds of spiritual, political and philosophical worldviews. Common contrasting personality traits that often go along with difficulties to collaborate up to the level of incompatibility include introverted and extraverted people, visioneers ('idealists' in the colloquial sense) and conservatives ('realists'), well-organised rationalists ('left-brainers') and chaotic artists ('right-brainers'), pragmatists and principalists. A complete psychological analysis of team dynamics would exceed the scope of this chapter. In accordance with the pragmatic approach of this book, the focus is put on aspects of teamwork that impact the integrity of the process of scientific knowledge generation and that potentially violate the codes of good scientific practice.

It is worthwhile having a short look at the dimensions of group performance (Figure 8.2). On the individual level, the crucial factor for successful work is attitude. Key individuals act with the self-understanding of a self-identity that resembles archetypical roles of different 'players' of a team (as in, for example, the Myers–Briggs temperament indicator): leaders, promoters, masterminds, generalists, communicators, teachers, and many more.[2-4] A team with two or more leaders might face conflicts in the delegation of tasks and competences. A team that lacks a good communicator has lower inner cohesion and a lower degree of trust. Within the team, the members' attitudes and personality types translate into behaviour. A productive and creative climate requires trust and openness, challenge and debate, involvement and support, freedom and space, goals and clear visions.[5] A team that has too much internal harmony and common-sense is less creative and, thus, less productive than a team in which a healthy discourse culture with constructive feedback channels and collegial challenge is maintained. As a specific form of trust, collegiality in research teams refers to treating each other in accordance with the rules and codes of conduct of science: as colleagues on equal footing, with respect, mutual support and understanding of the other's professional ties, duties and rights. Furthermore, for a team to be

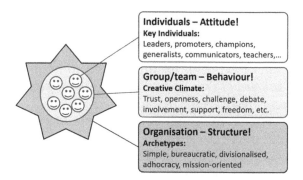

Individuals – Attitude!
Key Individuals:
Leaders, promoters, champions, generalists, communicators, teachers,...

Group/team – Behaviour!
Creative Climate:
Trust, openness, challenge, debate, involvement, support, freedom, etc.

Organisation – Structure!
Archetypes:
Simple, bureaucratic, divisionalised, adhocracy, mission-oriented

Figure 8.2 Three levels of success at work: individual attitude, group-internal behaviour, organisational structure.

successful, it should be made clear and transparent from the beginning who is held accountable for what and at what time. Accountability is different from responsibility in referring to the past (you are *now* accountable for what you decided, did or said in the past) while responsibility refers to the future (you are *now* responsible for the effects that your decisions, actions and statements might have in the future) (more details on this in Chapter 12).

The third level is the structure of the organisation in which the team is operating.[6] Universities are very different from corporations or public service agencies. Public institutions tend to be more bureaucratic, whereas companies have a rather divisionalised structure in which proper communication can lead to more flexible handling and management of inquiries. In any case, organisational support, the efficiency of work processes, and the local 'culture' (the way things are done around here) are crucial for a research team to yield productive output. An awareness of the importance of organisational embedment of research will help us understand the ethical conflicts in academia-industry collaborations better.

8.3.1 Potential Harm: Power

A source of ethical conflicts in professional teams is their hierarchic structure. Superiors (bosses, lab leaders, PIs, professors, senior scientists) have power over inferiors (employees, students, post-docs, technicians, junior scientists).[7] Discrepancies of influence and autonomy can also arise from cultural differences, familiarity with the environment, or financial dependence. In Chapter 6, we discussed the adverse effects that power and dependence have on whistleblowing. When something important is at stake for an inferior, for example, a grade, a promotion, a job, or emotional integrity, the interaction and communication with superiors may lack honesty, truthfulness, and scientificity. When a superior has conflicts of interests, poor leadership skills, or even a criminal intent (for example, discrimination, harassment, exploitation), his or her decisions and actions may damage the team performance and waste scientific capacity.[8]

Speaking up against misconduct in the form of bullying, discrimination and harassment is extremely important, not only for the personal integrity of the victim, but also for the mission of the research team. With a growing awareness of the existence and scale of ethical and legal issues arising from power misuse in recent years, institutional support has been improved. Public and private organisations have established ethics boards, anti-discrimination policies and offices that provide legal and institutional assistance in solving conflicts with an outcome that allows all involved parties to keep their face and reach their professional goals. In minor cases, pragmatic strategies for conflict resolution may be the better choice. Assuming that effective communication and argumentative discourse skills can and will always compensate poor attitude and flawed reason, the victim of a power misuse in a scientific research or innovation team may apply the scientific virtue rationale to bring forward valid and viable arguments: after all, professionalism and scientific integrity serve the goal of successful research and, thus, should be in the interest of the superior. Case 8.2 illustrates a situation in which a superior suggests something that the inferior judges as improper. It would be unlucky if the student would just accept the situation and do what he is told for the sake of his grade. The argument that everybody does it like this is invalid (a fallacy of relevance, see Chapter 4). Decisions like these should be made after a reasonable discussion of what serves scientific integrity in which the student is certainly empowered to make a point and should not accept what the more powerful party suggests.

8.3.2   Potential Harm: Conflicts of Interest

Research work is sometimes undermined by the interests of researchers that are not scientific in the sense of supporting the epistemic goals of the project (knowledge generation in academic science, innovation, and R&D in industry). These *conflicts of interest* (COI) are a very common ethical issue. First, we need to define what is understood as *conflict of interest*, as those conflicts arise in different forms in different co-operations. Following Shamoo and Resnik:[9]

An individual has a conflict of interest when he or she has personal,
financial, professional, political, or other interests that are likely to
undermine his or her ability to fulfil his or her primary professional,
ethical, or legal obligations.

What can these *personal, financial, professional, political, or other*
interests refer to in the case of scientists? Personal interests might
be opportunities of private and family life (*e.g.* moving to a particular location, helping one's spouse getting an affiliation, grant
or tenure position, taking revenge on an opponent or rival, *etc.*).
Financial interests arise when a scientist has stocks of a company
that he is collaborating with, therefore having an interest in the
company's success that is beyond his scientific interest. Professional interests are career opportunities, enlarging his research
group, increasing influence and power within one's institute or
organisation. Political interests refer to ideologies, promotion of
certain worldviews, impacting a certain legislative agenda, establishing one's own political views and beliefs. When all these interests collide with those professional, ethical and legal obligations
that we have framed by the virtues of good science conduct, then
we speak of *conflicts of interest*. Obviously, the abovementioned
interests might corrupt objectivity, the call for disinterestedness,
self-control, fairness, honesty, truthfulness – almost all the virtues
we outlined. There is just one problem with this definition of conflicts of interest (COIs): a scientist with an COI might be so biased
that he or she doesn't even notice that there is a COI. Therefore,
Shamoo and Resnik refine the definition:[9]

An individual has an apparent conflict of interest when he or she has
personal, financial, professional, political, or other interests that create
the perception of a conflict to a reasonable outside observer.

Sometimes, third parties (outside observers) with a more neutral
and unbiased stand can identify a COI much easier. In practice, this
is applied when institutions create independent offices or panels that
evaluate the chances for COIs of each and every project proposal and
collaboration agreement that is signed within the institute or organisation. This protects the institution and its individual members (the
scientists) from legal accusations and expensive lawsuits. Moreover, it
is important to note that not only individuals, but also entire institutions may have COIs, as expressed in a third definition:[9]

An institution has a conflict of interest when financial, political, or other interests of the institution or its leaders are likely to undermine the institution's ability to fulfil professional, legal, ethical, or social responsibilities.

A difference to the definition of COIs for individuals is that this one refers to undermining responsibilities rather than obligations. From this we should note that, legally, individuals in professional roles are attributed obligations to do their job well while institutions are, legally, attributed professional and social responsibilities in that they pay respect to their institutional justification built on the proper fulfilment of their tasks.

In academia-industry collaborations, financial COIs occur when scientists have stocks of the companies that they collaborate with, creating potential biases in their conduct of research. As mentioned above, some institutions implemented offices that check such COIs and eventually intervene in collaboration plans as the COIs of one of their co-workers have the potential to damage the institution's reputation or even cause legal problems. Although research bias is often deliberate, it may also operate at a subconscious level. People may not be aware of how financial or other interests are impacting their judgments and decisions. Studies have shown that financial or other interests can exert subconscious influences on thought processes. Additionally, even small interests can have an effect. People can be influenced by small gifts and small sums of money. Pharmaceutical companies give small gifts, such as pens, notepads, and free lunches, because they are trying to influence physicians' behaviours. Physicians often deny that they could be influenced by something as small as a pen, but studies indicate otherwise.

8.3.3 Potential Conflict: Intellectual Property

A third important field of potential conflicts that arise especially in scope of collaborations and co-operations is the protection of intellectual property (IP) rights.[10] We have already learned that data belongs to institutions, not to individual scientists. But who "owns" intellectual achievements such as research ideas, inventions, know-how, and so forth? There are several legal forms of IP rights and their protection:

- *Trade Secrets*: These are mostly relevant for companies and corporations that try not to reveal the basis of their economic success. The most common example is the recipe of CocaCola. What is granted here is the right to keep a particular information secret, not the secret itself. Once the secret is revealed, for example by accident or carelessness, it is no longer a secret, and nothing is left to be protected.

- *Trademarks*: Registered brand names, product names, process labels, and so on, are protected from being copied by others in the form of trademarks, often indicated by a small TM index, as in BrandXY™.
- *Copyrights*: Music, movies, any form of written work in print media, and computer software are protected by copyright regulations. When scientists publish their articles and essays, the words they write and the images they compile and design (photographs, diagrams, schemes, *etc.*) are protected by copyrights. A special form of copyright might be relevant for scientists when the outcome of the research is computer software (for example in IT sciences or theoretical physics/chemistry, *etc.*). Plagiarism is often associated with the violation of copyrights.
- *Patents*: This is by far the most important and most debated form of IP protection in the field of science. Patents are issued for technical artefacts and processes and their underlying theoretical foundations, often referred to as *inventions.* These inventions must fulfil certain criteria to have a chance of being granted a patent:
  o novelty (it must be the first time that someone came up with this idea).
  o non-obviousness (it must be the result of a visionary idea).
  o usefulness (it must be good for something).
  o enabling description (it must be possible to describe its features in technical terms in order to reproduce its fabrication and application).

It has been discussed whether it is possible or not to get a patent for natural elements, for example, chemical elements or DNA sequences. When geneticists succeeded in isolating DNA sequences that encode a specific protein, they applied for a patent, arguing that all the four abovementioned preconditions were fulfilled: it was new, it was not obvious (only after extensive research and with scientific knowledge), it was useful (for the directed synthesis of proteins), and it was perfectly scientifically and technically explainable and reproducible. However, the worry that excessive patenting of natural elements, even when engineered and artificially exploited, may lead to commercialisation of nature outweighed the technological arguments. Many expressed the worry of a *slippery slope*, a situation that occurs after one simple step is taken but in which the thus initiated drift into an undesirable direction cannot be stopped or controlled.

8.4 Chemistry in Academia

Scientific research in academia is organised around group structures.[11] Scholarly experts pursue very specialised and distinct epistemic goals in clearly defined academic niches. A professor in an organic chemistry institute may work on only a few reaction types including a defined set of substances and catalysts, for example. Group members share the interest in this particular research field and subscribe to the mission to advance expert knowledge. The head of a lab is expected to provide intellectual and behavioural guidance and mentorship. Occasionally, and to a growing extent, academic scientists collaborate in multi- and interdisciplinary settings that pose special challenges for the contributors. These two topics, mentorship and interdisciplinarity, are surveyed in this section.

8.4.1 Mentorship

All chemists, even those who work in industry and laboratories in the private sector, start their careers at university as science students. As Masters or PhD students, they have a supervisor, PI or senior researcher, usually a professor, as their mentor. The relationship between student and mentor is in many ways crucial for the young scientist's further career:[12,13]

- The mentor is an idol, both scientifically and personally. In many ways, the student will do things in the same way as his or her mentor did. This includes *scientific thinking*, the ways of identifying and specifying a scientific problem, the way they communicate with peers and handle critical issues, the style of writing research papers, interpersonal communication style, team leading and guiding, and much more.
- The mentor familiarises the student with methodologies and conduct of research, introduces him/her to techniques, devices, experimental setups, and so forth.
- The mentor introduces the student to the scientific community and the professional network. Here, the students form first bonds to academics outside of their home institutes, to increase job chances, find postdoctoral positions, and so on.
- Last but not least, the mentor also impacts the student's future job chances by grading courses and the thesis.

Mentors – all those who work with students – almost never learn how to be good mentors. This often leads to problems. Not all academic scholars are *natural born mentors* by personality. The bigger problem occurs when mentors even intentionally and consciously exploit their power in the unsymmetrical mentor-student-relationship. Cases of harassment and discrimination have been reported.[14,15] However, most ethical dilemmas in chemical education at university occur on a much more subtle level. Ethical challenges emerge during simply setting or marking a thesis. Being kind to a poor student has unintended consequences which are neither kind nor ethical. In fact, this kindness becomes less innocent when the lecturer's job or promotion depends on a good pass rate. Failing students limit funds available for promotion and cast aspersions on their professor's teaching abilities. But the upshot is less obvious: the beneficiaries of this easy pass are our future postgraduate students, teachers, academics, attorneys, political leaders, and experts in ethics. What is worse is that brilliant students are neglected or, at least, relatively downgraded. The challenge gets more complicated when the students enter the phase of their own research work: deciding how much assistance to give postgraduate students, estimating the difficulty of their research project, deciding on when and where to publish their work, all these are aspects that require tactic intuition.

Although the students' interests are clearly their successful graduation, fair treatment (in comparison with others) and a smooth start into their further career (*e.g.* being provided with the necessary skills, ideally placing a first publication in their field of interest), the PI's major focus is on funding, the management of research group resources, strategically well-timed and well-positioned publications, and a good reputation within and around their institute/faculty. Too high expectations and evaluation standards might scare away students, but when the degree can be obtained too fast and easy the PI risks a decline in credibility. The same factors that impact the current publication and funding practices also play a role at this stage of education: the potential quantity of publications in low-impact journals is the overriding consideration in designing research projects. In other words, the more traditional utility- or curiosity-driven research approach is career-limiting – a luxury that few can afford. Consequently, data collection replaces hypothesis-driven research because training operators instead of educating academics and scientists is more profitable and consumes fewer resources, and results are guaranteed. The responsibility for this situation is not primarily the individual scholar, but rather the systemic infrastructure manifested within the global scientific community. Interestingly, it has been shown that the abovementioned problems occur in almost every country and every cultural realm.[16,17]

What is a good mentor? First, a good mentor takes enough time for his or her students. A problem of our times is an often too big work group size, with some professors having more than 50 students to supervise. Then, a mentor should develop clear and transparent rules for the collaboration. This can build mutual trust and avoid conflicts. Case 8.1 depicts a professor (a real case known to the author) who told all the students who asked for joining his group that he does not allow love-relationships between two members of the group. At first, this sounds like an intrusion of the students' privacy. On the other hand, the professor has good reasons to establish this rule. The most important thing in terms of mentorship is that he made the rule clear and students can decide if they still want to join the group (accepting the rule) or look for another supervisor. Furthermore, a mentor should establish channels for confidential evaluation of the mentoring, which basically means that he or she is open for criticism and feedback. In an atmosphere of trust and co-operation, students are more likely to consult their "boss" directly whenever they feel that something does not go well. The mentor should also always protect whistle blowers, those who report when they observe that other group members apparently show hints for misconduct or cause other trouble in any way. All in all, a good mentor promotes a psychologically safe and non-discriminatory work environment with a diverse workforce, regardless of students' gender, cultural origin, political or religious views, and so on.

What can be done when the relationship between a student and a mentor is seriously disturbed or broken, or when the mentored feels mistreated or sees a clear violation of personal rights by the supervisor? In some cases, consulting a third party (an ethics board or office of research integrity at the university or institute) might be necessary. The first step should always be to establish an open and goal-oriented discourse with the mentor. As long as he or she is not entirely unreasonable (and from an adult person in an academic position we may expect a certain degree of reasonability), it should be possible with proper strategic communication to solve conflicting issues. The best argument wins. The virtues of science may help, here, too. It must be made clear that injustice, subjectivity (in treating team members differently, for example), emotion- or interest-driven biases are decreasing the scientific integrity and efficiency of the teamwork. In case the mentor violates fairness for the sake of his/her own benefits (authorship issues, for example), this can be pointed out using these guidelines of good scientific practice. This strategy will lead to better results than emotional pleas, begs, complaints, or threatening. There is a rationale underlying the science virtues that should have the power to convince a chemist in a mentor position. With this security backing up one's arguments, the affected student should enter such a conversation with confidence and affirmation. Even if the student doesn't succeed in solving the issue, the devotion to the science ethos will make him or her appear as the better scientist than the mentor, which will impress him/her and other team members that will then be on the student's side. On a side note, what is said about the student-mentor relationship applies in very similar ways to the employee-boss-relationship in any professional environment. Thus, mentorship guidelines can be extended towards leadership in general.

8.4.2 Multi-, Trans, Inter-disciplinary Collaborations

Academic sciences have become more and more interdisciplinary.[18] There is an obvious move away from the classical disciplines in natural sciences (physics, chemistry, biology) and engineering towards mission-oriented and contextual research areas like life sciences, computer sciences, or material sciences. Dissolving the borders and doing research at the interspace of disciplines yields viable and valuable insights that scientists within disciplinary constraints cannot elaborate. Yet, the sophistication level of our scientific knowledge is so high that scholarly generalists are the exception rather than the rule or trend. The best way, therefore, to maintain scientific research beyond disciplinary boundaries is to form collaborations among academic experts sharing their competences, knowledge, and equipment. This can be understood in several way, as Figure 8.3 illustrates.

The figure presents a few selected exemplary scientific or academic disciplines on the left side. Collaborations between disciplines in the same field (for example, natural sciences) count as *normal* or *common* collaborations, for example when physicists, chemists, biologist and perhaps engineers or researchers in investigative sciences like medical sciences work together, who all follow empirical-strategically thinking modes. The other disciplines in this overview differ significantly from the natural sciences in their specific methodology and scientific approach, like the semi-empirical social sciences, or the normative sciences jurisprudence and philosophy. When scholars from these fields collaborate, we have three possibilities:

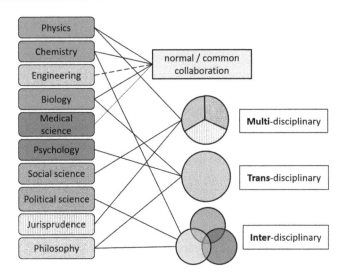

Figure 8.3 Conceptual differences between multi-, trans-, and inter-disciplinary collaborations.

1. *Multi-disciplinary Approaches*: The collaborating disciplines work in parallel and each within its own margin of methodological self-understanding. Here, an example connects physics with social science and jurisprudence. The early approaches to study the societal and environmental impact of nanotechnology (NT) worked like this. Nanotechnologists (mostly physicists) explained what is the state-of-the-art of NT research, social scientists estimated social acceptance of NT, law experts evaluated how current regulations deal with new developments and how they might be modified. They all work in parallel (mostly independent from each other), write their part of a report and *glue it together* as their final report.

2. *Trans-disciplinary Approaches*: The competences of various disciplines are combined to one new meta-discipline that is beyond the original ones. The shown example describes the discipline *neuroscience* as a combination of biology, psychology and philosophy (indeed, most neuroscientists are very eager in philosophical reasoning to justify their neuro-scientific worldview).

3. *Inter-disciplinary Approaches*: These want to explore what is between (*inter*) the disciplines. It is, therefore, attempted to create *synergies* by combining different competences and their strong points. Scientists of the collaborating disciplines look beyond the limits of their own expertise and elaborate on the matter together. In Figure 8.3, chemistry is connected with

political science and philosophy. This could be a project on the impact of new nano-scaled materials on the way that official agencies regulate the handling of them in order to ensure public health and safety (as performed for Registration, Evaluation, Authorisation and Restriction of Chemicals (REACH), the new chemical registry in the EU). The chemist (nano-scientist) contributes the knowledge on the physical and chemical properties of the compounds, the political scientist knows the existing policies and in which way they are insufficient for nano-compounds, the philosopher (ethicist) knows about demands for new definitions of *public health and safety* in terms of exposure to nano-materials. However, none of them can contribute anything meaningful to the debate when their insights are not informed by the others. Interdisciplinarity is characterised by mutual exchange, learning, and fruitful *epistemic dependence.*[19] Working in parallel (as in multidisciplinarity) is less efficient and productive.

Occasionally, interdisciplinary collaborations face some procedural and methodological difficulties.[20] First of all, there are *language problems*. People – even though speaking the same mother tongue – might use certain technical terms like *risk* or *law* in different ways according to the common meaning that it has within their discipline's standard terminology. Then, there is often a lack of understanding of other scientific disciplines' approaches to science. It is a common prejudice among empirical scientists like natural scientists that normative sciences and social sciences are non- or pseudo-scientific, whereas semi-empirical and normative scientists are convinced that the hard sciences need to be embedded into the epistemic frameworks of orientational knowledge in order to contribute anything meaningful to society. Here, the dialectical acceptance and respect between researchers from different disciplines is of crucial importance for fruitful collaboration. The abovementioned conflicts can and do occur here, often in more drastic forms: publication issues, authorship, IP, receiving credit in general. The claim that all participants are held accountable for the entire presented or published work, not only one's own contribution, is very important! The virtue of fairness asks researchers to represent collaborators' work correctly as such, including labelling it as their work and not as one's own. Conflicting interests concerning the type and use of the study results arise frequently: is it to serve as input for a political agency or office, or merely as an academic-intellectual contribution to advance knowledge? Who has how much access to which data and who has what authority to work with it? Priorities and expectations should be clearly communicated before the collaboration starts, and – if possible – written down in a binding project proposal and agreement.

8.5 Chemistry in the Private Sector

The conditions, the environment, and, thus, the professional-ethical requirements of private-sector R&D and innovation are very different from academic science. As outlined in Section 8.2, chemists in industry are embedded into a network of organisational and divisional competences, functions, and constraints, usually competing with other companies and corporations. While the goals of academic scientific inquiry are to advance human knowledge, educate students and fulfil the public mandate of providing public service, industrial research aims at producing goods and services, creating economic value (profitable innovation), increasing the competitiveness of the company, and maximising profits. Although the collaborative endeavour of academic science endorses openness and exchange of data and other assets, commercial R&D seeks to protect business information and proprietary interests. Important ideals in academic research are academic freedom, objectivity and constructive critical scrutiny, but industrial research is rather concerned about product quality enhancement and purpose-oriented innovation, thus regarding communication of results as a strategic, rather than an integral part of the work. Enhancing market share is the more or less openly communicated goal of research activities. For academic scientists, knowledge has an intrinsic value (for its own sake), whereas in industrial research it has instrumental value, can be managed and utilised, and is evaluated along the lines of its exploitation for practical and profitable applications. Although academic sciences, usually, do not compete (even though there is a competition of academic scholars for good jobs, fame, and epistemic authority), corporate and entrepreneurial R&D is operating under the conditions of a free market economy in which competition encourages or enforces companies to maintain high standards of product quality, to invest wisely, and to make effective and efficient strategic decisions. Finally, academic science is conducted under consideration of obligations towards students (including, perhaps, alumni), faculty, staff, government and other funders, the scientific and the public community, whereas corporations are indebted to their stockholders, customers, and workforce. Table 8.1 summarises this comparison between the characteristics of academic science and industrial R&D.

Table 8.1 Comparison of characteristics of academic science and private-sector R&D.

	Academia	Industry (private-sector R&D)
Main goal	Advance human knowledge, educate students, conduct public service	Maximise profits, produce goods and services, financial growth and stability
Data and idea dissipation	Openness, free exchange	Secrecy to protect confidential business information and proprietary interests
Research ideals	Academic freedom, free speech, free thought, honesty and objectivity	Research for specific purposes, restrictions on public communication, enhancing market share, ensuring quality of goods
Knowledge	For its own sake	Utilisation for the sake of profit or practical goals
Competition	No direct competition	Free market competition, must produce high quality, invest wisely, be effective and efficient
Obligations to	Students, faculty, staff, alumni, government, community	Stockholders, customers, employees, government, community

As mentioned in the introduction to this book (Chapter 1, Section 1.3) and in Chapter 5 (Section 5.5), professional codes of conduct and guidelines of good professional practice differ between the academic chemical profession and the private-sector industrial environment. Although a commitment to good scientific practice as outlined in Chapter 5 is, certainly, of great importance for the R&D-involved chemist, and in the context of interest- and purpose-driven research even greater than that of the academic scholar,[21] there are more ethical, legal and social norms to consider. As parts of enterprises, R&D departments are affected by matters of business ethics and of legal and organisational compliance. They are part of the company's *corporate social responsibility* (CSR) and *environment, health, and safety* commitments. We will come back to these aspects of chemical activity in Part 3 of this book, especially Chapter 12.

Despite the conceptual and paradigmatical differences between academic and corporate chemical research, co-operation is nonetheless commonly regarded as desirable and highly fruitful. Translational research, bridging the gap between basic (theoretical) science and applied (practical) science, is promoted as a key factor for economic competitiveness and driver of innovation. Companies and corporations have a genuine interest in collaborating with academic research groups for several plausible reasons:

- Gaining a strategic advantage from having access to first-hand scientific insight and trends;
- Securing profits and market share by increasing the chance of patenting innovations as the product of scientific progress;
- Having access to expensive technical and analytic equipment such as microscopes, spectrometers, reactors, and so forth, that are not profitable for a company but that universities finance with public budget or sponsorship;
- Attracting the best talents to join the company's workforce at an early stage (university level).

University institutions agree to join industry collaborations for different, but no less plausible, reasons:

- Technical know-how and commercialisation competences for effective progress that university labs do not have;
- Source of financial support;
- Higher visibility of research results and associated scientists' and institutions' names;
- The opportunity to 'place' alumni in famous and renowned companies as a form of self-advertisement.

The eras of bio- and nanotechnology and of other emerging techno-scientific systems saw a significant increase in public-private collaboration. It is widely associated with higher chances of progress and value creation.[22] Yet, ethical issues abound, especially in the form of conflicts of interest and of IP protection. The lure of financial prospect creates a bias that, potentially, blurs the chemist's scientific reasoning ability. This is apparent when the researcher possesses shares (for example, stocks) of the company and has a direct benefit from the company gaining advantages from academic support. To prevent COIs like these, laws have been implemented that ask full disclosure of all possible sources of COIs from contributors in private-public collaborative studies. As scandals have the power to ruin the reputation of both industrial partners and academic institutions sustainably, legal panels and ethics boards at both take COI issues extraordinarily seriously. IP rights and their protection, as expected in regard of Table 8.1, are a major source of conflict between academic and industrial partners. Although most academic scientists, including students, aim at journal publications as the most important credit for scholarly work, most companies wish to see their efforts culminating in patents. In order to achieve the latter, it is of strategic advantage not to share the acquired information in the form of the former. Unexperienced academic scientists may fall into traps, here. When negotiating the collaboration conditions before the project is started, publication opportunities including a timeline of envisioned outputs, authorship, and IP ownership should be openly discussed and the results be fixed in a written binding agreement.

Case 8.3 illustrates a typical example of the strict ties that academia-industry relations are operating with. Something apparently simple from the academic perspective, the use of a compound for a chemical reaction, causes an outrage on the corporate side. It may be acknowledged that the supervisor and his students communicated the case honestly and openly, even somewhat naively, to the involved companies. The loss of a financial source (company A) surely hurts the research group. From a purely scientific perspective, this is hard to understand as the scientific intuition that company B's ligand works well for student A's reaction and the subsequent proof that it does are uncritical in terms of the value of the chemical insight alone. Yet, the fact that confidentiality agreements have been signed and that company B has a plausible interest in third parties not knowing that they are after an application involving ligand B makes the academic partner appear careless and naïve.

Industrial funding creates a particular tension for academic institutions. As corporate funds continue to flow into a university, it may lose its commitment to basic research and knowledge for its own sake and may become more interested in applications and profits. The curriculum can also become transformed from one that emphasizes education in critical thinking and the liberal arts to one that emphasizes the training of students for careers in business, industry, or the professions. Talented faculty who might normally be involved in teaching may be encouraged to conduct research so the university can increase its revenues from contracts and grants. Corporate money can transform an academic institution into a private laboratory and technical training school. It is an ongoing debate how university faculties can position themselves and negotiate with industrial partners so that the balance between academic independence and the abovementioned benefits from translational and purpose-oriented research can be maintained.[23,24]

As in all other issues discussed in Part 2 of this book, the crucial question is a practical-pragmatic one: what can individual chemists do to prevent scientific misconduct? Here, we recall the three levels of discourse that the mindful chemist is engaged in: in an internal discourse, reflecting on own motivations and actions, the concerned chemist will identify and eliminate conflicts of interest in favour of unbiased scientific rationality. In a discourse with peers, colleagues and collaborators, the responsible chemist will apply the virtue approach to good scientific practice to argue for what this approach suggests as the *right* decision. As a member of a community, either of society as a whole or as a chemical researcher, the engaged chemist will try to contribute with effective argumentation and decision-making to the establishment of a healthy and beneficial environment in which scientific integrity is incentivised and professionalism is encouraged. These guidelines are not different for academic and industrial chemists, only their realisation must be adapted to the respective research environments.

8.6   The Impact of Policy and Governance

The role of science policy and governance for science consists partly of funding it and partly of providing oversight to the research enterprise to ensure that scientific inquiry proceeds in a high-quality, efficient, and ethical manner.[25] Public universities in most countries receive tax money to fulfil their tasks of doing research and educating students in academic and scientific thinking and practice.[26] However, in times of competing financial demands and interests among all kinds of social spheres, universities have to struggle to get their share of the public budget.[27] Legislations cut the resources for universities, forcing them to explore new channels (for example, industrial co-operations, see previous sections), or by channelling the money into specific innovation-oriented research and development programs, for example the National Nanotechnology Initiatives that many countries established. In this way, policy-making has influence on the particular research agenda (the decision of a research institute on what kind of research to do). Ideally, the political organisations represent the interests and expectations of the society, so they have an important control function for the public acceptance of science. In the other direction, science also plays a role for governance and regulation. Many political decisions are informed by knowledge that is generated by scientific means. Moreover, scientific and technological research enables innovation in the form of new technologies, so that the country benefits from increased economic competitiveness and life quality. At the same time, based on scientific insights, developments can be more sustainable for the society and the environment.

Current trends in funding – away from general funding of universities towards more directed financing of special research agendas and an increasing dependence on private-sector third-party funding – are expected to facilitate a trend towards less and less *basic research* and more and more *applied research*. This undermines academic freedom and inhibits the chance for revolutionary break-throughs as progress becomes predictable and confined to short-sighted margins.[28,29] The problem is this: when a chemist sees herself as someone who wants to do something useful and positively impacting the society very directly, she would probably not choose an academic career but one in industry. Then, academic freedom won't be a big issue for her. On the other hand, if her motivation to choose a career as an academic scientist at university or in other research institutions is driven by curiosity or the desire to understand the world and its features, then she would expect that her institution can provide a supportive research environment for that. If, then, it turns out that academic institutes subtly change into applied science centres, it may undermine her own understanding of scientific integrity.

At the same time, it causes a situation of increased pressure and stress for scientists. Imagine a government-funded research project with a tight time plan (3–5 years) and a clear research output (for example, materials for batteries with a high storage capacity and a low recharge time). The scientists – professors, senior researchers, postdocs, PhD students, and so on – always have in the back of their mind that they *have to be successful*. The prize for the generous funding of their research is a very high expectation on a particular output. "*It doesn't work*" is an inacceptable conclusion. Possibly, it even supports and facilitates fraud and misconduct in the form of benevolent and biased interpretation and communication of research results. The question for individual researchers is then: how can we maintain scientific integrity as described by the virtues of science (here especially *objectivity*, *disinterestedness*), even in a situation of financial dependencies and external pressure? On the one hand, this question has implications on the choice of research topics, but at the same time it may also imply that chemists need to stand up and raise critical voices (for example addressing political decision-makers) against current trends. On the basis of Merton's *communalism* virtue, it might be our responsibility to point out dangerous developments and trends that put the very idea of science as a social sphere at risk.

Last but not least, we revisit Case 8.2 one more time. Certainly, it is part of a chemical researchers' responsibility to use the resources they are entrusted with economically and for the purposes they are provided for. Ironically, Dr Lee's point cannot be refuted easily: the government grant won't be sufficient, in the long run, for her doing outstanding and impacting research. Thus, she could 'misuse' some of it to open a gate for more productive third-party funding. In this way, one of the grant's goals – scientific excellence – is achieved, just not in the way it was intended. To evaluate a case like this, it is important to know the underlying intentions and the particular uses of the budget. Here, there may be some leeway that allows the researcher to bumble through the financial limitations in order to reach a scientific milestone. Nothing in the description of this case indicates that any epistemic standard of scientific knowledge acquisition is violated.

A more drastic impact of the socio-political landscape on academic science is discrimination and the dogmatic or ideological restriction of what academic scientists are allowed to do and say in their professional roles as researchers, communicators, and teachers.[30] Academic freedom is doubtlessly essential for the mission and principles of scientific inquiry. Scientists who have to worry about being targeted for repression, job loss, or imprisonment when teaching or communicating ideas or facts that are inconvenient to external authorities in society and policy cannot do their job.[31] In some countries, oppression of the intellectual elite is obvious,[32,33] in other countries the limitation of academic freedom proceeds on a more subtle level.[34] Dealing with this problem is certainly beyond the virtues of science and a critical discourse attitude. Standing one's ground for justice and freedom is a virtue that not everybody is mentally or physically capable of. Basically, the affected scientist has two choices: go the pragmatic way and act within the boundaries as best as possible, or raise a voice and fight for academic freedom. Fighting this fight with the weapons of reason, bringing forward good arguments for academic freedom on the basis of the scientific virtues, may have a chance of success when the scientific community collaborates and collectively defends the integrity of science and research.

Box 8.1 Exercise: Defend academic freedom

*"**Academic freedom** is the conviction that the freedom of inquiry by faculty members is essential to the mission of the academy as well as the principles of academia, and that scholars should have freedom to teach or communicate ideas or facts (including those that are inconvenient to external political groups or to authorities) without being targeted for repression, job loss, or imprisonment."*[35]

(a) Identify and describe examples of victimization of scientists in the history of your country. Identify a scientist/researcher who may count as a role model in the active defence of academic freedom.

(b) How do current funding practices or other factors in the social, political, or economic environment of scientific research undermine academic freedom? Can you notice or feel that by yourself in your research context? What options do you have to maintain scientific integrity and act as a responsible professional and/or citizen?

Exercise Questions

1. What is the expected effect of current private (industry) funding practice?

   A: It supports only applied research and kills basic research.
   B: It puts too much focus on basic research.
   C: It guarantees a balance between applied research and basic research.
   D: It increases academic freedom because scientists don't need to worry about money anymore.

2. What may be stated about mentors in student-mentor relationships?

   A: Their behaviour is not important for the future of the student.
   B: The only important aspect is their scientific knowledge.
   C: They can affect the student's future professional success in many ways, for example by supporting publications, introducing him/her to the scientific community, grading the thesis, and so on.
   D: They cannot affect the student's future professional success because this will depend only on the student's skills.

3. Inter-disciplinarity refers to...

   A: ...the collaboration among different disciplines.
   B: ...the combination of different disciplines to create a new one.
   C: ...the exploration of the very nature of a discipline.
   D: ...the exploration of knowledge realms that lie between two or more disciplines.

4. The REACH regulations provide an example of a/an...

   A: ...multi-disciplinary approach.
   B: ...trans-disciplinary approach.
   C: ...inter-disciplinary approach.
   D: ...expert approach.

5. What should interdisciplinary work groups seek to do at one of their first meetings?

   A: The group should define key words (like "law" or "risk") in order to avoid complicated misunderstandings arising from differences in disciplinary jargon and different usage of such linguistic expressions.
   B: The members of the group should exchange the most important undergraduate textbooks so that each can study the basics of the other disciplines in order to be prepared for future meetings.
   C: Group members should provide their latest pay slips that show their salary so that financial conflicts of interest can be excluded, and trust can be built among the collaborators.
   D: The collaborators should elect a group leader who will then be given the authority to intervene in any conflict that diminishes the efficiency of the collaboration.

6. Which of the following individuals is likely to be the worst mentor?

   A: An idealist (working to make a better world).
   B: An introvert (who seeks solitude when considering solutions to problems).
   C: A chauvinist (who seeks to objectify women by denying that their scientific competence, in general, is equal to that of men).
   D: A bragger (who is continually showing off his own scientific achievements).

7. Which of the following factors is likely to lead to good mentorship?

   A: The mentor is the head of a huge work group with 30 research students and 4 postdocs.
   B: The mentor tells new group members very clearly about the rules and internal guidelines in his/her group.
   C: The mentor often travels to conferences and meetings.
   D: The mentor's office and the students' labs are located at different ends of the campus.

8. Which of the following should NOT provide a reason for chemists to collaborate with each other?

   A: Sharing expensive equipment saves resources and increases the productivity of the machines (spectroscopes, microscopes, *etc.*).
   B: Working in teams provides a first instance of "peer review" which would not be encountered when working in solitude.
   C: The collaborators share experience and knowledge. This is useful because two people always know more than one.
   D: Team collaborations usually result in publications with longer and, thus, more impressive author lists. Having two or more professors' names on a paper increases the chance of it being accepted and read.

9. Why was an interdisciplinary form of collaboration chosen for the REACH commission?

   A: Chemists had to supervise the regulators so that tax money was not wasted on another expensive but useless attempt to regulate chemistry.
   B: Only by combining the different competences from various fields of expertise was the commission able to work efficiently. Chemical background, regulatory framework, industrial needs and practices – all had to be combined to elaborate a new effective form of regulation.
   C: Originally, it was planned to delegate the task (elaborating a new chemical registry) to chemists alone, but representatives from industry worried that a too academic approach would ignore economic demands for profitability.
   D: The basic political paradigm underlying REACH is "sustainability". As this is a term relevant in many disciplines, it makes sense to equip a commission that works on the basis of this concept with people from various disciplines.

10. When problems arise between you (a student) and your mentor, what does it mean to "be prepared for solving them"?

    A: You have appropriate discourse and argumentation skills that enable you to explain your position, for example by referring to scientific virtues.
    B: You have a strong character that enables you to withstand a confrontation.
    C: You know the rhetoric tricks to "win" an argument against your mentor.
    D: You know where to find the right office, board or commission to accuse your mentor of improper mentorship and unethical conduct.

11. You work in an interdisciplinary team collaborating with a biologist, a doctor, and an ethicist. How should the group be organized?

    A: You (the chemist) should be the group leader, because chemistry is the most fundamental of these disciplines, and can serve as a model for the other involved "sciences".
    B: At the first meeting, it should be figured out who contributes the most to the project. That should be the leader and spokesperson of the group. The others are just "attachments".
    C: The collaboration doesn't need any leader. Everybody contributes his/her part on an equal footing, and exchange proceeds with equal respect for all involved competences and academic expertise.
    D: The scholar from the humanities (the ethicist) should be the group leader. He/she has the competence to "see the larger picture" whereas the others (including you, the chemist) have a rather narrow vision, limited mainly to expertise in their academic discipline.

12. What does the acronym COI that is often used throughout this chapter mean?

    A: Cause of incompetence.
    B: Chemical output investigation.
    C: Centre of Interdisciplinarity.
    D: Conflict of Interests.

13. Current funding practices are expected to result in which of the following? A shift...

    A: ...from basic to more applied research.
    B: ...from wasteful research to more fruitful research.
    C: ...towards less and less scientific research in general.
    D: ...from chemical research to biological research.

14. Direct funding of specific research agendas may lead to which of the following?

    A: More control and less fraud.
    B: Increased pressure to achieve positive results in a short time and, thus, an increased susceptibility for researchers to commit fraud.
    C: More publications.
    D: A better public reputation for chemistry.

15. Identify the statement that doesn't make sense according to the definition of "conflict of interest": Conflicts of interest in industry-academia collaborations arise from...

A: ...an academic scholar being a stockholder of the company he/she is collaborating with.

B: ...an entrepreneur (chief executive officer (CEO) of a start-up company) being the husband of the PI of the academic partner.

C: ...the company and the university being located in the same district of a city.

D: ...different goals concerning the results of such a collaboration (for example, patents *versus* journal publications).

16. Which of the following statements, concerning intellectual property rights, is the only one that is correct?

A: Data belong to the scientist who obtained them.

B: In order to use a diagram or graph from someone else's publication for a review article, one has to ask for permission from the copyright holder.

C: Everything scientists do is protected by copyrights.

D: Publishing of research results has no connection whatsoever with patenting so that the publication can proceed independently from the patent application.

17. Which of the following is NOT part of *academic freedom*?

A: Independence from political ideology and party politics.

B: Protection from suppression and programmatic limitation.

C: Freedom from being critically reviewed.

D: Self-organisation of academic institutions.

18. Which of the following statements concerning the roles of academia and industry is correct?

A: Academic research is intrinsically good, but industrial and economic thinking causes all the trouble!

B: Industrial research is good because it generates profit, whereas academic science is mostly a waste of resources and can, therefore, be regarded as negative.

C: All scientists and industrialists are, in the end, selfish and greedy. Thus, both fields are equally evil.

D: This question and the suggested answers A-C make no sense at all! Generalisations like this are invalid and a meaningless form of discourse!

19. Which of the following would not lead to a conflict of interest in the context of scientific integrity?

A: Aesthetic COI.

B: Financial COI.

C: Political COI.

D: Personal COI.

20. Which of the following differences in the interests/customs of academia and industry would not be considered likely to cause conflicts in collaborative ventures?

A: Different main goals (advancing human knowledge *versus* increasing financial growth and competitiveness).

B: Difference in working practices (often doing lab work until late night and at weekends *versus* a strict *9 to 18/Mo to Fr* schedule).

C: Difference in data and idea dissipation (openness and free exchange *versus* secrecy and confidentiality).

D: Difference in competition (no direct competition *versus* free market competition).

21. In which of the following cases should the ethics board of a faculty or university intervene in a collaboration?

A: When a professor has too many PhD students working on third-party funded projects.

B: When a professor is seen with products from a company he/she is collaborating with.

C: When a professor is going to get married to an employee from a company he/she is collaborating with.

D: When a professor "collaborates" with his/her own company.

22. Officially and publicly declaring not to have any COIs (for example in publications) is called...

A: ...a disclosure statement.

B: ...an acknowledgement.

C: ...a loyalty oath.

D: ...a declaration of independence (DOI).

23. Copyrights do NOT apply in the context of...

A: ...software.

B: ...images in scientific publications.

C: ...oral statements made in conference talks.

D: ...music and works of art.

24. Which of the following statements is correct? Trade secrets are...

A: ...forever safe in the hands of the company.

B: ...lost once they are revealed (deliberately or accidentally).

C: ...on the same legal intellectual property level as patents.

D: ...illegal and, wherever possible, uncovered by cartel offices and antitrust agencies.

25. What percentage (roughly) of research funds for academic science currently comes from the private sector (industry)?

A: <1%

B: ~20%

C: ~80%

D: >95%

References

1. OECD, *OECD Science, Technology and Industry Scoreboard 2015: Innovation for Growth and Society*, OECD, 2015, , DOI: 10.1787/sti_scoreboard-2015-en, ISBN 9789264239784.
2. F. Blackler, Knowledge, knowledge work and organizations, *Organ. Stud.*, 1995, **16**, 1021.
3. J. Sapsed, J. Bessant, D. Partington, D. Tranfield and M. Young, Teamworking and knowledge management: a review of converging themes, *Int. J. Manag. Rev.*, 2002, **4**, 71.
4. D. Duarte and N. Tennant Snyder, *Mastering Virtual Teams*, Jossey Bass, San Francisco, USA, 1999.
5. *Groups that Work (And Those that Don't): Creating Conditions for Effective Teamwork*, ed. J. R. Hackman, Jossey Bass, San Francisco, USA, 1990.
6. H. Mintzberg, *The Structuring of Organizations*, Prentice-Hall, Englewood Cliffs, USA, 1979.
7. *Power, Knowledge and the Academy*, ed. V. Gillies and H. Lucey, Palgrave Macmillan, Basingstoke, UK, 2007.
8. L. Jones, H. Riazuddin and J. Boesten, *Bullying and Harassment in Research and Innovation Environments: an Evidence Review*, UK Research and Innovation, Polaris House, Swindon, UK, 2019.
9. E. Shamoo and D. B. Resnik, *Responsible Conduct of Research*, Oxford University Press, Oxford, 3rd edn, 2015.
10. B. Andersen, Intellectual Property Rights, in *Encyclopedia of Applied Ethics*, ed. R. Chadwick (Chief), Elsevier, London, 2nd edn, 2012.
11. B. Bozeman and C. Boardman, *Research Collaboration and Team Science. A State-of-the-art Review and Agenda*, Springer, Cham, Switzerland, 2014.
12. V. Weil, Mentoring: Some ethical considerations, *Sci. Eng. Ethics*, 2001, **7**, 471.
13. C. Pfund, A. Byars-Winston, J. Branchaw, S. Hurtado and K. Eagan, Defining attributes and metrics of effective research mentoring relationships, *AIDS Behav.*, 2016, **20**, 238.
14. *Sexual Harassment at the Workplace*, ed. M. Stockdale, Sage Publisher, London, UK, 1996.
15. N. C. Cantalupo and W. C. Kidder, A systematic look at a serial problem: Sexual harassment of students by university faculty, *Utah Law Rev.*, 2018, **3**, 671.
16. *Chemistry as a Second Language: Chemical Education in a Globalized Society*, ed. C. F. Lovitt and P. Kelter, ACS Symposium Series No.1049, Washington, USA, 2010.
17. G. Mohamedbhai, The scourge of fraud and corruption in higher education, *Int. J. High. Educ.*, 2016, **84**, 11.
18. *Interdisciplinarity – Reconfigurations of the Social and Natural Sciences*, ed. A. Barry and G. Born, Routledge, Abingdon, UK, 2013.
19. H. Andersen, Collaboration, interdisciplinarity, and the epistemology of contemporary science, *Stud. Hist. Philos. Sci. A*, 2016, **56**, 1.
20. M. MacLeod, What makes interdisciplinarity difficult? Some consequences of domain specificity in interdisciplinary practice, *Synthese*, 2018, **195**, 697.
21. K. Shrader-Frechette, *Tainted. How Philosophy of Science Can Expose Bad Science*, Oxford University Press, New York, USA, 2014.
22. R. Zingg and M. Fischer, The rise of private–public collaboration in nanotechnology, *Nano Today*, 2019, **25**, 7.
23. T. Bollia, M. Olivares, A. Bonaccorsi, C. Daraio, A. G. Aracil and B. Lepori, The differential effects of competitive funding on the production frontier and the efficiency of universities, *Econ. Educ. Rev.*, 2016, **52**, 91.
24. C. Badelt, Private external funding of universities: Blind alley or new opening?, *Rev. Manage. Sci.*, 2020, **14**, 447.

25. A. D. Levine, Research governance, in *Encyclopedia of Applied Ethics*, ed. R. Chadwick (Chief), Elsevier, London, 2nd edn, 2012.
26. *European Higher Education Area: The Impact of Past and Future Policies*, ed. A. Curaj, L. Deca and R. Pricopie, SpringerOpen, Cham, Switzerland, 2018.
27. S. Calviac, Universities' funding: Evolution and challenges, *RFAP*, 2019, **169**, 51.
28. D. C. Poff, Research funding and academic freedom, in *Encyclopedia of Applied Ethics*, ed. R. Chadwick (Chief), Elsevier, London, 2nd edn, 2012.
29. K. D. Gariepy, *Power, Discourse, Ethics. A Policy Study of Academic Freedom*, Sense Publisher, Rotterdam, Netherlands, 2016.
30. J. Williams, *Academic Freedom in an Age of Conformity. Confronting the Fear of Knowledge*, Palgrave Macmillan, London, 2016.
31. T. Karran, K. Beiter and K. Appiagyei-Atua, Measuring academic freedom in Europe: a criterion referenced approach, *Pol. Rev. High. Educ.*, 2017, **1**, 209.
32. U. Özkirimli, How to liquidate a people? Academic freedom in Turkey and beyond, *Globalizations*, 2017, **14**, 851.
33. M. G. Hocevar, D. A. Gomez and N. J. Rivas, Threats to academic freedom in Venezuela: Legislative im-positions and patterns of discrimination towards University teachers and students, *Interdis. Pol. Stud.*, 2017, **3**, 145.
34. L. Allen, *Academic Freedom in the United Kingdom*, Academe, Fall 2019, vol. 105, Article No. 5.
35. https://en.wikipedia.org/wiki/Academic_freedom, accessed on August 26th 2020.

9 Animal Experiments

Overview

Summary: Admittedly, not all chemists face the situation of conducting animal experiments at any time in their career. Yet, those who do often struggle with ethical concerns about the justification of animal use in research and regulatory testing, see themselves confronted with public outrage, and face a load of paperwork that legislators request in order to limit animal sacrifice to a reasonable minimum. The bioethical considerations that inform the discourse on animal experiments easily exceed the competence of scientific researchers, toxicologists and analytic chemists in academia, industry and public service. At the same time, in practical terms – public debate and regulatory requirements – animal experiment practitioners can't escape the obligation to understand at least roughly what is at stake.

The goal of this chapter is to refine bioethical theory into an applicable overview of arguments that researchers can use as orientation for their daily discourses and experiments. The virtue ethics approach of the previous chapters is not sufficient for that. Instead, a brief introduction to bioethical reasoning strategies will give the reader a clearer sense of the conflict potential that the pro and the contra side face. Here, the image of ethics as a prism, as drawn in Chapter 1, is very powerful: we are not looking for *the right solution* as the result of an *ethical lens*, but rather attempt to refract the complexity of views into clear and plausible positions and their underlying justifications. It is also in this respect that *every* chemist, not only those who conduct animal experiments, may gain from paying careful attention to this chapter. In a propaedeutic sense, getting to know the approaches employed in an ethical *hot topic* like animal research supports ethical reasoning and argumentation competence and prepares for other seemingly intractable conflicts in the discourse on science and research.

Key Themes: Types of animals and purposes of animal experiments on vertebrates; means and ends of animal experiments; bioethics (benefits of outcomes (utilitarian); rights and duties (deontology); anthropocentrism and biocentrism); pro and contra arguments; regulation, 3R guidelines.

Good Chemistry: Methodological, Ethical, and Social Dimensions
By Jan Mehlich
© Jan Mehlich 2021
Published by the Royal Society of Chemistry, www.rsc.org

> **Learning Objectives:**
>
> Chemists might be affected by the ethical debate on animal experiments in two ways: they find themselves attacked or criticised by opponents of animal testing (sometimes unjustified or unreasonably), or they are asked to follow legal and ethical guidelines for animal experimentation. Thus, in this chapter, two competences are acquired:
>
> - Responding to objections and verbal attacks with proper and plausible arguments, so that credibility is maintained, and argumentation is reasonable and convincing.
> - Understanding the ethical background of regulations and fulfilling formal requirements (the necessary paperwork before and after animal experiments) professionally and satisfyingly.

9.1 Introductory Cases

Case 9.1: Bad Publicity

A medium-sized company produces paints and lacquers as consumer products. Innovative new products need approval in terms of dermatological innocuousness and environmental safety. As regulations require, the former aspect was tested in cooperation with an analytic institute that conducted animal studies as part of an EHS assessment (environment, health, and safety). The company became the focus of a local animal rights activist group that hired a reporter to publish a not so benevolent article on the animal cruelty case in this company. The next day, a crowd of upset protesters gathers in front of the manufactory facility and confronts the head of the unit, a chemist by training, with accusations of animal torture. The heated atmosphere threatens to escalate.

Case 9.2: Standing Ground

A popular TV talk show hosts a discussion panel on animal testing. The debaters are:

- A toxicologist conducting animal experiments for basic research.
- A chief executive officer (CEO) of a cosmetics company (known for testing their products on animals).
- An animal rights activist protesting against any form of animal experiments.

The scientist faces aggressive opposition from the activist, but gets support from the CEO. At the same time, he feels sympathy with the activist's concerns and despises excessive animal use for regulatory testing of luxury consumer products which, for him, is different from the reasonable occasional conduct of animal experiments for the purpose of scientific insight on species ecology and biomedicine.

Case 9.3: Evaluating Alternatives

A graduate student is asked by her principle investigator (PI) to order 15 mice in order to perform a series of experiments investigating the possible mutagenic effects of a newly synthesized class of compounds. Although quite common in her discipline, having never considered non-human animal experimentation necessary to her research, the student suggested that perhaps these experiments wouldn't be necessary. The PI emphatically disagreed and, in response, explain that although non-human animals may potentially experience suffering throughout the study, this was an essential and inevitable part of scientific research. It would be more important, the PI reasoned, to safeguard the health of humans who would later come into contact with these compounds in the future than those animals that were bred specifically for this type of research. Given a few days to design an experiment that would minimise the suffering of the mice, the student finds alternatives in the literature. Machine learning of toxicological big data from the Registration, Evaluation, Authorisation and Restriction of Chemicals (REACH) database has been used to predict and verify the hazards of tens of thousands of known molecules. Also, there are protocols in place that use cell cultures rather than mice. The student remains conflicted. As suggested by her PI, non-human animal experimentation seems to be the fastest and most well-established methodology to investigate these new compounds, while the alternatives carry their own sets of risks and consequences: unknown transferability, highly-specialized skills, longer experiments and greater cost.

9.2 Some Facts and Numbers

Animal experiments and animal testing are controversial topics, as the realistic example in Case 9.1 illustrates. Sometimes, the atmosphere gets so heated up that violence and vandalism are the result. Despite the seriousness and ethical relevance of the discourse on animal testing, this form of militant activism and criminal form of protest must be condemned.[1] In this chapter, we attempt to respond to the emotional debate with fact-based *is*-premises and well-reasoned *ought*-premises (remember the ethical argument structure in Section 1.4). Yet, as we will see, this strategy is complicated by the fact that there is no simple agreement on what the starting point of the ethical reasoning is. This will be enlightened and disentangled in Section 9.3. First of all, however, we start with a few numbers from which we can derive the is-premises, as without facts there is no basis to discuss, after all.

9.2.1 Number and Purposes of Animal Experiments

The discussion revolves exclusively around vertebrate animals. This does not include invertebrates like insects (for example, fruit flies and cockroaches). It is estimated that more than 100 million vertebrate animals are used worldwide per year for experiments and testing procedures.[2] Available reports are eager in pointing out that statistical data has to be treated with care as different countries around the globe treat animal testing very differently. The reporting is extremely inconsistent, and estimates are most likely much higher than the reported numbers. In the following, numbers for the European Union for the year 2017 are presented. They are taken from a report of the European Commission to the European Parliament.[3] According to this report, 9.39 million vertebrate animals were used in animal research, which is a slight reduction compared to 2015. More than 60% of those were mice, 13% fish (mostly zebra fish), 12% rats, 8% other mammals (most of them guinea-pigs, rabbits and pigs), 6% birds (mostly domestic fowl), 0.3% amphibians, reptiles and cephalopods, and approximately 8000 monkeys and non-human primates (7200 of which were macaques (*cynomolgus monkeys*)) (see Figure 9.1). No great apes were used for animal experiments in the EU. Species of *"particular public concern"* – cats, dogs and non-human primates – represented less than 0.3% of the total number.

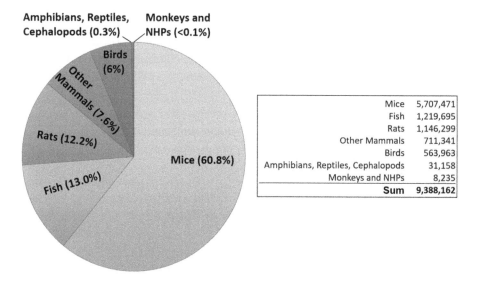

Figure 9.1 Number of animals by type used in the EU in 2017.[3]

In addition to the number of animals used for experiments and testing, we need to look at the purposes for which they are used. The report names five categories with the following details (see Figure 9.2A). The order of detailed purposes in a category represents the ranking by number (highest to lowest):

- *Basic research* (45%) on: nervous system, immune system, oncology, ethology/animal behaviour/animal biology, cardiovascular blood and lymphatic system, multisystemic, endocrine system/metabolism, gastrointestinal system including liver, urogenital/reproductive system, musculoskeletal system, respiratory system, and sensory organs (skin, eyes, ears).
- *Translational and applied research* (23%) on: human cancer, human nervous and mental disorders, human infectious disorders, animal diseases and disorders, diagnosis of diseases, human endocrine/metabolism disorders, non-regulatory toxicology and ecotoxicology, human immune disorders, human cardiovascular disorders, animal welfare, human sensory organ disorders, human gastrointestinal disorders including liver, human musculoskeletal disorders, human urogenital/reproductive disorders, and plant diseases.
- *Regulatory purposes* (23%) related to batch potency testing (by far the most), efficacy and tolerance testing, reproductive toxicity, batch safety testing, repeated dose, pharmaco-dynamics, developmental toxicity, ecotoxicity, acute, kinetics, skin sensitisation, safety testing in the food and feed area, pyrogenicity testing, carcinogenicity, genotoxicity, target animal safety, skin irritation/corrosion, neurotoxicity, eye irritation/corrosion, and phototoxicity.
- *Routine production* (5%) of blood based products, monoclonal antibodies using the mouse ascites method, and other product types.
- *Other* (4%): namely higher education or training, protection of the natural environment, preservation of species, and forensic enquiries.

51% of the uses were assessed as mild or less than that, 32% as moderate, 11% as severe and 6% as non-recovery, which is defined as a *"procedure that has been performed entirely under general anaesthesia from which the animal has not recovered consciousness"* (Figure 9.2B). These values indicate nothing about the survival of the animals. Most lab mice and rats are euthanised after the study is completed.

Another insightful overview is that of specific sector legislations that require the satisfaction of its guidelines and regulations. 61% of the uses occurred for medicinal products for humans, 15% for veterinary

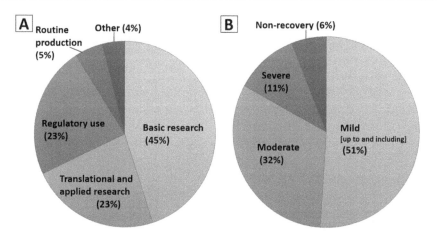

Figure 9.2 (A) Purposes of animal experiments in the EU in 2017. (B) Severity of the experiments in the EU in 2017. Data obtained from ref. 3.

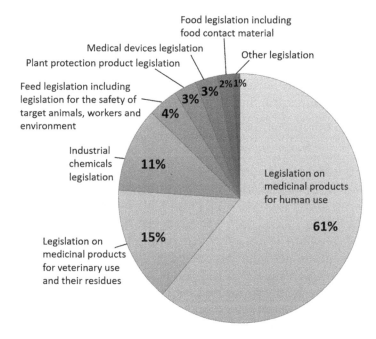

Figure 9.3 Regulatory testing as required by different legislations in the EU in 2017. Data obtained from ref. 3.

medicinal products, and 11% for industrial chemicals (Figure 9.3). No uses were reported under the cosmetics legislation.

These numbers will be important when we come to talk about relationships between the purposes and goals of animal testing and the justification of sacrificing animals for those goals. Only a very small fraction of research on and with animals benefits non-human entities, for example the animal species that is studied, biodiversity, the natural environment, or plants. A considerable fraction of the experiments with animals require chemical competences (toxicology, analytic chemistry, study of new materials and compounds). Although cosmetic testing and experiments with great apes are banned (or: legally framed so that they are factually banned), it remains questionable whether these regulations only shift the problem to other countries, for example in South-America and Asia, where regulations are less strict or insufficiently enforced.[4]

9.2.2   History and Present Day Animal Research and its Opposition

Animal testing dates back very far in history.[5–7] References to animal experimentation were found in the writings of Greek-philosopher-physicians of the third and fourth centuries BC, for example Aristotle who performed dissections to study the inside of animals, or Erasistratus who, reportedly, was the first to do experiments on living animals. Medieval clerical scholar van Galen (Germany) began to dissect goats and pigs and compared his findings to what he knew about humans, earning him the title 'father of vivisection'.[8] In the early Renaissance, the Cartesian worldview regarded animals as some kind of biomechanical machine and endorsed treating animals as objects and examining them in the cruellest of ways, for example, nailing dogs onto boards and cutting them open to observe the beating of a heart.[9] When the scientific experimenting with animals turned into a more systematic method on a large scale (in the 19th century), the first movements of resistance formed, the *anti-vivisectionists*. Today, organised activism against animal research is widespread. Animal rights movements like that organised by PETA are well known. Occasionally, the protest even turns violent against individual scientists and/or their institutions, for example in the form of attempts to set research labs on fire or threatening researchers at their home.

Why is animal testing such a hot topic? Those supporting research on animals defend their approach by referring to the huge benefits for mankind, for example safety and efficiency of new drugs and cosmetics, anatomic and other medical knowledge, understanding of natural (biological) processes, and so forth, for a relatively low price – *only* animals – avoiding human suffering. The contra arguments are much more diverse and focus on different aspects of animals and their usage for research. Indeed, some activists put a strong focus on emotion and aesthetics, pointing at the suffering and despair of animals, and the misuse of *beautiful*, *cute*, and *lovely* animals. This is, understandably, the target of criticism, as personal preferences and aesthetic standards cannot be applied in a philosophical debate.[10-12] In the following section, it is attempted to survey the arguments in a concise systematic manner. The aim is not a complete bioethical elaboration of arguments, but an understanding of the main positions in the debate and their roots. Two goals are followed: first, chemists conducting animal research should be able to respond with rationally plausible arguments to activists and other critiques, which involves understanding the logic of the counterarguments and the formulation of a response that has the potential to convince the other side. It is assumed that most scientists working with and on animals are very critical themselves with the issue rather than *heartless cruel monsters* as activists often depict them.[13] This chapter shall help these critical and reflected researchers to sharpen their argumentation skills. Second, chemists who are affected by the regulations concerning animal experimentation (for example, REACH guidelines) should understand why they are in place, under what rationale they have been made, and what implications they have for the chemist's lab practice.

9.3 Ethical Perspectives

The starting point for our reflection is the relationship between *means* and *ends*. We apply means in order to achieve certain ends in almost all our actions. Ends are our goals, purposes, and objectives. Usually, we aim at something which we believe is somehow a benefit for us and/or others, for example an increased quality of life or the fulfilment of a need or desire. Then we exploit certain means to reach this goal, or at least to take steps to move towards the realisation of the goal. Sometimes we take other people as means (for example, a bus driver) to realise ends (for example, being transported from A to B). We may think of animal experimentation as a means to achieve certain ends. What are these? In Figures 9.4–9.8, the size of the arrow (the means) and the circle (the ends that are to be achieved with the help of those means) illustrate the significance, importance or value of the respective means and ends.

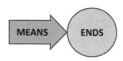

Figure 9.4 Means-ends-relationship.

First, we may ask whether the ends are *justified*? We can illustrate the justification, the legitimacy, or the value of certain ends by the size of the circle representing the ends in the scheme. With regards to animal testing, the ends are either scientific knowledge (as we have seen: almost 70% of animal experiments are for basic and applied research) and its dissipation (education), or efficient and safe drugs (through medical research), or efficient and safe consumer products. Many people see a strong necessity to perform drug testing (big ends), but cosmetic products are considered a *luxury good* that animals shouldn't suffer for (small ends) (see Figure 9.5).

A variation of the question of the legitimacy of ends is whether they justify the means. Are these achievements really so important that torturing and killing animals is an acceptable strategy? We can also ask about the size of the means: is the sacrifice of animals for science and research really such a big issue (Figure 9.6A), or are these *only* animals (Figure 9.6B)? Does it make a difference if we use chimpanzees, mice or fruit flies, or should we generally treat all animals the same? Wouldn't it otherwise be speciesism?

In addition to the *weight* of different means and ends, specific problems in the means-ends-relationships in animal testing have been pointed out.[14] First, the means-ends-relationship here is highly asymmetric. The ones who benefit from the ends (predominantly humans) are not the ones who are affected by the means. Or, in other words: the ones exploited as means (animals) have no benefits at all from the ends. Secondly, by referring to several empirical studies, it has been argued that the means are not effective in achieving the

Figure 9.5 Evaluation of ends. (A) Important, valuable, desirable ends, and (B) less important ends.

Figure 9.6 Evaluation of means. (A) The means have a high value, and (B) the means are less significant.

Figure 9.7 Mismatch between the means and ends.

ends (Figure 9.7). Animal models may not be adequate to relate to the conditions in human physiology, so that medical testing on animals may not be potent enough to provide valuable and reliable insights. Additionally, it is often sufficient to perform *in vitro* experiments (on tissue samples) rather than *in vivo* experiments (on living organisms). Literature on alternatives to animal experiments is abundant and has been growing in recent years.[15–17]

9.3.1 Consequentialist Arguments

The most common arguments in favour of and against animal testing follow a parlance of benefits and costs.[18] This resembles the paradigm of utilitarianism, which is a form of consequentialism, an ethical theory that evaluates the goodness of an action by the consequences: *Good* is what maximises the benefit of the largest possible number of affected entities.[19] We can regard this as an equation or drawing the balance sheet: we have a positive side (benefits, gains, increase of pleasure or satisfaction of needs – our ends), and a negative side (what has to be invested, sacrificed, and has negative side effects – here: the means), and we weigh these against each other. Arguments like *"Animal testing saves life!"*, or *"Animal testing is necessary to achieve the desired benefits!"*, are such a Utilitarian evaluation. The benefits outweigh the harm. We invest little and gain a lot.

Let's assume this situation: we start from the possibility of performing toxicity tests on cell cultures. With these simple and uncritical experiments, the insights allow estimations on toxicity but without reliable evidence. When we aim at acquiring more insightful knowledge, we need to change to animal models, so we substitute the cell cultures for mice, which allows reliable conclusions. Using the scheme, we would start from a situation with a small arrow (uncritical means: cell culture) and small circle (humble ends: hints for toxicity), and end up with a slightly bigger means (mice instead of cells) but a massively increased benefit (exploitable and regulation-relevant toxicological knowledge) (Figure 9.8).

Consequentialist considerations have been challenged. For ethicists like Peter Singer,[20] appraising the benefits for mankind without

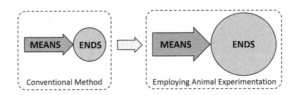

Figure 9.8 Higher gain (bigger ends) with a bit more means?

granting animals their place in the equation is not justified. Animals as ethically relevant entities have little to no benefit from those experiments (confirmed by the data presented in Section 9.1.1). Their suffering diminishes the value of the benefits so massively that it is impossible to regard such experiments as morally acceptable. There are two things we have to check at this point: Is consequentialism an appropriate theory to apply here, and are animals justifiably part of the equation?

9.3.2   The Trolley Problem

A famous thought experiment to question the validity of consequentialist reasoning is the Trolley problem.[21] Imagine the following situation (originally described by Philippa Foot):[22] A trolley (a tram car) is out of control and rolling down a slope, threatening to kill five workers on the tracks who don't recognise it approaching (or five otherwise *trapped* people). You stand at a lever that can switch the track to another one on which the trolley would kill only one person. If you do nothing, five people die. If you switch the lever, only one person dies (see Figure 9.9). Will you pull the lever?

The big majority of people exposed to this problem (~90%, varying in different empirical studies[24]) answer "yes". The simple consequentialist consideration is: one person dead is better than five people dead. Now we can introduce some special variations: what if the one person is a 2 year-old boy and the five people on the other track are all over 90 years old and terminally ill cancer patients? What if the one person is a member of your family and the five are all strangers? What if the choice is to be made between one of your family members and 1000 strangers? As we can see, these considerations often reach limits in which simple numerical calculation of benefits (or consequences in general) becomes impossible or insufficient. Here is a variation on the thought experiment: again, the trolley is approaching the five people on the track, certainly killing them if nothing stops it. You are on a bridge above the track, together with a very fat man. It is clear that if

Figure 9.9 The Trolley Problem, variation 1: Switching the lever. Image by David Navarrot, Reproduced from ref. 23 http://www.ciencia-cognitiva.org/?p=1147 under the terms of the CC BY 3.0 licence https://creativecommons.org/licenses/by/3.0/deed.es_ES.

you push down that fat man, he would block the trolley, saving the five people, but dying himself (Figure 9.10). Is it permissible to push the fat man down from the bridge to save the other five people?

In essence, the outcome is the same: one person dying instead of five. However, in this case, less than 10% of people respond that they would push down that man.[24] There is a clear difference between the first and the second variation: in the first case, all deaths can be regarded as an accident, an unlucky chain of events. Either the one person or the five are just unlucky. In the second case, however, pushing down the fat man (instead of simply pulling a lever) is an act of killing – murder. Whereas in variation 1 the death of the one person is *accepted* or *condoned*, the death of the fat man in variation 2 is *induced* or *caused* with an intention – a distinction that plays an important role in jurisdiction. Intuitively, for most people, reducing unavoidable harm to a minimum follows a rationale that is different from interfering directly in the life of possibly innocent people (bystanders that would not be affected by the harm without your intervention). Therefore, calculating benefits alone, obviously, cannot determine the goodness or rightness of a decision. Again, we can vary this scenario slightly: what if the fat man is a villain who put those five poor people onto the track? Would it be different? Under these circumstances, would it be permissible to push him down to save his victims and take his own death into account? Most people answer "yes".

The trolley episode tells us that another consideration that is beyond the mere consequentialist calculation of benefits is important: The rights of people and the ethical duties that stand behind actions and decisions to act.

Figure 9.10   The Trolley Problem, variation 2: pushing the fat man down the bridge. Image by David Navarrot. Reproduced from ref. 23 http://www.cienciacognitiva.org/?p=1147 under the terms of the CC BY 3.0 licence https://creativecommons.org/licenses/by/3.0/deed.es_ES.

9.3.3   Deontological Arguments

In Section 9.3.1, the basic consideration that motivates the idea of looking at the outcome of an action as the ethically relevant factor was the benefits (in terms of well-being, satisfaction, utility, for example) for human beings or other entities. Philosophers remarked that *well-being* is too vague a concept, very subjective, corrupt, or hedonistic. Another idea has many supporters in many cultural realms and civilisations during the last few centuries: rational beings like humans do good out of their duty to respect the rights and freedom of others. This is expressed, for example, in the *Golden rule* that is found in many cultures and societies and their philosophies around the globe: *don't do to others what you don't want to be done to you!* A famous variation is Immanuel Kant's *categorical imperative*: *Act only according to that maxim whereby you can at the same time will that it should become a universal law.* In practice, when you consider a certain action you have to ask yourself: if the motivation (*maxim*) for this action would be a law (or allowed by law), would that be a world (or society, or state) I'd like to live in, or a world that would be *reasonable*? For example: is *lying* a moral or immoral act? Can we imagine a society in which *lying as a strategy to achieve personal goals* is allowed by law that works reasonably well? A rational person with a sufficiently clear mind would surely find that a society of liars could never work out well. Therefore, we conclude that lying is unethical. That means, as the ultimate conclusion, from our rationality (especially: the rational insight that we *have to do good* in order to maintain social harmony) and our free will (especially: the ability to decide what to do) we derive the *duty to do good*. This idea of duty ethics – the moral obligation to act morally – is known as *deontology*, with Immanuel Kant as its most prominent advocate in the West.[25]

An alternative formulation of the categorical imperative serves the purpose of understanding arguments concerning animal testing better: *never treat **a moral agent** merely as **means**, but always also as **ends**!* Kant did not use the term moral agent as he referred solely to human beings. The term *moral agent* is introduced, here, as a response to bioethical considerations in contemporary applied ethics. This way of expressing the golden rule is applied by most animal rights activists – those who are not utilitarian (see above) but who argue that animals are moral agents that need moral protection. In animal experiments, they argue, animals are treated merely as means for ends that they have nothing to do with or that are not for themselves, which makes it unethical to perform such experiments. Illustrated as a means-ends-scheme, as in Figure 9.4, the three alternatives are that animals are, in any case, moral agents so that the means become big (a morally heavy investment); or they are not moral agents so that the means are small; or there is a gradual ascription of moral agency so that some species deserve physical integrity and freedom from harm while other species' exploitation for experiments is permissible. For a deontologist who grants rights to animals, the ends wouldn't even play any role: the instrumentalization of even just one animal for the safety of thousands of people is not morally justified.

For both utilitarians and deontologists the question whether and to what extent animals are *moral agents* is crucial to the result of their argument. The utilitarian needs to clarify whether animals play a role in the equation that assesses benefits and costs, while deontologists need to decide whether animals have non-instrumental value and, thus, rights *for their own sake*.

9.3.4 Centrisms in Bioethics

There are several positions in bioethics on who or what *counts* in ethical considerations. These are so called *centrisms*, based on what the respective worldview puts into the centre of its morality.[26,27] The most confined viewpoint is *"I only care about myself!"*. This is known as egocentrism and certainly not helpful for an ethical discussion. The next level would be a kind of *sociocentrism*: members of a society (for example a family, a clan, a village, a nation) care about the members of this particular society and give them ethical relevance. A variety of that, ethnocentrism, would define the group that carries moral status by ethnicity. It is obvious that these selective criteria are discriminative and may amount to fascism or racism. Therefore, reasonable philosophical approaches start at the *mankind* level (see Table 9.1).

Some might say, only the human sphere is important for us humans. We can reflect on ethical aspects only for mankind because other spheres (like animals, mountains, planets) cannot participate in the discourse, can't express what they want, and wouldn't understand what we conclude. Therefore, our only option is to conduct ethical debates from the human point of view and, as a consequence, ascribe moral relevance, intrinsic value, or moral standing (ethicists differ in their wording here) only to humankind. The rest is beyond our capability and responsibility. This is called *anthropocentrism*. It does not mean that animals, plants, the planet, and the cosmos don't matter to us! But if they do, we must understand it as a human interest! We must treat dogs nicely for the sake of our own good, not for the sake of the dog! Kant, for example, pointed out that mistreatment of animals has negative impact on the personal character disposition. Violence against animals just forms a bad character. The suffering of the dog is not a point at all.

Table 9.1 The centrisms used in bioethics to describe who/what is granted moral status.

Who/what has moral status?	Bioethical terminology	Ethical relevance
Myself	Egocentrism	None (or: unethical)
My community/society/ culture	Sociocentrism	
My ethnic group	Ethnocentrism	
All humans (mankind)	Anthropocentrism	Other entities may have instrumental value.
Every sentient being (able to feel pain)	Pathocentrism	Includes only some species (speciesism), depends on human knowledge.
The biosphere	Biocentrism	Includes all lifeforms, gradual ascription possible.
The entire planet (eco-system)	Ecocentrism	May include landscapes and weather/climate.
The universe ('everything')	Cosmocentrism/holism	May amount to ascribing no moral value to anything at all.

The next level of ethical relevance is a *pathocentric* viewpoint: every sentient being, at least every being that obviously feels pain and tries to escape from it, has a moral interest for its own sake. For some, that might sound somehow incomplete. Why only those organisms that feel pain? What about the border cases in which we just don't recognise the expression of pain because our knowledge of animal species is not elaborated enough? Taking the feeling of pain as a criterion bears the danger of *speciesism*, the discrimination of species by putting some above others in the same way as anthropocentrism puts mankind above other animals.

Biocentrist go further and grant all living organisms (including animals and plants, and maybe life forms of other planets) the status of *ethical entities* that have an intrinsic ethical value *for their own sake*. Some people argue that this is still incomplete: geological formations, mountains, landscapes, the entire planet with its sophisticated ecosystem and its rich biodiversity has an ethical value as such. This is an *ecocentric* viewpoint. It is also possible to extend the ethical realm to the universe, either as the entirety of its elements and parts (*cosmocentrism*) or the whole as such (*holism*): everything counts! Holism endorses the view that everything in the universe is connected and everything has its place. The harmony of *the whole* is, therefore, the fundamental value of all existence.

All these considerations come into play when we reflect on the environmental impact of science and technology. Do we only care about the well-being of human individuals or mankind? Do we want to protect the environment because we understand that we, as the human race, need an environment that maintains its integrity and function? Or do we want to protect it because we believe that *nature* has a value independent of mankind and human value ascription? Is it acceptable to take the pollution of a river into account as the price for a few more jobs in a nearby factory? May we shoot polluted materials (like, for example, radioactive waste) to the Moon, to Mars or into the sun to get rid of them, but risk polluting extra-terrestrial places? We will come back to questions like these in Chapter 11 on sustainability. Here, we confront two views that play a role in understanding arguments concerning animal testing:

- Anthropocentrist positions grant human rights but no animal rights. Animals may have instrumental or functional value, though.
- Biocentrists hold that animals have a moral value for their own sake, independent from human interests.

9.3.5   Overview of Positions

Now, we want to see how the two different ethics approaches (consequentialism and deontology) combine with the two bioethical centrisms. Within all four of the possible positions, we will find both pro and contra arguments concerning animal experiments (with one exception).

First, we consider an anthropocentric consequentialist. Such a person will assess benefits and risks that affect humankind and set them into perspective. An argument in favour of animal testing would be that the benefits for humans justify the relatively small sacrifice of animal lives. Yet – and this might turn it into a contra argument – the acceptability of the means (animal testing) strongly depends on the justification of the ends. Therefore, it matters significantly whether the animals are sacrificed for drug safety or for cosmetic products – not because the means suddenly have to be evaluated 'bigger', but the ends are 'smaller' (or less beneficial).

Then, we may meet a biocentric consequentialist who insists on making animals an impacting element of the risk-benefit equation: it is likely that this person will promote the *contra*-argument, that the suffering and death of animals shifts the balance so much to the negative side that human interests and benefits don't have sufficient justification power. Yet, there is a chance to formulate a *pro*-argument for a biocentric utilitarian: in cases in which the animals themselves, or the species, or the biodiversity of our ecosystem benefit from our experiment, animal research is ethically acceptable, because this increases the beneficial ends for those who suffer the adverse consequences as means to exactly these ends.

On the other side, we have deontologists. For them, a lot depends on whether animals are seen as moral agents or not. Anthropocentric deontologists mostly argue in favour of animal testing. Human safety and integrity – here understood as human rights – clearly outweigh animal rights (if they even have any). The ends justify the means. Possible counter arguments from this group of people might be that mistreatment of animals on a large scale leads to a serious form of alienation from nature and environment, or to character decay (as Kant argued concerning mistreatment of dogs). In case of biocentric deontologists, an argument in favour of animal experiments cannot be formulated. It seems, with this ethical conviction, one must in any case argue against animal testing. With this viewpoint, animals have rights for their own sake, experimentation on them is not justified at all.

Animals should never be instrumentalised as mere means. Table 9.2 summarises all these positions.

Equipped with this overview, its application can be practised with Cases 9.1 and 9.2. A confrontation with militant activists is not the right occasion for a philosophical debate. Yet, it should be expected from an educated chemist to try, at least, to de-escalate the situation with calmness and diplomatic reason. First, we try to understand the protesters' argument. As they claim to be *animal rights* activists, their view can be identified as deontological and biocentric: animals deserve full moral protection for their own sake. Rejecting their argument and endorsing the benefits for humans (product safety, reduction of health risks) will not reach them at all as they are not arguing on consequentialist grounds. Rather, their biocentric commitment should be embraced and turned into a counter-argument: the activists' protest is misdirected as: (a) the company itself doesn't perform animal testing (which can be proven by inviting representatives of the group to inspect the facility); (b) the agency doing the testing is fulfilling all the necessary guidelines to ensure the animals' welfare (see next section for details); and (c) these guidelines and regulations are rightfully in place because the legal organs (and, of course, the company and the chemist) agree with the activists that animal life is precious

Table 9.2 The pro- and contra-arguments of biocentric and anthropocentric consequentialists and deontologists.

	Consequentialist	Deontologist
Anthropocentrist	*Pro*: The benefits for humans justify the relatively small sacrifice of animal lives.	*Pro*: Human safety and integrity (as rights) outweigh animal rights. The ends justify the means!
	Contra: The acceptability of the means (animal testing) depends on the justification of the ends (purposes).	*Contra*: Mistreatment of animals might lead to alienation from nature and to character decay.
Biocentrist	*Pro*: Animal experiments are acceptable when they benefit the animals, the species, biodiversity, and so forth.	*Pro*: —
	Contra: The suffering and death of animals shifts the balance so much to the negative side that human interests and benefits don't have sufficient justification power.	*Contra*: Animals have rights for their own sake! Animal experiments are not justified at all! Animals shouldn't be treated as means. That would be an unacceptable instrumentalization!

and deserves our sympathy. Details of the animal welfare measures in the testing agency should be provided to prove the point. Certainly, some activists will not accept these counterarguments as they aim to stop all animal experiments. Yet, signalling that their point is not dismissed as do-gooder nonsense, but taken serious in the company's *corporate social responsibility* (CSR) concept will surely contribute to a de-escalation of the conflict and a more rational discourse.

In the context of a TV debate as described in Case 9.2, the toxicologist has a better opportunity to exhibit scientific integrity by going deeper into the bioethical considerations as proof that animal researchers are taking the issue seriously and trying everything in their power to reduce animal suffering. What are the positions of the stereotypical debate participants? The activist will likely represent the biocentric deontologist position, as in the previous case. The CEO, committed to legislation, CSR, product quality, and the commercial success of his company's product, will probably argue on anthropocentric consequentialist grounds: only unreasonable dreamers would assume that product safety and life quality come at no cost, and the sacrifice of animals, bred for exactly that purpose, is the inevitable price for this life quality. Nobody can seriously want the products to be tested on humans instead! As an instance of reason and scientificity on this public panel, the toxicologist is well advised to treat both co-debaters with respect and critical distance at the same time. That means, in both responses, he would first point out to which part of the other's argument he agrees, and then, what alternatives are that the opponent could, potentially, understand and agree with. As suggested above, the activist may be countered with biocentric consequentialism that embraces the view that animals have intrinsic value but, at the same time, allows the view that animal experiments are, under certain conditions, permissible. If it is the case that the cosmetics company that the CEO represents is performing unnecessary animal tests – of course, the scientist will enter the debate prepared and with this kind of knowledge – then both the biocentric consequentialist and the anthropocentric deontological view can be employed for expressing the critical stance towards the company's practices. In the former view, the acclaimed ends can be reached with other means (*in vitro* tests, machine learning based statistical analysis, for example) so that they don't have sufficient power to justify the sacrifice of animals. In the latter view, the rash but unnecessary consideration of animal testing as a tool for the company to reach its goals just shows the alienation of the entrepreneurial attitude from our environmental roots in which more respect for other species is due. This argumentative logic, at the same time, gives the toxicologist the plausible leeway to argue for his own engagement with animal experimentation. With guidelines followed and animal welfare assured, the careful planning of experiments and the reduction of animals subjected to research makes basic research involving animals an important good that serves both human and animal interests.

In addition to these discourse-related aspects, the arguments presented in this section play an important role in policymaking and regulation on animal protection. Chemists (toxicologists, for example) are requested to work with these regulations. We may say that ethical evaluation of animal experimentation is far beyond the core competences of chemists, so that interpreting and applying ethical guidelines like these is very challenging. On the other hand, the regulations in place define a clear framework and orientation (including technical tools) for dealing with animal testing issues, so that chemists are supported in making ethically sound and acceptable decisions.

9.4 Regulations on Animal Experiments

The considerations in Section 9.3 found their way into current regulations and guidelines on animal testing. The EU implemented rules and laws on how to conduct animal experiments.[28] The underlying principle is known as the 3R-guideline that claims that the following methods should be applied in animal experiment planning and conduct:[29]

- *Replacement*: If it is possible to answer a research question without using an animal, replace the animal with a methodology that does not use animals, such as cell studies or computer modelling. If it is possible to answer a scientific question using a morally "lower" species of animal, replace the "higher" species with a lower one.
- *Reduction*: If it is possible to answer a research question using a smaller number of animals, reduce the number of animals used.
- *Refinement*: Wherever possible, refine research methods, techniques, concepts, and tools, for example statistical analysis, to reduce the need for animals in research and to reduce harm to animals.

The EU Directive 2010/63/EU spells out the 3R principles and make them a legal requirement in the context of animal experiments for scientific purposes in which this term covers basic, translation and applied research, regulatory testing and production, as well as education and training. This regulation illustrates the attempt to mediate between animal welfare (a deontological animal rights approach) and scientific knowledge acquisition for human and non-human benefits such as the understanding of diseases, the safety of products, or the protection of biodiversity (a consequentialist approach that is widely interpreted as anthropocentric). The 3Rs aim at minimizing harm to animals under the assumption that the research that is done using them is motivated by justified ends, creating either scientific, medical, or social value. To ensure this, Shamoo and Resnik suggest a fourth R:[30]

- *Relevance*: Research protocols that use animals should address questions that have some scientific, medical, or social relevance; all risks to animals need to be balanced against benefits to humans and animals.

They add a fifth R that plays a key role in U.S. animal research regulations:

- *Redundancy Avoidance*: Avoid redundancy in animal research whenever possible—make sure to do a thorough literature search to ensure that the experiment has not already been done. If it has already been done, provide a good justification for repeating the work.

The 3R principle and its further advancements are not seen uncritically. The concept may be outdated, originally intended as a guideline in the behavioural sciences, and, thus, does not meet the requirements of regulating current animal research practices and purposes.[31] Yet, they constitute a helpful and pragmatic link between the bioethical considerations concerning animal research and the daily lab practice of those doing (or perhaps, having to do) animal experiments. On the basis of these principles and the legal regulations, practical guidelines for researchers are in place to support the ethical conduct of animal experiments[32] and the animals' well-being and welfare.[33] Researchers who want to conduct an animal experiment must provide a protocol that explains why the experiment is necessary. In other words, they have to apply to use animals and give good reasons for it. Such an application needs to address the following points:

- *Scientific Necessity*: It must be shown that the experiment delivers new insights (not repeating former experiments) and that the obtained knowledge is valuable.
- *The Appropriateness of the Animal Model*: It must be demonstrated how the information gathered from animals can be transferred to the human model (in medical research). Researchers are obliged to prove by thorough literature study that their chosen animal model serves the respective purpose.
- *Number of Animals Used*: The researcher must find the right balance between the principle of reduction (use as few animals as possible) and the statistical significance of the findings (as many as necessary). Good statistical design and ethical practice are intertwined, here.
- *Promote Animal Welfare and Reduce Animal Harm*: Researchers must make sure that the animals are treated with a minimum of suffering, pain, and torture, including appropriate use of analgesia, anaesthesia and euthanasia. Also, lab animals deserve proper living conditions and care.
- *Alternatives to the Animal Model*: The principle of replacement implies that researchers should check whenever it is possible to substitute animal experiments by other methods of knowledge acquisition, for example cell and tissue cultures or computer simulations.

This list can hopefully show that ethical research conducted in the context of animal experiments is not facilitated by ethical expertise (like the bioethical considerations above, or philosophical textbook knowledge), but by our chemical core expertise: judgments on experimental and methodological design, statistical models, heuristic and conceptual refinement of studies on the basis of scientific reasoning. After all, as this topic shows illustratively,[34] applied ethics disciplines like science ethics, research ethics, bio- and environmental ethics, as well as all ethical considerations concerning professionalism and responsible conduct in one's job, are always closer to the respective fields of competence and practice than to moral philosophy.

We are now competent to advise the graduate student in Case 9.3. First, she should study the experimental details and explanatory power of the alternative methods. For example, if she considers machine learning techniques with the input of toxicological big data,[35] it would be necessary to analyse the feasibility of this option in view of available computation competences and equipment, the quality of existing relevant data sets, and the validity of the models to enable viable interpretation of the results. With this homework done, the PI can be approached and challenged with a chance of success. It is important to keep the discussion at a research-related content level that focuses on the desired inquiry result, and not on sympathy with the animals. Avoiding the animal tests is possible with scientific means, not out of emotive or ideological reasons. The exercise in Box 9.1 provides a step-by-step guide for the preparation of such a conversation.

Box 9.1 Exercise: Arguing against Animal Experiments

Suppose you are the graduate student in Case 9.3 and meet with your advisor to discuss experimental design. To prepare for the meeting, use the matrix of Table 9.2 to map out the arguments to better understand your advisor's perspective.

Step 1: Identify the likely perspective (*e.g.* biocentric consequentialist, anthropocentric deontologist, *etc.*) of the graduate student's PI. Is the PI pro or contra?
Step 2: List as many possible arguments that you believe support the position of the PI. Although you may or may not agree, what reasoning is the PI using to understand the need for non-human animal experiments?
Step 3: Which arguments exist for the opposing position (pro or contra)? For example, if you believe that the PI is an anthropocentric utilitarian who is 'pro' the proposed experiments, what would an anthropocentric utilitarian 'con' position sound like?
Step 4: How could the details of the alternative methods (*i.e.* computations and cell studies) inform, support, or alter the position of the PI (Step 2) and the position opposing that of the PI (Step 3)?

Guiding questions:

- Do animals have the same rights as humans to live without being subjected to suffering?
- In the case study, substitute the mice for another species (*e.g.* pigs, fish, rabbits, dogs, *etc.*). Does your argument change in regard of the species? If so, why?
- Should access to time or resources influence whether non-human animal studies are appropriate?

Exercise Questions

1. With the discussion limited to animal experimentation on vertebrates, which of the following is not an issue in this context?

A: Dolphin
B: Fruit fly
C: Ape
D: Mouse

2. Which of the following positions is not accepted and, thus, not found in a bioethical discussion?

A: Biocentrism
B: Pathocentrism

C: Egocentrism
D: Anthropocentrism

3. Which animal is by far the most commonly used species in animal experimentation?

   A: Mouse/rat
   B: Monkey
   C: Fish
   D: Frog

4. Which experimentation area of research is unlikely to involve animal experimentation?

   A: Breeding.
   B: Veterinary medicine.
   C: Medical research.
   D: Basic research on atomic structures.

5. In the past, it was not unusual for dogs to be dissected alive in order to observe the beating heart. Some people strongly opposed such cruel experimentation. What term is used to describe these early animal rights activists?

   A: Dog police
   B: Antivivisectionists
   C: Luddites
   D: Caninophiles

6. What does the word vivisection mean?

   A: Operation on a living body without anaesthetics.
   B: Cutting something in live shows.
   C: The part of life that a person spends in public.
   D: An animated film.

7. Who of the following would not grant animals rights "for their own sake"?

   A: An ecocentrist
   B: A pathocentrist
   C: An anthropocentrist
   D: A biocentrist

8. Who or what is a stakeholder?

   A: Someone or something that is at stake.
   B: Someone for whom something is at stake.
   C: Someone who bought stocks of a company.
   D: Someone with a piece of meat at a BBQ party.

9. "Sentient" means...

   A: ...able to perceive and express feelings.
   B: ...knowledgeable.
   C: ...in a love affair.
   D: ...watching over something.

10. Which of the following is not among the "3Rs" as communicated by the REACH commission?

    A: Reduction
    B: Refinement
    C: Reproduction
    D: Replacement

11. In addition to the "3Rs", two other principles starting with the letter R have been proposed. What are they?

    A: Restriction and retribution.
    B: Rats and rabbits (as the only animals acceptable for research).
    C: Reanimation and resurrection.
    D: Relevance and redundancy avoidance.

12. For consequentialists, the main factor that determines whether an action is ethically favourable or not is...

    A: ...the outcome of the action.
    B: ...the educational background of the agent.
    C: ...the intention of the agent.
    D: ...the success rate with which the agent reaches a goal.

13. As the word suggests, deontology is concerned with...

    A: ...tooth ache.
    B: ...duty.
    C: ...stupidity.
    D: ...metaphysics.

14. A researcher is required to take care of the living condition for laboratory animals. Which of the following is likely to describe his duties?

    A: Releasing as many of them, as possible, into freedom once they are no longer needed for experiments.
    B: Offering them opportunities to reproduce (with the side effect that more specimens are produced for free).
    C: Providing food, shelter, and safety.
    D: Playing soft music that drowns out the scary background noises of the laboratory.

15. How does ECHA work to support the reduction of animals used for testing?

    A: ECHA sells the lab animals directly to researchers who need them.
    B: ECHA provides frequently updated reports on methodologies and practices that serve as valuable alternatives for animal testing.
    C: ECHA allows animal testing only for drugs, but not for cosmetics.
    D: ECHA collaborates with PETA and Greenpeace on public campaigns against animal testing.

16. Which one of the following statements makes sense?

    A: All anthropocentrists are necessarily deontologists.
    B: Biocentrists only look at the outcome of an action.
    C: Anthropocentrists and consequentialists frequently get into fights with each other.
    D: A biocentrist following deontological ethics would claim that animals, themselves, have rights.

17. The word "end" has frequently been used in this chapter. What does it mean?

    A: A stop to animal experimentation.
    B: A goal or purpose that is achieved using certain means.
    C: The last step of a development.
    D: A basic minimal agreement about something.

18. Which of the following is not a justification for animal experimentation?

    A: Scientific necessity.
    B: Appropriateness of the suggested animal model.
    C: Price of animals (*e.g.* "Rabbits are cheaper than monkeys or human research subjects.").
    D: Statistical significance as a reason for the number of animals used.

19. Which of the following is not of concern to "bioethics"?

    A: Animal treatment.
    B: Synthetic biology.
    C: The moral status of embryos.
    D: The unacceptably high level of unemployment among biology PhDs.

20. Which of the following describes the main conclusion from the Trolley experiment?

    A: Evaluating the ethical implications of alternative actions simply in numerical terms is counter-intuitive. Our morality does not work as simply as Utilitarians suggest!
    B: Beware of trains!
    C: People are very susceptible to slippery-slope arguments!
    D: Letting 5 people die "by accident" is still better than actively killing 1 person.

References

1. P. M. Conn and J. V. Parker, *The Animal Research War*, Palgrave Macmillan, Basingstoke, UK, 2008.
2. K. Taylor, N. Gordon, G. Langley and W. Higgins, Estimates for Worldwide Laboratory Animal Use in 2005, *ATLA, Altern. Lab. Anim.*, 2008, **36**, 327.
3. European Commission, 2019 report on the statistics on the use of animals for scientific purposes in the Member States of the European Union in 2015–2017, Report from the Commission to the European Parliament and the Council, COM(2020) 16 final, February 5th 2020.

4. Chemical Watch, *Feature: Animal testing in Asia – Harmonised Approach Unlikely*, October 23rd 2019, https://chemicalwatch.com/83443/feature-animal-testing-in-asia-harmonised-approach-unlikely accessed on August 16th 2020.

5. B. J. Cohen and F. M. Loew, *Laboratory Animal Medicine: Historical Perspectives in Laboratory Animal Medicine*, Academic Press, Orlando, USA, 1984.

6. A.-H. Maehle and U. Tröhler, Animal experimentation from antiquity to the end of the eighteenth century: attitudes and arguments, in *Vivisection in Historical Perspective*, ed. N. A. Rupke, Croom Helm, London, UK, 1987.

7. N. H. Franco, Animal Experiments in Biomedical Research: A Historical Perspective, *Animals*, 2013, **3**, 238.

8. C. R. Greek and J. S. Greek, *Sacred Cows and Golden Geese*, The Continuum International Publishing Group, New York, USA, 2000.

9. W. Harvey, *On the Motion of the Heart and Blood in Animals*, Part 3. The Harvard Classics, P.F. Collier & Son, New York, USA, 1909–1914 (original: 1628), Bartleby.com, 2001, **vol. XXXVIII**, www.bartleby.com/38/3/ accessed August 16th 2020.

10. I. A. S. Olsson and P. Sandøe, Animal Research, in *Encyclopedia of Applied Ethics*, ed. R. Chadwick (Chief), Elsevier, London, 2nd edn, 2012.

11. J. R. Garrett, *The Ethics of Animal Research: Exploring the Controversy*, MIT Press, Boston, USA, 2012.

12. D. Newton, *The Animal Experimentation Debate*, ABC-CLIO, Santa Barbara, USA, 2013.

13. G. Guitchounts, *Are Animal Experiments Justified?*, Nautilus, Issue 72, Quandary, 2019, http://nautil.us/issue/72/quandary/are-animal-experiments-justified accessed on August 16th 2020.

14. *Part 5: Animal Experimentation. The Animal Ethics Reader*, ed. S. J. Armstrong and R. G. Botzler, 3rd edn, Routledge, London, UK, 2017.

15. *Alternatives to Animal Testing*, ed. R. E. Hester and R. M. Harrison, Royal Society of Chemistry, Cambridge, UK, 2006.

16. *Methods in Bioengineering: Alternative Technologies to Animal Testing*, ed. T. Maguire and E. Novik, Artech House, Norwood, USA, 2010.

17. *Alternatives to Animal Testing*, ed. H. Kojima, T. Seidle and H. Spielmann, Springer, Singapore, 2018.

18. A. Knight, *The Costs and Benefits of Animal Experiments*, Palgrave Macmillan, Basingstoke, UK, 2011.

19. W. Sinnott-Armstrong, Consequentialism, in *Stanford Encyclopedia of Philosophy*, ed. E. N. Zalta, Summer edn, 2019, https://plato.stanford.edu/entries/consequentialism/ accessed August 16th 2020.

20. P. Singer, *Animal Liberation*, Avon Books, New York, USA, 1975.

21. F. M. Kamm, *The Trolley Problem Mysteries*, Oxford University Press, Oxford, UK, 2016.

22. P. Foot, The Problem of Abortion and the Doctrine of the Double Effect, *Oxford Rev.*, 1967, **5**, 5.

23. J. D. Corey and A. Costa, The foreign language effect on moral decisions, *Ciencia Cognitiva*, 2015, **9**, 57.

24. S. Bruers and J. Braeckman, A Review and Systematization of the Trolley Problem, *Philosophia*, 2014, **42**, 251.

25. L. Alexander and M. Moore, Deontological Ethics, in *Stanford Encyclopedia of Philosophy*, ed. E. N. Zalta, Winter edn, 2016, https://plato.stanford.edu/entries/ethics-deontological/ accessed August 16th 2020.

26. J. B. Callicott, Environmental ethics: Overview, in *Encyclopedia of Bioethics*, ed. W. T. Reich, Simon & Schuster Macmillan, New York, USA, Revised edn., 1995, pp. 676–687.

27. D. Jamieson, *Ethics and the Environment: An Introduction*, Cambridge University Press, Cambridge, UK, 2008.

28. European Commission, Animals used for scientific purposes, https://ec.europa.eu/environment/chemicals/lab_animals/index_en.htm accessed on August 16th 2020.

29. W. Russell and R. Birch, *Principles of Humane Animal Experimentation*, Charles C. Thomas, Springfield, USA, 1959.

30. E. Shamoo and D. B. Resnik, *Responsible Conduct of Research*, Oxford University Press, Oxford, 3rd edn, 2015.

31. J. Lauwereyns, *Rethinking the Three R's in Animal Research: Replacement, Reduction, Refinement, Palgrave Pivot*, Springer International Publisher, Cham, Switzerland, 2018.

32. *Principles of Animal Research for Graduate and Undergraduate Students*, ed. M. A. Suckow and K. Stewart, Academic Press, London, UK, 2017.

33. R. C. Hubrecht, *The Welfare of Animals Used in Research: Practice and Ethics*, Wiley-Blackwell, Chichester, UK, 2014.

34. T. L. Beauchamp and D. DeGrazia, *Principles of Animal Research Ethics*, Oxford University Press, New York, USA, 2020.

35. T. Luechtefeld, D. Marsh, C. Rowlands and T. Hartung, Machine Learning of Toxicological Big Data Enables Read-Across Structure Activity Relationships (RASAR) Outperforming Animal Test Reproducibility, *Toxicol. Sci.*, 2018, **165**, 198.

Part 3: Chemistry and Society

So far, we have looked at aspects of *good chemistry* from the perspective of methodology and of professionalism. Both concern the daily professional practice of a chemist – here understood as a scientist or researcher in academia or private sector research and design (R&D) – and his or her conduct that is expected to follow guidelines of scientific integrity. It has been argued that the obligation to maintain a high degree of integrity arises from the fact that chemical research is legitimized by social acceptance and the pay-off of its progress in society and its spheres (culture, policy, economy, etc.). However, whereas methodology is entirely a *profession-internal* topic, the research-ethical considerations in Part 2 are also about *technically doing a good job.* In this part of the book, the focus shifts towards social and environmental implications of chemistry and its professional actors. First, it will be justified that these considerations are of practical significance for chemists (Chapter 10). After understanding that chemistry is embedded in social values and interests, sustainability as a workable framework for the evaluation, protection and promotion of such values is introduced and explained in Chapter 11. The external responsibility of chemists as important agents who, among other agents, affect these values and their manifestation, is analysed and exemplified in Chapter 12. In practice, chemists contribute to public discourse on scientific and technological (S&T) progress in the form of risk assessment and management (Chapter 13) as part of national and international S&T governance strategies (surveyed in Chapter 14). A special form of outreach of chemical science and research is communication with the public or, more generally, chemical laymen, posing difficulties that are addressed in Chapter 15.

10 Chemistry and Values

Overview

Summary: In this Part 3 of the book, we attempt to understand the impacts that chemical activity has on society and the environment. With the same application-oriented approach of the other two parts, we will focus on the role that chemical professionals play in the network of actors and decision-makers. Yet, first, it is necessary to show in which way chemistry is not per definition neutral or value-free. It is not difficult to see how the chemical industry, private sector research and development (R&D) and innovation, but also regulation and governance of chemical progress, have social and environmental implications that can be assessed and *evaluated* (thus, necessarily, implying *values*). For chemistry as academic science and research it is not that obvious, though. Therefore, we will outline the ties between chemical science, technology development, innovation, and how progress is embedded in the social and cultural lifeworld of the people that it effects. Moreover, the chapter will introduce the contemporarily predominant social constructivist view of science and technology (S&T) progress by a short historical comparison with earlier paradigms. This will help us understand why reflecting on normative dimensions of scientific and innovation activity is not trivial or a waste of time, but an important element of research on how to make S&T progress sustainable and beneficial.

Key Themes: Chemistry as social sphere; chemistry as science and/or technology driver; neutrality theses; dual use; technological determinism *versus* social constructivism.

Good Chemistry: Methodological, Ethical, and Social Dimensions
By Jan Mehlich
© Jan Mehlich 2021
Published by the Royal Society of Chemistry, www.rsc.org

> **Learning Objectives:**
>
> This chapter shall convince you that:
>
> - Scientific activity is not neutral or value free but embedded in social practices and normative frameworks.
> - Science is a main driver and facilitator of technological development, and as such the subject of the same ethical considerations.
> - An ethical evaluation can't start at the application level in which it has a visible impact on society and the environment, but must start at the early development level (scientific research) in order to identify and push trajectories of development that are desirable and beneficial.

10.1 Introductory Cases

Case 10.1: Manhattan Project

[Real case] The standard textbook example for social responsibility of scientists is the Manhattan project, including Einstein's letter to US president Roosevelt, and the role of basic researchers like Oppenheimer. Since the discovery of nuclear fission by Otto Hahn and Liese Meitner in Germany in 1938, its scientific investigation and subsequent application was firmly connected to the historical political situation. Almost immediately after it became clear what incredible forces it could unleash, scientists (!) expressed concerns about the chance that Nazi-Germany could develop a weapon that exploits the devastating effects of nuclear fission. Under the leadership of the US government, thousands of scientists and staff worked on winning the race of being the first to have a nuclear bomb. Three weeks after the first successful test, two nuclear bombs were dropped on Hiroshima (August 6th 1945) and Nagasaki (August 9th 1945), killing 180 000 people instantly and probably up to 300 000 in the following years through radioactive contamination. Otto Hahn considered suicide after hearing the News from Hiroshima. The scientific leader of the Manhattan project, Robert Oppenheimer, quit the project right after the bombs have been dropped on Japan and fought against the further development of nuclear bombs, especially the hydrogen bomb. He was then accused of disloyalty by the US government.

Case 10.2: Agent Orange

[Real case] Arthur Galston discovered a chemical compound, 2,3,5-triiodobenzoic acid (TIBA), that inhibits the growth of leaves and, instead, increases the number of floral buds on soybean plants. In higher concentrations, though, it leads to defoliation (abscission of leaves) and the death of the plant. He learned later that the US army, interested in this chemical's defoliation effects, produced derivates of this compound and used the most powerful ones (known as *Agents Orange*, *White* and *Blue*) in the Vietnam war, having disastrous effects on the ecosystem, food chains, and the aboriginal culture and lifestyle. Galston remained extremely concerned about the implications of his work, not shying away from confrontation with the US government. He became a public voice of science that reminded fellow scientists of the inherent dual use potential of every scientific discovery. *"The only recourse for a scientist concerned about the social consequences of his work is to remain involved with it to the end!"*

Case 10.3: Boom!

[Real case] Prof. K. from M. is doing research on nitrogen-rich compounds that have promising potential as explosives. In papers and presentations, he explains their application in mining, construction industry (for example, tunnels) and space rockets. A closer look at his profile reveals that most of his work is funded by the army. It is obvious that most of his compounds are of interest in warfare products (missiles, bombs, landmines, *etc.*). Confronted with the dual use potential of his work, he states: *"See, I am just the scientist. I only deliver knowledge about how things work. I am not the one to tell others what to do with it. Of course, I envision that my explosives will make somebody's life easier, for example the miner or the engineer. I can't judge whether a war is justified or not, and I also can't see how my chemical compounds would have any implication for that question."*

10.2 Chemists' External Responsibility?

The standard textbook example for social responsibility of scientists is the Manhattan project (Case 10.1), including Einstein's letter to US president Roosevelt and the role of basic researchers like Oppenheimer.[1] It is employed to illustrate the historicity of science, the integral role of even the most basic science in technological application and, thus, in the inscription of values and worldviews in technological designs, the experiences and emotions of scientists who learn that their achievements are used for something they did not intend, and the power of science and technology in our society. It is acknowledged that the physicists who recognised the destructive potential of the newly discovered phenomenon of nuclear fission understood it as their responsibility to warn political authorities in the 'free world' of the disastrous possibility that the Nazis in Germany could develop a nuclear weapon. Firmly convinced that such a weapon in the hands of good people would serve only good ends (civil use like energy production, as envisioned by Fermi; or the sole purpose of deterrence but not dropping a nuclear bomb, as expected by Oppenheimer), scientists engaged wholeheartedly in the Manhattan project, only to find later that they are significant elements in the causal chain of decisions and actions that led to the death of hundreds of thousands.[2,3]

Historians of science agree that this incident has changed how we think of science, its impact, the way we perceive of and discuss technological progress, and the role of scientists in this discourse.[4] Yet, it is not clear how a deeper reflection on this case generates helpful insights for contemporary chemists and the current generation of students. The times have changed and the conditions and concepts of S&T-related progress and its assessment are different – possibly, partly thanks to the Manhattan project. As we will see later, a major turning point in paradigm emerged during the 1960s and 70s. Since then, S&T discourse arenas have been established and improved. Offices of technology assessment have been installed by governments and parliaments in the USA and in Europe, advising policymakers on issues of scientific and technological progress. In recent decades, S&T-accompanying research on *ethical, legal and social implications* (ELSI) and on *responsible research and innovation* (RRI) has become obligatory for many large-scale research projects in the EU and its member states. In this fashions, scientists – including basic researchers, like academic university scholars – engage in roundtables and other forms of exchange in order to facilitate a more sustainable development of *socio-techno-scientific systems* like nanotechnology, biotechnology, cognitive and information sciences (and the convergence of all of these). The term indicates that the borders between the classical disciplines of (academic) science and technology (with engineering) have become blurry. Today, scientists don't need to be prominent (like Einstein) and write letters to presidents, but can easily communicate concerns on risks and other adverse effects of scientific progress in their field to other stakeholders that use this information for S&T assessment and sustainable decision-making.[5]

An influential proponent of social responsibility of all those with power and impact as professionals – including chemists and other scientists – was Hans Jonas, well-known for his book *"The Imperative of Responsibility"*.[6] This imperative – *"Act so that the effects of your action are compatible with the permanence of genuine human life"* – is explicitly demanded from scientists and researchers. We will elaborate in greater detail in Chapter 12 what exactly the responsibility claim on chemists is (or: may legitimately be). Yet, we face an important question here: is scientific activity always and, by definition, free from any restrictions, from value ascriptions and from normative judgment, so that it would be invalid and unjustified to claim social responsibility from scientists? Proponents of scientists' social responsibility often argue *via* the link between science and technological progress. Apparently, it is easier to spell out the connection between technology and ethical/social responsibilities. Therefore, we would need to identify the strong ties between science and technology, so that we can understand that the ethical considerations in technology ethics are also applicable and justifiable in the context of scientific activity. Three approaches are imaginable:

- *Congruence*: 'Science is just like technology'. Thus, what is ethically significant for engineers and technology designers applies in the same way to scientific researchers with academic degrees.
- *Linear relation*: 'Science as basic research precedes technology as the result of applied research'. Thus, science plays a causal role in the emergence of technology. Scientists have at least partial responsibility for the real-world manifestation of their scientific insights.
- *Complex network relation*: 'Science and technology are, together with other social spheres, parts of a complex network of potentials, influences, and competences'. Disentangling these complicated interrelations is tedious but worth the effort because normative claims can be plausibly and convincingly set in relation with different actors' particular competences, roles, and professional activities.

Those scientists who intuitively (or, perhaps, with a good reason) reject even the notion of 'S&T' – that seems to imply that there is no distinction between the two – mistake it to mean *congruence*. To make this clear: discussing 'S&T' in Chapters 10 to 15 of this book does not intend to ignore or reject the useful distinction between scientific and technological agency. The two spheres require different competences, different education, different epistemologies, and different evaluations concerning societal contexts and impacts. Thus, the first of the three approaches above is rejected. Yet, it would also be too simple to assume a clear border between what researchers (science) and engineers (technology) do. As explained in Chapter 1, the notion that science has only research ethics implications while technology is solely an issue for engineering ethics is not tenable. Thus, the second approach is rejected, too. This urges us to take the more difficult path and try to figure out the exact roles and impacts that different chemical professions have in the wider context of technological development and the real-world manifestation of epistemic and technical scientific progress.[7] This is in line with the pragmatic approach followed throughout the book: All that can be normatively required from chemical experts must be directly rooted in their particular competence, knowledge, and expertise.[8]

Whereas the chapters in Part 1 of this book were informed by philosophy and history of science, epistemology, and science theory, and Part 2 was built on insights in applied ethics, the elaborations in Part 3 have accounts in the sociology of science and science governance. We won't dive deeply into the theories of those disciplines. This chapter introduces basic ideas insofar as they are necessary to understand the practical implications on the competences it wants to convey: fulfilling the obligations that arise from plausible external responsibility ascriptions to chemists with solid judgment-ability, critical thinking, awareness of ethically relevant chemistry-related issues, and argumentative skills that empower the chemical expert to participate effectively in fruitful S&T-related discourses.[9] First, we need to debunk the claim of the value-neutrality of science (Section 10.3). Once it is accepted that scientific activity is, on all levels, pervaded by normativity, its role in technological progress can be assessed (Section 10.4). The ethical call for doing something about the impact of chemistry on our lifeworld (the ought-premise of that claim) is circumstantiated in Section 10.5.

10.3   The Neutrality Thesis in Science and Technology

We have learned that scientific reasoning follows certain rules that make it scientific. When these rules are undermined by fallacies and biases, the scientificity of the inquiry is at risk. This led many scientists to claim that scientific research must be value-neutral as normative judgments would introduce unscientific elements into the reasoning process and, thus, create biases. The underlying image is that of a good scientist focusing on *how things are*, and a poor scientist following interests, desires, and preferences (*how things should be*) that blur the clear reason and logic rationality of the inquiry. This confrontation of cognitive clarity as *scientific* and value judgments as *bias* is a misunderstanding. Neutrality, understood in analogy to the virtues of *objectivity* and *dedicated disinterestedness*, is undoubtedly important in the methodological sense.[10] Controversy arises when scientific actors confuse this methodological neutrality either with the claim that science has no non-scientific presuppositions or with the conviction that scientific output (knowledge) is value-neutral.

10.3.1 Neutrality as 'No Presuppositions'

For many natural scientists, the paradigm of naturalism implies the view that scientific universalism can only be achieved when scientific inquiry is stripped off all presuppositions. The scientific researcher is committed to nothing but the truth about the world. This amounts to the claim that scientific inquiry should rest entirely on so-called epistemic values and no other (moral, social, economic, political, *etc.*) values. A version of this distinguishes scientific neutrality as the opposite of dogma or ideology. Nothing of the latter should play any role in doing science.

Critics argue that this is impossible for the human mind.[11-15] All our reason, and even the fact that we apply reason, is affected by environment, upbringing, socialisation, culturalization, and the inevitable goal-orientedness of all activity. The scientist who plans and conducts a sophisticated experiment or a whole study makes many normative choices by selecting topics, setting parameters for deciding on the direction of the inquiry, or by formulating goals and criteria for the decision when the goal is successfully reached (or not). The simple fact that our perception is selective introduces presuppositions for the conduct of science.[16] Even more basic: the inductive character of most scientific reasoning requires normative factors for the interpretation and comprehension of data to be meaningful. This embeddedness of meaning construction in our social and cultural lifeworld does not stop at laboratory and institute doors. Identifying, embracing and working with these presuppositions may be more effective for an epistemic advancement of scientific agency than the denial of their existence.[17,18]

10.3.2 Neutrality as 'Only Means'

An often-heard claim is that *"science only produces knowledge!"*. It can be understood in two ways. First, it could mean that *science is only means*. It implies that scientific inquiry is a mere tool for knowledge generation. This form of instrumentalism, say the critics, is an untenable understatement of scientists diminishing the social influence and knowledge authority of the sciences. Chapter 2 has clearly shown that science is not a professional island detached from its social or cultural roots. As such, it serves ends and is clearly driven and motivated by it. Second, the claim could mean that *knowledge is only means*. An alternative way of expressing this is to say that *science (or: knowledge produced by scientific inquiry) only tells us what we can do, but never what we should do*. It is probably intended to see normative responsibilities only in those people and instances that make a decision on the latter, but that the mere potential of application does not justify any normative claims (thus, making it neutral).[19-21] This claim can be refuted easily.[22]

Scientific and technological progress create action potentials. It means, that *qua* the invention of new (types of) artefacts, processes, techniques, and so on, but also *qua* the generation of new knowledge and scientific insight, people have actions to choose from that they would not have without this progress. In its most elaborated form, in Bruno Latour's *actor-network theory*,[23] knowledge (same as artefacts or technologies as such) has the status of an actor that pulls argumentative equilibria for decision-making into a particular direction, the same as, for example, an influential human discourse partner would. Thus, also knowledge itself becomes a factor in making normative choices and choosing options for actions. As such, its potential is not a mere possibility, but already manifested upon the time of its *coming-into-existence*. Moreover, turning the abovementioned claim around, *should* always implies *can*: We can only choose from options that are possible. Therefore, whenever decisions are made on *should*-questions (for example, in S&T policy and governance, in R&D processes, even in academia, as explicated before), we need to clarify first what we *can* do and what the available options each imply. Here, scientific purpose is of great significance, if we don't want to leave the field to dogmatists, ideologists, post-factualists, or others who base their claims on non-scientific decision factors.

Many developments (technological artefacts, procedures, techniques, but also scientific insight and knowledge) are ambivalent in the sense that they can be exploited for something *good* and something *bad* (whatever that means). This is known as *dual use*.[24] This is easier to understand for technical artefacts than for knowledge bits. Thus, the first examples chosen are technical devices: it is said that washing machines are good because they facilitate doing the laundry and make life (here: housework) easier, whereas the guillotine is bad because it chops off heads and is an instrument for killing. Proponents of a neutrality thesis might argue that these artefacts are, strictly speaking, neutral, because you could also do other things with them (killing cats with washing machines, or chopping cabbage to feed the poor with guillotines), and that the normative judgment of the rightness or wrongness of the actual act is directed at the intention of the actor rather than the device. As a counter argument, the actions described first are the intended purposes that are *inscribed* by design into those artefacts. Such an attribution is more difficult for artefacts like kitchen knives or the internet, as these are conceptualised, designed and produced for multiple purposes. The more complex a machine or tool the greater is the ambivalence of good and bad application potentials. Whereas in the first case (washing machine, guillotine) the attributes *good* and *bad* are related to their intended means-ends-relations and intended purposes and applications, in the second case (knives, internet) such attributions either refer to the success rate with which ethically unacceptable unintended means-ends-relations are suppressed and disabled or to the expectation on the (side-) effects of the artefact on the life quality of current and future generations.

Now, we transfer these insights onto the chemical knowledge that chemical researchers produce.[25] Examples that come to mind, here, could be research on explosives (for mining, tunnel construction, space rockets, or for missiles and other weapons, see Case 10.3), or persistent organic pollutants (POPs) (enabling efficient and convenient devices like fridges, but harming the environment and humans after disposal). *Good chemical knowledge* does not refer to that knowledge that enables exclusively beneficial applications and developments, or in other words: knowledge that prevents dual use. This is impossible. Instead, good chemical knowledge is that which points out application potentials and risks rather than being silent on a neutral stance or hoping that knowledge becomes manifested in exclusively beneficial applications. A researcher on explosives, for examples, may choose funding sources or collaboration partners wisely (for example ESA or construction tech firms, but not the army or arms industry), control the application of explosives by patenting them, or creating public pressure by pointing out publicly what his/her research is intended for and what not. Of course, he/she could insist that only basic research into nitrogen-rich organic compounds is performed, but that would be blue-eyed and not very credible. Indeed, decisions are made by others beyond the control of the scientist (regulators, companies, engineers, *etc.*), but when it is true that science is the most valuable source for factual knowledge, it is also upon the scientists to point out all relevant information that affects the desirability or harmfulness of potential applications. For chemists, this is especially significant in terms of toxicological properties of substances and materials. This does not mean that it is the scientists' responsibility to prevent *bad use* (the negative side of dual use), but to be critical and communicate warnings whenever they identify potentials for misuse or potential side effects of chemical knowledge. Robert Oppenheimer (Case 10.1) and Arthur Galston (Case 10.2) are two historical figures who have demonstrated concern over the footprint they left with their scientific achievements. At their times, their activism was extraordinarily difficult and burdensome – Oppenheimer, for example, being accused of disloyalty by the US government. Today, being involved in public discourse about scientific development and its manifestation in the social lifeworld is easier and more welcome than ever before.[26]

10.4   Chemistry in Science, Technology, and Innovation

The diversity of chemical professions makes it difficult to address all of these in one book about *social implications of chemistry*. Certainly, some academic chemistry in the realm of basic research might have very little to no touching points with society besides its being published in scientific journals. Yet, most chemists with a university degree – chemical engineers included – pursue a career in which their chemical competence and expertise has an impact in the form of either knowledge (*episteme*) or things in the widest sense (resulting from a crafting process, *techne*). The theoretical and practical knowledge of matter and its properties makes chemistry a creative science.[27,28] Chemists, thus, play a vital role in science and innovative R&D. A third career opportunity for chemists is in public service, such as agencies, patent offices, regulatory bodies, and so on. Even in this field, decision-making informed by chemical expertise has an influence on the direction that S&T progress takes (for example, issuing patents, allocating research funds, setting threshold for pollutant concentrations, approving innovation strategies, *etc.*). When in the following the focus seems to shift away from chemistry as a science towards chemistry as driver and facilitator of technological progress, it does not mean that suddenly the book addresses chemical engineers and product developers but no longer chemical scientists. Aware of the conceptual and practical differences between science and technology, this section is based on a conception of *technology* that is broad enough to involve basic researchers in academia and chemists in public service as enactors and stakeholders in its wide network of agents.

Technology is best understood as a system of things, actions and knowledge that is embedded into complex social, cultural, political, and historical contexts.[29] Here, we are not so much concerned about particular artefacts (for example, the ethics and social impact of cars), but about implications of technological development that affects the individual and social life of people and the environment. Figure 10.1 presents a simplistic overview of the elements of technological development that also illustrates how complex this social sphere really is.

The arrow indicates that technological development proceeds in steps with starting conditions or initial phases, and with an output. This must not be understood in the temporal, but in a conceptual sense. There are elements that play an important role in the early phase of development while others are concerned with later stages. However, all elements impact the development at all times, not only at the beginning, the middle or the end.

On the input side, the most evident element of technology is knowledge (know-that, know-how, technical expertise). Without that, no artefact and no process can be invented. This knowledge can be scientific, but not necessarily. Artefacts are usually invented by non-scientists with creative minds and practical goals. The steam engine is a prominent example. Sparks of innovation can come from engineering knowledge, business and market knowledge, or simply lifeworld knowledge of a common problem that can be solved by technical means. In addition, thinking of technology as a societally embedded system, and also sociological (and related) knowledge, might be of importance for the success or failure of technological progress.

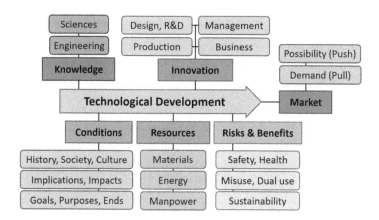

Figure 10.1 Elements of technological development.

Technological development needs more than knowledge to be initiated, facilitated and manifested in real-world items and processes. The environmental conditions (here not in the sense of natural, but social and cultural) must be *right* to serve as fertilizing ground on which technology can flourish. As seen above, artefacts and techniques are developed and applied as means to achieve certain ends. Manifold factors (society, culture, Zeitgeist, politics, history, *etc.*) determine what kind of ends are regarded as desirable or urgent and by what type of means they can be realised. Public acceptance and societal embedment of technologies follows complex culture-dependent mechanisms. Knowing these is a strong strategic advantage in innovation management.

On the other end, the output end, we have the factors that motivate the development: innovations are inventions that are established successfully on a market, either serving a specific demand (understood as *pulling* the development) or solving someone's problem with a novel possibility (as *pushing* the development). The difference is that a development can either be the response to specific articulated interests and desires (the demand for a technological solution for something), or the possibility-driven occupation of market niches that exist without prior awareness among customers and users.

Innovation itself is commonly located in the realm of industrial and corporate R&D. This ranges from start-ups with few people and resources, but great creativity and flexibility, to established large firms with huge resources but a slow and bureaucratic organisational structure. A diverse range of people have impact on the innovation process: developers and designers (sometimes engineers), producers and manufacturers, managers and executives, marketeers and traders. All of these professions could be occupied by chemists (here: people with a university degree in chemistry or related). It must be acknowledged, though, that in recent years more and more innovation and application-driven research is funded and carried out by public institutions and bodies. This can be observed in the bio- and nanotechnology eras since the late 1990s and is at a peak, currently, in energy-related research, the healthcare sector, mobility and transportation, and sciences involved in information technology, artificial intelligence, machine learning, cloud computing, big data, internet of things, digitalisation, and *industry 4.0*. All these mentioned fields of science, research and innovation are sophisticated entanglements of scientific and technological drivers in academia, industry and society. Here, we speak of *socio-techno-scientific* systems rather than technologies.

Every technology consumes resources throughout its life cycle from invention to production, dissemination, application to disposal or recycling. It requires energy, materials, and people. It means, every technological innovation comes at a price. It is possible, of course, that a progressive technology substitutes a less efficient or more wasteful previous technology, making processes or applications more sustainable. Yet, nothing is for free. As we will see in Chapter 11, chemists play an important role in global material flow management, optimisation processes, resource stewardship, disposal and recycling, and the environmental impact of human technology.

The efficiency of all these elements determines if the intended goals are reached and purposes are met. In addition to the successful or failed fulfilment of these ends, technology may also have various unintended side-effects and impacts. Generally, the effects of technology on social and environmental stakeholders are either *goods* – then we speak of *benefits* – or *evils* – then we speak of *risks* or harms. In all stages, the conditions, implications, and purposes of technology play an important role for the performance and role expectations of designers, producers, managers, and appliers. In order to minimise risks and maximise benefits, the most important considerations in each phase are related to safety and health aspects, to the possibilities of misuse or dual use, and to aspects of sustainability.

Some of these elements of technological progress are rather abstract. It is not clear who embodies them. Chemists as knowledge producers and innovators play their role in this development, but who are the other actors and stakeholders? The network illustration in Chapter 8 (Figure 8.1) provided some insights: academic chemists do their job in collaboration with many other scientists and, possibly, private-sector partners. Corporate chemists work in teams that are embedded into more or less complex organisational structures. Legislators and regulators, democratically legitimated by public mandate, provide legal frameworks and facilitate progress with funding and subsidies. In addition to these epistemic, economic, and political agents, four other stakeholder groups play an important role for technological development:

- **Consumers, Appliers, Users:** All technology – as knowledge-how, artefact, technique, or procedure – is produced for someone to exploit its intended functions and effects. Some technology (production machinery, reactors, industrial tools, *etc.*) advances industrial production, manufacturing, transportation, resource harvesting, agriculture, and other professional spheres. Other directions of innovation provide consumer products, ranging from daily life items to luxury goods. It can be stated confidently that every human on this planet is affected by technology in their job and in private life. The capabilities, intents, preferences and requirements of consumers and appliers of technologies constitute an important driving force of their design and assessment.

- **Third Parties:** There are entities that are affected by technological progress rather indirectly and without their choice or consent. These include marginal groups and minorities, future generations, non-human stakeholders like animals and plants, and those who reject the usage of particular technologies but can't escape the effect that these technologies have on society and their lifeworld.

- **Environment, Eco-system:** The conversion of materials and energy that drives technological progress exploits natural resources. The environmental impact of human activity is so massive that the current geological age is referred to as the Anthropocene. While severe adverse effects are ubiquitous in both real world and in mass media (pollution, destruction of habitats, change of landscape, loss of biodiversity, climate change, migration, to name a few), attempts to make technological processes more sustainable increase along with growing knowledge and the availability of alternatives. The environment with its sophisticated eco-system plays a double role as a stakeholder of technology: it supplies the process with materials on the input side, and it is affected by the application and operation of technologies on the output side. In both roles, the environment shares the characteristics of third party stakeholders: it has no own voice in the discourse (like future generations or animals), it has no own benefits from the ends for which it provides the means, and it never gave consent.

- **Evaluators, Assessors:** In recent years, another group of stakeholders became more and more important and involved: those who do not *do* (produce, apply) technology but *talk about it.* Evaluators and technology assessors analyse and study technological progress in order to understand its mechanisms, driving forces, and effects. Among them are social scientists and – for the evaluative part – ethicists. Their insights feed the decision-making processes in R&D, innovation management, technology governance and innovation policy.

All stakeholders, from scientists to designers, industrialist, entrepreneurs, workers, consumers, ethicists, regulators, and environment, have one collective interest: keeping the risks low and increasing the benefits. What exactly is understood as a risk or benefit is the subject of endless debate and conflict. The first step towards more progress in this discourse is the awareness that all these other stakeholders exist and operate towards the same goal. The goal of this section was to outline the range of players that the chemist in academia and in industry is cooperating with (here in a wider sense, not in a procedural sense). Chemical activity has an impact on the lifeworld of people, some of which is not intended and not desired. It has a firm connection with the environment and eco-system. Moreover, it is under critical scrutiny by those who assess and evaluate S&T progress. The practical implications of all these ties are the subject of Chapters 11 to 15. Yet, before we get there, one more missing link in the argument needs to be explained: what roles do values and norms play in technological progress, why would that matter to chemical actors, and what legitimises the role of technology assessors and evaluators that scrutinise chemical activity? We need to consult the sociology of science and technology, often labelled STS (which, depending on the country or the academic context, means *Science, Technology, and Society*, or *Science and Technology Studies*).[30]

10.5 Science, Technology, and Society

The understanding of the connection between S&T progress and society follows two very different paradigms in the recent history of technology studies (the analytical look from outside at technology). In the simplest possible way of putting it, one says that *technology shapes society*, and the other one says that *society shapes technology*. The former is termed *technological determinism* and holds that technological progress is inevitable, unstoppable and somehow *happening to us*. In contrast, the latter is called *social constructivism*, believing that technologies are embedded in a network of social demands, technical possibilities, cultural identifications, and various forms of social practices.[31]

Both positions find examples to substantiate their viewpoint. Determinists refer to historical examples like the hand mill that brought about the feudal system of the medieval ages with its lords and peasants and the steam mill that changed the society into an industrial one with bosses and workers (the separation of capital and labour). Constructivists present pathways of technological development that prove the dependence of technology on the knowledge of the time, the social acceptance and the economic mechanisms of demand and market potentials. The two viewpoints may be illustrated in this way: determinism takes the technological progress as a given constant continuous process following contingent trajectories, while the social development follows stepwise with more or less radical changes triggered by technological milestone innovations. For example, after the invention of book printing the society was a different one, the invention of the steam engine changed it again, the automobile did, and most recently the internet. In contrast, constructivism takes social development as the underlying contingent progress along historical and cultural lines, but the technologies come and go dependent on the stage and level of society (see Figure 10.2).

Technological determinism (not to be confused with the determinism debate in physics, challenged by quantum theory) was the predominant view until the late 1960s. Influential philosophers, sociologists, psychologists and cybernetics scientists at that time popularised and elaborated constructivism which gained much more influence. Today, it is hard to find any serious determinist. Constructivism is by far the most accepted conceptual framework for S&T development and its study.

This has significant consequences! The change from deterministic to constructivist technology models initiated a postmodern and post-positivistic understanding of technology. Most importantly, this implies that S&T progress is regarded as assessable, controllable, debatable, and designable. It does not simply *happen to us*, but we *make it happen* and are in control of it.[32] Moreover, S&T is not value-neutral, but its effects are matters of responsibility, justice, and fairness. Only on this basis (controllability, debatability) can S&T development become a public and political endeavour. In particular, it becomes

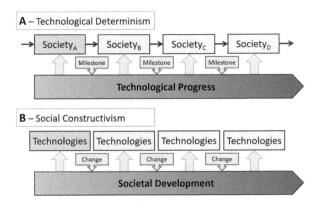

Figure 10.2 Paradigms of technological progress. (A) Technological determinism. (B) Social constructivism.

a subject of ethical discourse in the form of risk assessment, technology foresight and technology assessment (TA). Additionally, it is believed that technological progress is, in any case, more sustainable when the public is participating in its discourse, and not just various experts. Under the constructivist paradigm we see *the larger picture* of S&T (Figure 10.3).[33]

Figure 10.3 illustrates a development in several respects. First, it depicts a ranking of different assessment tools and concepts along the lines of levels of complexity. Second, it also describes advancements in the development of S&T-accompanying studies over time, namely the past three to four decades. Third, it outlines a connection between these concepts and the depth of the intended technology governance effect. The simplest and oldest form of dealing with S&T impact in scientific ways is empirical risk assessment that identifies hazards (for example, toxicity of substances), studies the exposure (how much? for whom? where? when?) and characterizes the risk on the basis of these findings. This risk is communicated and managed as best as possible. This approach responds to the risk perception and awareness of public or other stakeholders of a technological development. It turned out that this is not sufficient. In order to gain *public acceptance* – a basis for the success of a new technology – clear definitions of standards concerning environmental, health and safety issues (EHS) are necessary. Many companies defined these standards for their internal safety and quality measures, both for worker protection (labour rights) and for appearing trustworthy to the public (corporate social responsibility). In science and research, an increasing awareness of the importance of reflecting ethical, legal, and social implications (ELSI) of scientific activity could be observed. National or EU-funded research programs implemented work packages on ELSI. Some even talked about this

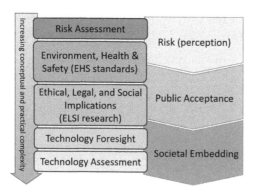

Figure 10.3 The larger picture of scientific and technological progress.

trend as *elsification of science*:[34] All scientific activity, academic and corporate, public and private, is the subject of public scrutiny and evaluation! Whereas these approaches still somehow separated the institutions *technology* and *public* – here the technology enactors that work on progress, there the public that has to accept the output – latest approaches aim at *societal embedding* of S&T progress. *Technology Foresight* as a sociological method aims to guide the development with proper methodologies like scenario analysis, modelling, assessments, and so forth. TA goes one step further and aims at enriching developments in interdisciplinary and transdisciplinary discourse on S&T and institutionalising the ELSI reflection as a governance tool.

Now that we understand the paradigmatic conditions for evaluating S&T progress, we can elaborate further on the relationship between science, technology, and other disciplines that have something to contribute to it. The common belief and, perhaps, plausible intuition are that science comes first and produces the necessary knowledge that, in the next step, is applied and exploited for the design and engineering of technical artefacts. First basic research, then applied research, then innovation. This view is contested by empirical research on the history of S&T.[35] The steam machine, for example, was developed by craftsmen (James Watt, Thomas Newcomen) who had no background in physics or other sciences. The practical problems and flaws of the steam machine that occurred in the years after its invention triggered a more systematic scientific study of thermodynamics and mechanics. In this respect, we can say, a technological challenge that engineers and craftsmen had to face inspired scientists to study it and provide solutions. Technology precedes science in most of the cases. Moreover, undoubtedly, man created artefacts long before the elaboration of a scientific methodology (Figure 10.4).

Then, as described above, for many centuries, there was the idea that technological progress is somehow inevitable and unstoppable, shaping the society that it is having an impact on. Common examples

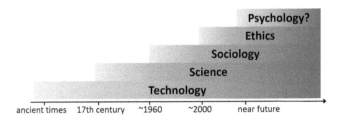

Figure 10.4 From technology to socio-techno-scientific systems and beyond.

(citing Karl Marx) are the windmill bringing about the feudal system and the steam mill inducing the transition to an industrial society. Around the 1960s, the paradigm shift towards *social constructivism* moved technological development and its risks and uncertainties more and more into the focus of social sciences. At the same time, in the face of nuclear threat and increasing environmental destruction, deterministic thinking (*"There's nothing we can do about it, anyway!"*) was replaced by constructivist thinking (*"We can intervene in the construction process!"*). There was big optimism that technology governance informed by empirical methodology-based scientific studies on S&T progress could reduce risks and choose future trajectories that maximise the (intended) benefits.

Around the 1990s and early 2000s, significantly triggered by the rise of biotechnology and nanotechnology, many scientific disciplines (the natural sciences, sociology, political sciences, philosophy, the Humanities) recognised the need for more profound reflection on ethical issues of S&T. The widespread, irrational, and in parts aggressive, opposition of the public against genetic engineering surprised the enactors of this S&T field and left them hamstrung. Great prospects (envisioned by the scientists, medical practitioners, politicians) were juxtaposed with significant moral challenges and imagined threats for humankind. The rise of nanosciences and nanotechnologies, efficiently supported by governmental and private-sector funding and cooperation, saw a growing interest in the ethical and social implications of such techno-scientific systems, almost as if the strategic mistakes of the genetics and biotechnology wave had to be avoided by all means. The major concern in nanoethics at that time arose from *unclear risks*: expecting risks without knowing what the particular risks would be, how strongly they impact and who is exposed. The challenge of established and previously unquestioned normative frameworks posed by technological innovation led to normative uncertainty and gave rise to a call for ethical analysis, as the common tools and reasoning strategies proved inefficient in light of the new emerging conflict potentials.

Now, another 20 years later, the normative discourse on S&T-related issues has moved out of the background and become an integral part of innovation and its governance. Experts still do their jobs and apply their specific competences and skills, but the daily professional conduct has become more interdisciplinary.[36] Chemists don't have to be ethical experts, but they interact to an increasing extent with actors outside of their core competence, engaging in normative debates concerning the impact of the scientific progress that they are part of.

Practical experiences have proven that this form of collaboration is the most effective and fruitful alternative to prominent figures writing letters to presidents (see example Case 10.1). The goal is to set the course for sustainable and socially sound progress as early as possible in socio-techno-scientific systems. The high degree of interdisciplinarity is demanded by the nature of this endeavour in which many epistemic agents depend on each other's sharing of knowledge and competence.

Note in Figure 10.4 that the borders between the disciplines contributing to shaping technological progress get more and more blurry towards the right side. While it is difficult, today, to draw a clear line between scientific and technological agencies, the social study of S&T progress has become part of the progress itself. Now, normative technology assessment is growing into a mature element of innovative processes. Figure 10.4 contains an element that indicates a future vision: it is quite possible that human demands, intents, capabilities, and preferences come into the focus of technology assessments. What matters to the people and why? Are the experts' estimations on what matters to the public always realistic and appropriate? What makes people purchase, use or reject a certain technology (besides sociological answers to this question)? How do people construct meaning from the existence and availability of technological artefacts? Currently, the psychological aspects of the social construction of technology are dealt with in the same way that ethical issues have been treated for a long time: somehow *in the background*, without granting it the level of *expertise* that it deserves. It was believed that *everybody can do ethics*. In the face of intractable conflicts, it turned out to be crucial, however, to include professional ethical expertise. Currently, it is the psychological aspects that are given only a marginal importance (for example, environmental psychology that studies the effect of wind turbines visible at the horizon on the perceived life quality of residents). Maybe soon in the future we will include social, environmental and cognitive psychologists in our S&T assessments.

Why does this matter for chemists? In accordance with the characterization of interdisciplinarity in Chapter 8, chemists do not need to worry about having to operate outside of their competence. What is more important is an awareness of the complexity of the contemporary assessment of S&T. They are sophisticated and elaborated enough to pay attention to more and more factors without losing the scientific character and normative plausibility that is necessary to efficiently influence and facilitate sustainable S&T progress. It is in this respect that values are important in chemical professions. In a collaborative effort, the values that are at stake are made visible. Then, the chemist takes the insights into the lab and aligns professional decision-making with the goals and purposes.[8]

From here, we will proceed in four steps: first, we need to define an operable normative framework for value discourses in S&T (sustainability, Chapter 11). With this in place, we can specify the particular responsibility attributions to chemists and their options to fulfil them (Chapter 12). Then, we need to define levels of risk discourse and what role chemists play in it (Chapter 13). Finally, it must be explicated where these discourses take place and how chemists with their competence and expertise can contribute to the more sustainable and socially sound progress of S&T (Chapter 14).

Exercise Questions

1. Which of the following is NOT stated by the neutrality thesis?

   A: Science only produces knowledge.
   B: Science is concerned about means, but not about ends.
   C: There is no "wrong" in science. Everything that scientists say is somehow justified.
   D: Science tells us what we can do, not what we should do.

2. "Manhattan project" is the name for...

   A: ...the project that developed the first nuclear bomb in USA during WWII.
   B: ...the first animal experiment with a monkey.
   C: ...a material science project that was looking for new materials for the construction of stable skyscrapers.
   D: ...a famous case of fraud at New York University.

3. Which of the following is stated by Hans Jonas in his *Imperative of Responsibility*?

   A: YOU are responsible for everything you do and say!
   B: Act so that the effects of your action are compatible with the permanence of genuine human life!
   C: Take responsibility, even for something that is not your business!
   D: Scientists have to make human life better, because that is what they are paid for!

4. Which of the following describes the connection between science and technology?

   A: Science is first, and its results lead to the invention of new technologies.
   B: Science is one key element of technological development (among others).
   C: They are distinct fields of professionalism: scientists do science, engineers work on technology.
   D: Science is theory whereas technology is practice (applied science).

5. In a social science context, the field of nanosciences and nanotechnologies is often referred to as...

   A: ...a political campaign.
   B: ...an economy boost.
   C: ...a grey-goo scenario.
   D: ...a techno-scientific system.

6. Which of the following is implied by the term "techno-scientific system"?

   A: It is difficult, in contemporary developments, to identify a clear border between (basic) research and technological progress.
   B: Science would not be able to achieve anything without support from technology (*e.g.* spectroscopes and other lab equipment).
   C: Contemporary societies live in technocracies in which their members are slaves of an intellectual elite building a cold engineered technical world.
   D: Education of scientists and engineers, which was once separated between universities and technical universities, has now become so similar that they may all have the same lectures and classes.

7. Which of the following is NOT considered a key component of technological development?

   A: Science and engineering.
   B: Natural resources and manpower.
   C: Production and management.
   D: Communism and socialism.

8. Which of the following statements is not tenable?

   A: Stakeholders are interested in keeping technology's risks as low as possible.
   B: Scientists contribute to the development of technology.
   C: Technology ethics is only a matter of engineering ethics.
   D: Designers are involved in technological developments.

9. Which of the following views is NOT employed to refute the neutrality thesis?

   A: Scientific progress creates action potentials.
   B: The expensive endeavour 'science' is built on social justification and, should therefore, be devoted to the public good.
   C: It is more beneficial for everyone when the scientific elite wields power than power lying entirely in the hands of greedy chief executive officers!
   D: The more knowledge we have, the more efficiently we can decide upon normative questions. Science increases this knowledge level so that it facilitates sustainable decision-making.

10. What is technological determinism? Choose one false statement from the following.

   A: The view that technology develops inevitably along certain trajectories while societies change accordingly.
   B: The view that technological progress is unstoppable and uncontrollable.
   C: The view that technology determines the course of history.
   D: The view that technologies will eventually eradicate mankind from this planet.

11. What is social constructivism (in the context of technology)? Choose one false statement from the following.

   A: The view that society shapes technology.
   B: The view that all infrastructure should be constructed with participation from the general public.
   C: The view that technological progress is the result of complex social influences like desires, demands, needs, and values.
   D: The view that technological progress is controllable, designable, and debatable.

12. Which of the following academic disciplines is NOT directly associated with or concerned about technological progress?

   A: Natural sciences
   B: Sociology
   C: Ethics
   D: Linguistics

13. Which of the following does STS stand for, in the context of this chapter?

   A: Science, Technology and Society (a subdiscipline of sociology).
   B: Search for Terrestrial Safety (a global program to develop a defence system against asteroid impacts on Earth).
   C: Sustainable Technical System (a label awarded to very eco-friendly home appliance devices like TVs and fridges).
   D: Self-organised Thin Superlayers (a nano-scaled innovative coating applied in house paint to keep the façade clean for many decades).

14. When did the paradigmatic move from technological determinism towards social constructivism occur?

   A: Around 1900.
   B: In the 1960s and 70s.
   C: Just recently (around the year 2000).
   D: It didn't. It is a utopian future vision.

15. Which of the following describes what it means to "see the larger picture" of S&T progress?

   A: A scientist is not narrowly focused on his/her scientific discipline.
   B: No technical innovation must be made without reflecting on the philosophy of being human and on the spiritual purpose of life.
   C: A science museum exhibits a large map with scientific disciplines, their interconnection with technological inventions, all illustrated along a timeline.
   D: S&T progress is not only evaluated technically in classical risk assessments, but also in terms of public acceptance, ultimately aiming at the societal embedding of developments.

16. Which of the following statements on scientific activity and societal responsibility is the only tenable one in view of this chapter?

   A: Scientific activity is always by definition free from any restrictions.
   B: Considerations in technology ethics are not applicable to scientific activity.

C: Scientists have a societal responsibility.

D: There are strict borders between science and technology.

17. In the context of technological progress, which of the following would NOT be of interest to psychologists?

    A: Can everybody do ethics?

    B: What matters to people and why?

    C: What is the consumer ethics of technology?

    D: How do people construct meaning from the existence and availability of technologies?

18. Who is the main driver of technological progress?

    A: Politicians

    B: Consumers

    C: Scientists and engineers

    D: The complexity of the issue doesn't allow a simple answer.

19. Which of the following is NOT part of a common definition of technology?

    A: Knowledge (know-how, know-what).

    B: Artefacts (devices, machines, *etc.*).

    C: Malfunctioning (everything that can be out of order).

    D: Procedures and techniques (*e.g.* industrial processes).

20. Which of the following is the only tenable statement?

    A: New nitrogen-rich compounds are good because they can be used for mining and space rockets!

    B: New nitrogen-rich compounds are bad because they can be used for bombs and missiles!

    C: Explosives are neutral. The person who uses them is acting either morally right or wrong.

    C: *Good chemical knowledge* on the synthesis of explosives refers to a chemist's competence to identify, describe, and prevent dual use potentials.

References

1. P. J. Gilmer and M. DuBois, Teaching social responsibility: The Manhattan project, *Sci. Eng. Ethics*, 2002, 8, 206.
2. *Oppenheimer and the Manhattan Project*, ed. C. C. Kelly, World Scientific, Singapore, 2005.
3. *The Manhattan Project: The Birth of the Atomic Bomb in the Words of its Creators, Eyewitnesses, and Historians.* Revised edition. ed. C. C. Kelly, Black Dog & Leventhal, New York, 2020.
4. *Science, Technology, and Society: Emerging Relationships. Papers from Science, 1949–1988*, ed. R. Chalk, American Association for the Advancement of Science, Washington, USA, 1988.
5. *Worldviews, Science and Us. Redemarcating Knowledge and its Social and Ethical Implications*, ed. D. Aeerts, B. D'Hooghe and N. Note, World Scientific, Singapore, 2005.

6. H. Jonas, *The Imperative of Responsibility: In Search of an Ethics for the Technological Age*, Universitye of Chicago Press, Chicago, USA, 1984.
7. *Science, Worldviews, and Education*, ed. M. R. Matthews, Springer, Dordrecht, Netherlands, 2009.
8. *Ethics on the Laboratory Floor*, ed. S. van der Burg and T. Swierstra, Palgrave Macmillan, Basingstoke, UK, 2013.
9. J. Mehlich, Chemistry and Dual Use: From Scientific Integrity to Social Responsibility, *Helv. Chim. Acta*, 2018, **101**, e1800098.
10. J. Reiss and J. Sprenger, Scientific Objectivity, in *Stanford Encyclopedia of Philosophy*, ed. E. N. Zalta, Winter edn, 2017, https://plato.stanford.edu/entries/scientific-objectivity/, accessed August 18 2020.
11. D. R. Mandel and P. E. Tetlock, Debunking the Myth of Value-Neutral Virginity: Toward Truth in Scientific Advertising, *Front. Psychol.*, 2016, **7**, 451.
12. S. P. Restivo, *Science, Society, and Values: Toward a Sociology of Objectivity*, Lehigh University Press, Bethlehem, 1994.
13. *The Dynamics of Science and Technology: Social Values, Technical Norms and Scientific Criteria in the Development of Knowledge*, ed. W. Krohn, E. T. Layton Jr. and P. Weingart, D. Reidel Publishing, Dordrecht, Netherlands, 1978.
14. L. Laudan, *Science and Values: The Aims of Science and Their Role in Scientific Debate*, University of California Press, Berkeley, USA, 1984.
15. *Science, Values, and Objectivity*, ed. P. Machamer and G. Wolters, University of Pittsburgh Press, Pittsburgh, USA, 2004.
16. *The Cognitive Basis of Science*, ed. P. Carruthers, S. Stich and M. Siegal, Cambridge University Press, 2002.
17. D. B. Resnik and K. C. Elliott, Value-entanglement and the integrity of scientific research, *Stud. Hist. Philos. Sci. A*, 2019, **75**, 1.
18. L. G. Christophorou, *Place of Science in a World of Values and Facts*, Kluwer Academic Publishing, New York, 2001.
19. K. C. Elliott, *A Tapestry of Values: An Introduction to Values in Science*, Oxford University Press, New York, 2017.
20. L. G. Christophorou, *Emerging Dynamics: Science, Energy, Society and Values*, Springer International Publisher, Cham, Switzerland, 2018.
21. *Current Controversies in Values and Science*, ed. K. C. Elliott and D. Steel, Routledge, Abingdon, UK, 2017.
22. H. Lacey, *Is Science Value Free? Values and Scientific Understanding*, Routledge, London, 1999.
23. B. Latour, *Reassembling the Social. An Introduction to Actor-Network Theory*, Oxford University Press, Oxford, 2005.
24. *Innovation, Dual Use, and Security: Managing the Risks of Emerging Biological and Chemical Technologies*, ed. J. B. Tucker, MIT Press, Cambridge, 2012.
25. M. Develaki, Social and ethical dimension of the natural sciences, complex problems of the age, interdisciplinarity, and the contribution of education, *Sci. Educ.*, 2008, **17**, 873.
26. *Why Trust Science?*, ed. N. Oreskes, Princeton University Press, Princeton, USA, 2019.
27. L. G. O. Dolino, Chemistry as a creative science, *Found. Chem.*, 2018, **20**, 3.
28. S. A. Matlin, G. Mehta, H. Hopf and A. Krief, One-world chemistry and systems thinking, *Nat. Chem.*, 2016, **8**, 393.
29. *A Companion to the Philosophy of Technology*, ed. J. K. B. Olsen, S. A. Pedersen and V. F. Hendricks, Blackwell Publishing, Chichester, 2009.
30. *The Handbook of Science and Technology Studies*, ed. E. J. Hackett, O. Amsterdamska, M. Lynch and J. Wajcman, MIT Press, London, 3rd edn, 2008.
31. W. E. Bijker, Social Construction of Technology, in *A Companion to the Philosophy of Technology*, ed. J. K. B. Olsen, S. A. Pedersen and V. F. Hendricks, Blackwell Publishing, Chichester, 2009.

32. *Technology and Society, Building Our Sociotechnical Future*, ed. D. G. Johnson and J. M. Wetmore, MIT Press, Cambridge, USA, 2009.
33. *Shaping Emerging Technologies: Governance, Innovation, Discourse*, ed. K. Konrad, C. Coenen, A. Dijkstra, C. Milburn and H. van Lente, IOS Press, Netherlands, 2013.
34. D. G. Stein, ELSI: Creating Bureaucracy for Fun and Profit, *AJOB Neurosci.*, 2010, **1**, 21.
35. B. Gremmen, The Interplay between Science and Technology, in *A Companion to the Philosophy of Technology*, ed. J. K. B. Olsen, S. A. Pedersen and V. F. Hendricks, Blackwell Publishing, Chichester, 2009.
36. *Interdisciplinarity – Reconfigurations of the Social and Natural Sciences*, ed. A. Barry and G. Born, Routledge, Abingdon, 2013.

11 Sustainability

Overview

Summary: In the previous chapter, we outlined the role of values and societal factors in and for scientific and other chemical activity (research, innovation, industry, business). We have seen that the complexity of the debated issues can overwhelm professionals and practitioners with chemical expertise to a point that any ethical or social consideration is rejected and delegated to other actors and stakeholders. The overall goal of this chapter is to systematise and, thus, *tame* the discourse on values implicit in science and technology (S&T) development. The most prominent, most widespread and best accepted approach is *sustainability*.

In recent years, sustainability became a key concept in environmental and social politics, both at the local and global levels. However, it is often not clear what exactly people mean when they use this term and what it implies in particular. Let's try to bring a bit of light into it. Please keep in mind that we do this because we want to understand it as the reason for reflecting on ethical aspects of S&T progress. The principles of sustainability will give us a framework for the evaluation of risks and benefits of S&T.

This chapter sets the scene for the following chapters. It is necessary to understand that evaluations of risks, responsibilities, desirable or undesired developments of science and technology proceed in professional realms (governance, commissions, academic and economic decision-making) in discourses among stakeholders on the basis of plausible principles of justice and fairness. Generally, the question is *"How do we want to live, and how can we make sure that future generations also have the freedom to ask this question and decide upon it?"*. This is the idea of sustainability.

Key Themes: Concept and definition of sustainability; sustainability as a normative framework for S&T discourse; sustainable chemistry; sustainability and ethics (values in the *sustainable chemistry* discourse; sustainability as value co-creation; needs and necessities).

Good Chemistry: Methodological, Ethical, and Social Dimensions
By Jan Mehlich
© Jan Mehlich 2021
Published by the Royal Society of Chemistry, www.rsc.org

Learning Objectives:

Upon completion of this chapter:

- You as an actor in S&T development – here: a chemical professional – will understand what sustainability implies and what it means in practical terms for your job.
- You will be able to evaluate various stakeholders' interests, identify overlaps and conflict potentials, and mediate between different values applying principles of justice and fairness in your decision-making.
- You will have the skill to analyse the consequences of your decisions in terms of sustainability, so that related processes (in research and development (R&D), in industry, in economy) become, indeed, sustainable.

11.1   Introductory Cases

Case 11.1: Competing Interests

A metal-working firm produces metal parts that come out of the production process with a greasy film. Before shipping them to their clients, they must be cleaned using a solvent. This solvent is purchased from a chemical supplier. The amount of used solvent is rather large owing to inefficient cleaning procedures, spillages and workers' incompetence in carrying out the cleaning procedure. The supplier, however, seems reluctant to support the metal-working firm in solvent handling as an increase in efficiency would go along with selling less solvent. The result is more chemical usage than is necessary, higher costs, higher risk in transportation, storage and disposal of the solvent, and a higher environmental burden. Is there any way to reduce the solvent usage without violating any actors' interest (including the supplier's)?

Case 11.2: Chemical Complexity

The result of synthetic chemistry as a productive creative process is a large and ever-growing number of new compounds and substances with potential applications in industrial and academic R&D. Safety regulations demand that all chemicals that workers, users and third parties possibly get in contact with must be properly characterised and handled. Yet, once a chemical registry is maintained, it is already outdated and incomplete. How can the gains and objectives of a chemical registry – safety, completeness, correctness, feasibility – be weighed against the exorbitant requirements on its maintenance? How can a reasonable load of scientific and administrative work power best be invested in creating a chemical data base that supports industry, R&D, business AND the wider public?

Case 11.3: Whose Value?

Geologists have located a large amount of a rare earth element rich ore. Unfortunately, its mining would immensely affect a national park that is one of the last protected natural habitats of an endangered animal species. The mined resources are industrially extremely important, used worldwide in IT consumer goods, generally believed to increase life quality and global equality. How can the economic and social values be compared to or weighed with the environmental values (biodiversity, health of the ecosystem, *etc.*)? What role can and do chemists play in the discourse on such problems and their practical solutions?

11.2 The Basic Idea

In recent years, sustainability has become a key concept in environmental and social politics, both on local and global levels. However, it is often not clear what is exactly meant by it and what it particularly implies. In the following attempt to define the concept and meaning of sustainability, it should be kept in mind that we do this because we want to understand it as the reason for reflecting on ethical aspects of chemical progress. The principles of sustainability will give us a framework for the evaluation of the risks and benefits of chemistry as an important element in S&T.

As the word sustainability suggests, it describes the ability of something (an event, an incidence, a development, a process, a phenomenon) to *sustain*, that means to keep itself proceeding smoothly and continuously without exhausting itself and without drifting towards decay or catastrophe. One of the oldest records of sustainability – without calling it as such – might be a statement by the ancient Chinese sage Mengzi (or 'Mencius' in Romanised transcription):[1]

> 不違農時, 穀不可勝食也; 數罟不入洿池, 魚鱉不可勝食也; 斧斤以時入山林, 材木不可勝用也. 穀與魚鱉不可勝食, 材木不可勝用, 是使民養生喪死無憾也. 養生喪死無憾, 王道之始也. [If the seasons of husbandry be not interfered with, the grain will be more than can be eaten. If close nets are not allowed to enter the pools and ponds, the fishes and turtles will be more than can be consumed. If the axes and bills enter the hills and forests only at the proper time, the wood will be more than can be used. When the grain and fish and turtles are more than can be eaten, and there is more wood than can be used, this enables the people to nourish their living

and mourn for their dead, without any feeling against any. This condition is the first step of royal government.] (Mencius 1A3)

Mengzi's three examples (agriculture, fishing, forestry) are crucial for society and its survival. In this advice for a leader, he recognises that the most important factor for long-term social stability is to make sure that the applied techniques to generate food and exploit natural resources enable a steady recovery of resources so that they don't run out. Natural balance and harmony – in good accordance with Chinese cosmology (Yijing), Daoism and Confucius' teachings – must not be messed up as are the basis for human life. In other words: agriculture, fishing and forestry can only *sustain* when they are done *in the right way*. His three examples can serve as an illustration of three different notions of technology:

- Husbandry (or agriculture): This represents a system of knowledge as a guideline for action, in other words: science and technology as *know-how* and *know-that.*
- Fishing nets: Here, technology means *technical artefacts.*
- Axes and bills: As particular ways to cut wood in the forest, these represent technology understood as *techniques* or *procedural methods.*

Science and technology, therefore, can be sustainable or not in view of their application and impact. Know-how (if not ignored) has the potential to manage activities in a way that they become more sustainable. Technical artefacts can be sustainable or not through their particular design and usage. Techniques and methods can be sustainable or not through the impact they have on the environment in which they are applied (landscapes, societies, social (sub-)systems, *etc.*).

Mengzi also gives a reason for the importance of sustainability: *"This enables the people to nourish their living"*. For him as a follower of Confucius, that means above all that they have time to cultivate and follow their rituals (*li*). Only when their existential needs are fulfilled through knowingly sustainable production methods, they will have the temporal, mental and spiritual capacity to actively make their life quality better. This could be taken as an example for an anthropocentric view on sustainability. Grain, fish, turtles and the forest are not of interest for their own sake, but in view of their importance for human well-being. For modern approaches to sustainability aspects of S&T development and innovation, we must keep in mind that progress is made and evaluated by humans for human purposes. This, however, does not make its assessment necessarily anthropocentric! Scientific and technological advancements clearly have an impact on all the above mentioned spheres (society, environment, economy), either as direct or indirect effects, which can both be either intended or totally unexpected. Whether these effects are evaluated as positive or negative depends on how we apply ethical reasoning (which we will do later).

Finally, Mengzi also explains where (in which realm) sustainability is debated: *"The first step of royal government"*. Sustainability is a political task. Stakeholders in the discourse can be manifold from all kinds of institutions and organisations, but in the end, it must be politically manifested in the form of decisions, action plans, agendas, regulations or laws. The underlying ideal is that politics' priority is the governance of different interests and viewpoints in order to promote and support the well-being of the entire society and its environmental foundations. Politics is also the arena in which ethics leaves the academic surrounding of pure theoretical reflection and enters the stage of real-life relevance and pragmatic application.

11.3 Contemporary Definitions

In the Western world, the history of sustainability as an important principle is much shorter.[2] After many centuries of a paradigm that suggested endless resources and unlimited exploitation of *God's creation* (Earth as a gift of God to mankind), deeper systematic and scientific insights into nature and society facilitated a higher awareness of the vulnerability of the environment and the limits of exploitation. The first record of sustainability (in its German word, *Nachhaltigkeit*) is from a German forestry regulation in the middle of the 19th century. It explains how to maintain a forest in a *sustainable* way by finding the balance between cutting wood and letting trees grow (note the similarity to Mengzi's advice).

In 1980, the *World Conservation Strategy* was produced by the International Union for the Conservation of Nature, the World Wide Fund for Nature, and the United Nations Environment Programme. This first international attempt to apply sustainability on a political scale was driven by non-governmental organisations. Soon, this pressure led to political action: in 1987, the *World Commission on Environment and Development* (also known as *Brundtland Commission*, named after its chairman) produced the report *Our Common Future* with strategies for global environmental balance and a call for action in the face of progressing environmental destruction and pollution. Another milestone was the *United Nations Conference on Environment and Development* (UNCED) – the Earth Summit – in Rio de Janeiro in 1992. Here, conventions on climate change and bio-diversity, guidelines of forest principles, a declaration on Environment and Development, and an extensive international agenda on action towards sustainable development for the twenty-first century (Agenda 21) was presented as a major output. Sustainability was also a topic at the *United Nation's Millennium Summit* in 2000 in New York, but with a slightly different focus than the others: major outputs were more focused on *human affairs* rather than the environment, such as the Millennium Development Goals concerning eradication of poverty and hunger, universal education, gender equality, child and maternal health, combating HIV and AIDS, and global partnership. Worth mentioning here is also the *World Summit on Sustainable Development* (WSSD) in Johannesburg in 2002. Its report gives more details on what is understood as sustainability. The participants agreed upon a declaration that committed to "*a collective responsibility to advance and strengthen the interdependent and mutually reinforcing pillars of sustainable development – economic development, social development, and environmental protection – at local, national and global levels*".

This definition – sustainability as a matter of society, environment and economy – can help us understand sustainability as a regime between different interests of different *stakeholders* – instances or people for which something is *at stake* (Figure 11.1).

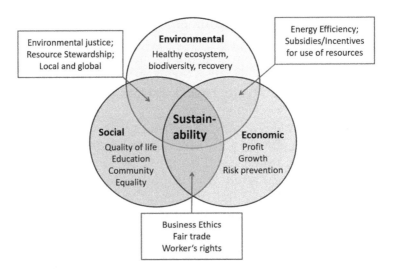

Figure 11.1   The triple-bottom-line definition of sustainability.[3,4]

There is a society with its members' interest in a good quality of life (whatever that means), in education, community (family and friend life), and equal chances for everybody. Then, there is an economic system as a special field of human activity with its interest in profit and growth, but also in risk prevention and smoothly running processes. In addition, there is an environment with its interest (as far as we can state that) in a healthy ecosystem, biodiversity and the steady chance of recovery from impacts. All these spheres may have internal conflicts of interest, so that we can speak of *social sustainability* when all interests within society are met, and analogously so for the other spheres. Our larger idea of sustainability, however, starts where the interests of different spheres overlap. The quality of human life, obviously, depends strongly on the health of the ecosystem, so that both instances (society and environment) have an interest in a balanced ecosystem. This can be achieved by following principles of environmental justice, for example resource stewardship (which is more than just *resource management*) on a local and global scale. Society's and economy's interests overlap in the call for business ethics, a healthy work force, protecting worker's rights, and fair trade. Economy and environment share the demand for energy efficiency (not wasting sources of energy) and a profitable use of resources. Ideally, human activity and respective decision-making on it take into consideration as many interests as possible and identify those options as the best ones that meet all three spheres' interests. These options, according to this *triple-bottom-line* definition of sustainability, constitute the overlap area of actions and decisions that deserve the label sustainable. In summary, we may express the principle of sustainability like this:

Sustainability means to govern human activities and decision-making in a way that current interests are met while the freedom of choices of future generations concerning the fulfilment of their needs is not diminished.

It is important to realise the two dimensions in which S&T have to do with sustainability: on the one hand, obviously, S&T progress enabled human activities that can now be identified as *unsustainable*, for example in the form of environmental destruction or social disruptions: S&T development as the problem. On the other hand, we are only able to understand what processes may be sustainable by scientific means, that means by research into environmental, social and economic processes. Moreover, many of these processes can be made sustainable by technological means, for example the improvement of existing technology or the invention of new techniques, artefacts and procedures: S&T development as the solution!

The most recent noteworthy UN event was the 2015 Sustainable Development Summit in New York at which Agenda 2030 with its 17 *sustainable development goals* (SDGs) was introduced (Figure 11.2).

The 17 SDGs in Figure 11.2 specify fields in which action towards sustainability is of special importance. For each of them, various instances in politics, economy, S&T, and the global public are requested to work together in order to achieve the desired goals. Here, we are especially interested in those goals in which chemical activity (in academia and industry) is of special significance.[7] The following eight SDGs may be the most obvious:

6. *Clean Water and Sanitation*: Owing to the global basic demand for clean water supply and healthy hygiene conditions, this goal is of special importance. New materials for filter technologies and improved analytical testing methods for the monitoring of contaminations are examples of chemical R&D facilitating progress in this field.
7. *Affordable and Clean Energy*: The future success of renewable energy depends to a large degree on the efficiency of respective technologies (solar and photovoltaic (PV) panels, wind turbines, storage devices, *etc.*). Chemists develop novel materials with a

Figure 11.2 The 17 sustainable development goals of the United Nations summit in New York, 2015.[5,6]

higher energy conversion efficiency, increased stability and durability, and better economic performance from supply to application to recycling.

8. *Decent Work and Economic Growth*: Without doubt the chemical industry is one of the major employers around the globe. Its sustainability is of major interest for many countries, societies and individuals, as well as for the environment. If it does not want to face the same fate as the 19th and early 20th century coal mining industry, with its decline resulting in high unemployment, as visible in the Ruhr area in Germany, the chemical industry is well advised to maintain innovative and dynamic conditions that enable it to respond to trends and progress. Then, job security and profitability may go hand in hand with creating benefits for the environment and society.

9. *Industry, Innovation and Infrastructure*: New compounds and materials, as well as novel fabrication and synthesis processes bring about new options and possibilities for innovative products and infrastructural changes, for example in roads and traffic systems, global and local communication infrastructure, industrial manufacturing and supply chain management, irrigation, energy supply, and so on. Chemicals that support the development and application of products with sustainable life cycles (from innovation to application to recycling and disposal) are highly desirable.

11. *Sustainable Cities and Communities*: This SDG concerns the maintenance and efficiency of urban systems like transportation, communication, energy, waste management, housing, and basic services (health care, education, food supply, *etc.*) in terms of life quality and social as well as economic advancement. As for SDG No.9, novel solutions and beneficial developments may be facilitated by chemical means (materials, processes, technologies).

12. *Responsible Consumption and Production*: Chemical progress may support the development of alternative materials and compounds in consumer goods such as household items, electronics, packages, and so on, and their production and disposal (or, desirably, recycling). Moreover, chemicals in agriculture play an important role for sustainable food consumption. Especially in contexts in which a reduction in the amount of consumption is difficult or impossible, substitution of critical compounds by more sustainable ones – a task for chemists – might be the more viable solution.

14. *Life below Water*: Chemicals in the global water cycle, for example persistent organic pollutants (POPs) like polychlorinated biphenyl (PCB) constitute a major threat for marine life and those who come into contact with it, for example the consumers of fish and other seafood. Facilitating the development and usage of eco-friendly chemicals in industrial production and agriculture (just to mention the most important fields) helps to protect life below water, while the removal of existing pollutants may be a second major task that involves the expertise and knowledge of chemists.

15. *Life on Land*: As for SDG 14 above, the substitution of chemical pollutants by safer and eco-friendlier substances can be achieved by chemical R&D. Here, too, existing threats can be reduced or removed by chemical means (purification methods like sequestering, precipitation, extraction, *etc.*).

11.4   Sustainable Chemistry?

In recent years, *sustainable chemistry* and *green chemistry* have become popular concepts in academic and industrial chemistry. Curricula with respective names, theoretical and practical courses, but also numerous journals, conferences and funding programs illustrate the culmination of a 60 year history of environmentalism and awareness of the chemical impact on the environment and society into a well-accepted discipline. Albini and Protti describe sustainable chemistry as "*a broad concept that deals with all aspects of making and using materials and chemical compounds in the man-built world, including safety and risk policy, remediation technologies, water purification, alternative energy, and, obviously, green chemistry*".[8] The latter is involved in the optimization of synthesis, in the use of renewable rather than non-renewable resources (both chemicals and energy) and in the qualitative and quantitative control of the artificial materials employed and produced (as well as of the accompanying waste). The concept, if applied and realised aptly, empowers chemical scientists and engineers "*to protect and benefit the economy, people and the planet by finding creative and innovative ways to reduce waste, conserve energy, and discover replacements for hazardous substances*".[9] Clark writes: "*The challenge for chemists and others is to develop new products, processes and services that achieve the societal, economic and environmental benefits that are now required. This requires a new approach which sets out to reduce the materials and energy intensity of chemical processes and products, minimize or eliminate the dispersion of harmful chemicals in the environment, maximize the use of renewable resources and extend the durability and recyclability of products in a way which increases industrial competitiveness*.[10] As a consequence, the goal of sustainable chemistry is not only to control chemical toxicity but to facilitate *"energy conservation, waste reduction, and life cycle considerations such as the use of more sustainable or renewable feedstock and designing for end of life or the final disposal of the product"*.[11]

Anastas and Warner developed a concise list of *12 principles of Green Chemistry*[12,13] that have been prominently promoted on the website of the American Chemical Society (ACS).[14] These guidelines for decision-making and action in chemical R&D contexts are:

1. *Prevention*: It is better to prevent waste formation than to treat or clean up waste after it has been created.
2. *Atom Economy*: Synthetic methods should be designed to maximize the incorporation of all materials used in the process into the final products.
3. *Less Hazardous Chemical Syntheses*: Wherever practicable, synthetic methods should be designed to use and generate substances that possess little or no toxicity to human health and the environment.
4. *Designing Safer Chemicals*: Chemical products should be designed to maintain their desired function while minimizing their toxicity.
5. *Safer Solvents and Auxiliaries*: The use of auxiliary substances (*e.g.*, solvents, separation agents, *etc.*) should be made unnecessary wherever possible and innocuous when used.
6. *Design for Energy Efficiency*: Energy requirements of chemical processes should be recognized for their environmental and economic impacts and should be minimized. If possible, synthetic methods should be conducted at ambient temperature and pressure.
7. *Use of Renewable Feedstock*: A raw material or feedstock should be renewable rather than depleting whenever technically and economically practicable.
8. *Reduce Derivatives*: Unnecessary derivatization (use of blocking groups, protection and deprotection, temporary modification of physical and chemical processes) should be minimized or avoided if possible, because such steps require additional reagents and can generate waste.
9. *Catalysis*: Catalytic reagents (as selective as possible) are superior to stoichiometric reagents.

10. *Design for Degradation*: Chemical products should be designed so that at the end of their function they break down into innocuous degradation products and do not persist in the environment.
11. *Real-time Analysis for Pollution Prevention*: Analytical methodologies need to be further developed to allow for real-time, in-process monitoring and control prior to the formation of hazardous substances.
12. *Inherently Safer Chemistry for Accident Prevention*: Substances used in a chemical process should be chosen to minimize the potential for chemical accidents, including release, explosion, and fire.

These themes seem plausible in the context of risk reduction, safety, protection of the environment, workforce and public. Yet, critics have expressed that these *green chemistry* guidelines do not meet the overall idea of sustainability, or at best touch the considerations of it (for example in the claim for renewable feedstock). Even though the principles address interests of stakeholders (society, economy, environment in the terminology of the triple-bottom-line), they hardly give orientation for identifying the overlap of interests and how to decide on an action that targets this overlap.

Cases 11.1 and 11.2 illustrate that sustainability and chemistry meet in ways beyond the idea of green chemistry as characterised above. Whereas the abovementioned guidelines predominantly address chemists in (academic) science and research, the two cases set their scenes in the private sector (industry) and in public service.

In Case 11.1, a metal-working firm buys solvent to clean specimens. The seller is interested in selling as much solvent as possible. Therefore, he benefits from the metal workers' lack of competence in the cleaning procedure because he will need more solvent than necessary. In this traditional fashion, the seller wants to sell as much solvent as possible, while the buyer wants to decrease the consumption to a minimum (but may not expect efficient support from the seller). The two actors, thus, have differing interests and 'pull in opposite directions'. As a result, more solvent will need to be produced, transported, applied and disposed.

An incentive to increase the efficiency of the cleaning and to reduce the volume of the required solvent is a change in the business model: According to a concept called *chemical leasing*, the solvent producer does not sell solvent but a *cleaning service* (or *clean metal parts*). The buyer pays for the amount of specimen being cleaned. Now, the business for the seller is better the less solvent is applied or wasted. His interest shifts to higher cleaning efficiency. Moreover, instead of transporting solvent, it could be an option to ship the specimen (less risk on the road). This is an example for a model in which a simple incentive changes the interests of a stakeholder so that all pull in the same direction: towards higher sustainability.[15]

Case 11.2, as the reader might have noticed after discussing the collaboration approach of interdisciplinarity in Chapter 8, describes exactly what is the challenge for the new EU chemical registry REACH.[16,17] Recall that REACH stands for *Registration, Evaluation, Authorisation and Restriction of Chemicals*. It entered into force on June 1st 2007 and has been extended and developed constantly. The European Chemical Agency (ECHA) is in charge of enacting the regulations and supervising their execution and fulfilment. On their webpage, they state:[18]

> *"REACH is a regulation of the European Union, adopted to improve the protection of **human** health and the **environment** from the risks that can be posed by chemicals, while enhancing the competitiveness of the EU chemicals **industry**."*

A core element of REACH is that it provides one coherent framework for new and existing chemicals. It finds a balance between a reasonable workload of updating the data base and a flood of new compounds and substances, especially in view of countless nano-scaled compounds that have different properties at different particle sizes. The idea is that the registry puts a priority on the reporting of those substances that are of relevance for industry and applications (sorted by various steps of production/consumption in tons per year). This form of organisation only works on the premise of good communication along the supply chain, from manufacturers, importers, distributors, to downstream users. By this, there is a shift away from public authorities towards industry. What may look like a burden for industrial stakeholders (and analytical chemistry) at first, is in fact an increased efficiency in regulation. Given the vast amount of new materials from chemistry and chemical engineering, no agency or other public authority would be able to manage the testing and regulation of all these chemicals in a feasible way. By shared competences and a clever combination of them, REACH is now one of the most sophisticated, but also most helpful, chemical registries worldwide.

11.5   Sustainability and Ethics

We have learned that sustainability is an empty concept when it is not filled with ethical arguments. Sustainability itself has no concrete ethical claims to make, besides providing a referential framework for justice and fairness as the major concepts. Sustainable is what appears *just* and/or *fair* to all involved stakeholders (including the environment, and future generations). It connects normative parameters like *quality of life*, *environmental justice* or *risk* with the actual consequences of human activity (like business or technological progress). We still have to *do ethics* in order to determine what *good quality of life* means, what *justice* is, or how to evaluate certain risk situations.[19,20] The crucial question is, again, towards whom one has an obligation to be just or fair and in respect of which values.

11.5.1   Values in the *Sustainable Chemistry* Discourse

These reflections lead us to this situation: we are looking for a value basis from which we can derive directives and guidelines for action in a deductive fashion. We want actors to realise by means of logic and reason that acting sustainably would be the best thing to do. Unfortunately, experiences across time and space show that people (and organisations, in some respect) act in self-interest and in view of expected utility. Moreover, changes in established routines and processes are notoriously difficult, facing oppression and resistance. How, under these conditions, can we induce (politically, or societally) a change in human practices towards greater sustainability? Generally, we may say, we need a convincing set of foundational values and use logic and reason to make everybody see clearly what is *the right thing to do* as the consistent consequence of those value commitments. More specifically, we need to provide incentives for actors that make them act sustainably *voluntarily* rather than by force. A good example for that is the *chemical leasing* business model introduced above.

A well-established strategy is to define a set of values that covers what is valued by technology enactors and stakeholders including the wider public and the environment ("third parties"). The ethical principles (like freedom, autonomy, privacy, *etc.*) are then defined by how they relate to those values. The most famous set of values is the 'Oktogon' by the VDI (German Engineers' Association).[21] Originally consisting of eight items (therefore the name), two are summarised in one so that there are seven boxes now (Figure 11.3).

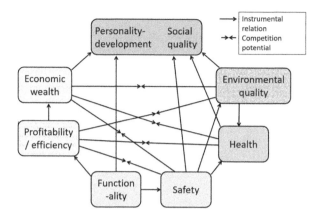

Figure 11.3 The VDI Oktogon of values in science and engineering practice.[21]

First, technology is evaluated in regards to its *functionality*, that is, how well it meets its proclaimed means-ends-relations. In addition, aspects of *safety* play an important role. From the economic side, significant values are *profitability* (or *efficiency*, depending on how to translate the German term *Wirtschaftlichkeit*) and economic *wealth*. The members of the society are interested in personal *health* and *environmental quality*. Moreover, the quality of life is affected by *social quality* (or balance) and options for *personality development*.

These items are connected in two ways. The first (indicated by a one-headed arrow) is an instrumental relationship: one supports (or stronger: is necessary for) the other. For example, the functionality of an item determines its safety and its efficiency. More safety means (usually) better health. Higher environmental quality also has a positive effect on health and social quality, and so on. The other type of connection is a competitive one (indicated by two colliding arrows), indicating that one cannot be supported without diminishing the other. Profitability often conflicts with safety aspects and health effects. An increase in economic wealth usually goes along with a decrease in environmental quality. With this scheme, it is then possible to classify arguments by supporting or neglecting one or more of those values, and to identify their support of or conflict with other arguments in favour of or against other values.

There are a few difficulties with this approach. It does not provide any orientation on how to prioritise or hierarchise these values. Exactly that, however, is necessary in a conflict situation, so that the value set does not bring the ethical discourse any further besides a few clarifications. Moreover, it is by far not clear why the VDI value list should be in any way complete or consistent. Alternative sets have been proposed, for example one by Hentig: *1. Life, 2. Freedom, 3. Peace, 4. Peace of mind, 5. Justice, 6. Solidarity, 7. Truth, 8. Education, 9. Love, 10. Health, 11. Honour, 12. Beauty.*[22] Of course, there are relationships between the VDI values and these ones here. Some (love, honour, beauty) are not represented as such by the VDI Oktogon. This shows that the value approach appears helpful in some respects, but still lacks ethical consistency and doesn't settle any of the most urgent ethical debates. Another criticism was expressed concerning the argumentative justification of those values. One option is a coherent-reconstructive reasoning that derives values from tradition and obvious social consent. *"We have always done it like this!"*, or *"For 200 years the society X is based on these values!"*. This approach is not very ethical but rather descriptive and, therefore, lacks ethical justification. Another option is intuitionism, a holistic conceptualisation of a coherent value system. Again, it is highly questionable whether this theoretical reflection has any justifiable foundations. A third option draws conclusions from reflexive reasoning from discourses or reflexive action.

11.5.2   Sustainability as Value Co-creation

The triple-bottom-line concept of sustainability provides a normative framework based on principles of justice and fairness. It facilitates a more integrative multi-stakeholder view and, above all, gives silent stakeholders like environmental entities and future generations a voice in the discourse. The goal of establishing such a framework is not primarily the protection of real (physical) goods and things, but the protection of values. Yet, the meaning of value and what it refers to is seen differently by different stakeholders and respective scholarly disciplines representing them. Moreover, in practice, the conceptual distinction between the spheres of environment, society and economy turned out to be rather an obstacle than a helpful orientation. Thus, it is suggested to redefine sustainability in terms of value creation rather than in terms of interest protection.[23]

The VDI Oktogon may serve as a starting point. It may be summarized as representing the value ascriptions from the viewpoints of engineering, economy, society at large, individual people (insightfully represented by psychology), and – as an outstanding discipline covering all areas – normative considerations such as ethics and law. The engineering value of a product is seen in its utility that results from its function-by-design, while the economic value is its price and profitability. For individuals, a product's value is mirrored in the satisfaction level of the user applying it. On the social level, satisfaction is not sufficient as a representation of the value of things (physical and abstract). Instead, we may look at the fulfilment of the interest in integrity (physical, mental, ethical). Last but not least, in normative sciences, the value of something is seen in it being *good/bad* or *right/wrong*. The design, production and dissemination of goods deserves the label sustainable when value is created in all five fields, that means when functionality serves certain purposes and desires, when economic profit is generated, when consumers are satisfied, when the societal integrity is maintained, and when ethical or legal norms are supported and/or protected (Figure 11.4).[24-26]

Why would this approach to sustainability be beneficial in practice? Let's take Case 11.3 as an example of a meaningful discourse on what to do and what chemists contribute to it. The mined ore contains an element that serves important functions in electronic devices. From an engineering perspective, the availability of that material enables the design and application of parts with desirable functionality. From an entrepreneur's economic perspective, the market share is the most important consideration. Having access to considerable amounts of a natural resource (perhaps in closer vicinity and with comparably low mining effort) supports a competitive price and, thus, grants a greater revenue. Consumers desire the utility they gain from exploiting the products' functionality. Thus, for them, mining the resource is a legitimate means for this end. Yet, the negative implications (destruction of unique habitat, loss of biodiversity, pollution) may have a drawback on societal integrity in the form of loss of life quality and, perhaps, immense follow-up costs for repairing the damage or reinstating the lost biotopes. From an environmental or bioethical perspective, it needs to be assessed whether the exploitation of natural and ecological means for human purposes is justified.

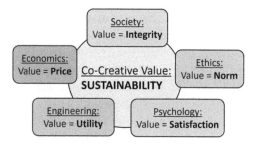

Figure 11.4 Sustainability understood as value co-creation.[23]

The special competence of chemists allows – or, perhaps, obliges – them to contribute important expert knowledge to the increase in value creation. Possibly, based on chemists' basic research, there are alternative materials that gain the same or even greater functionality without requiring mining. The chemical entrepreneur running a business somewhere in the down- or upstream industry (supplier, manufacturer, transportation, consumer goods, retail, recycling) may consider business models that grant a revenue or market share without relying on the cheap exploitation of depletable resources. Toxicologists and environmental chemists help understand the pathways, impacts and removal of pollutants in ecosystems so that this knowledge can facilitate the development of better mining and refinement methods. Chemical lifecycle assessments highlight how ecosystems, but also the human sphere as such, are affected by the exploitation and processing of natural resources. Last but not least, also the normative considerations in environmental and bioethics require a careful analysis of the postulated is-premises (see Chapter 1). Again, we meet the common theme: chemists are not more empowered than anybody else to tell us how we should want to live, but they, indeed, are empowered to inform our decision-making with valuable expert knowledge on the characterisation and impact of chemical compounds, their processing, and their application.

11.5.3   Needs and Necessities

There remains the question of how to prioritise or rank values in cases of conflicts that require a weighting. An approach inspired by Maslow's pyramid of needs is discussed in the following (Figure 11.5).[27,28] He distinguishes three levels of human needs, manifested in five steps of particular interests. The *basic needs* are the most fundamental physiological needs (enough food and water, sufficient warmth and the chance to rest) and safety needs (being free from harm and danger). Then, there are psychological needs such as belongingness and love (having relationships, family, friends) and esteem needs (feeling productive and being merited for one's accomplishments). Finally, people have self-fulfilment or self-actualisation needs (having hobbies, being creative, expressing and satisfying one's inner states).

This pyramid can be read in various ways. For example, the suggested hierarchy may be understood as an order of development of both human civilisation as a whole and individual human beings in particular. Another reading that is of relevance for our topic is the relationship between those needs and the granting of human

Figure 11.5 Maslow's pyramid of needs.[27,28]

rights. The more basic a need, the more we are inclined to grant the satisfaction of that need as a human right. It seems plausible to understand the physiological and safety needs as *more urgent* than, for example the need to have a hobby or a job. This hierarchy is also mirrored in international agreements on human rights protection and manifested in actual law-and-order systems. Third, there is an ethical reading in the pyramid, even though it is not clear whether Maslow or others who apply this illustration would think of it in this way: this pyramid may serve as an orientation for a hierarchy of rights. When two need-based rights collide, the one further down in the pyramid is to be prioritised over the one further up. Ignoring the more basic need or right in favour of one closer to the top is, then, considered unethical.

Meanwhile, this simple taxonomy has been commented, criticised, and refuted with empirical evidence that real social processes and organisation do not follow this model.[29] Yet, it may be saved by advancing it into a concept that has more depth and complexity. An additional or even supplementary *pyramid of necessities* may be postulated (Figure 11.6). The pyramid of needs does not say anything about the sources for the satisfaction of those needs. What must be given for a certain life quality? How can that be prioritised or hierarchised in order to come to insights that can serve as orientations for actions and decisions (such as sustainable S&T governance)?

The basic necessity for survival is environmental stability. Embedded into an ecosystem, human beings cannot survive without it. If the fine-tuned environmental balance is disrupted, the whole system will be affected, for example through changes in biodiversity,

Figure 11.6 Mehlich's pyramid of necessities.

food chains, climate, chemical constitution of the atmosphere, and so on. Environmental health is the basis for our food sources, for access to fresh water, for breathable air and the ecological niche of the human race. All anthropogenic activity (including system formation such as society, culture, economy, money, *etc.*) is dependent on it and, therefore, secondary to it. Second, human needs can only be satisfied when there is a certain level of social stability. In extreme cases (war, riots, anarchy, violence), this can affect the survival chances. In a more moderate sense, political stability provides autonomy and grants rights to the citizen that it is governing, thus enabling integrity. Here, integrity means inviolacy and the ability to act at all. However, it gradually (in the pyramid upwards) takes up the meaning of righteousness (ethical integrity) when the levels further down are taken care of.

The third level that corresponds to Maslow's belongingness and love needs is labelled ethical stability. This is understood as an atmosphere of trust and co-operation among family members, neighbours, colleagues, and peers (those in direct vicinity of one's life). Only in that kind of surrounding can people start building close ties and rely on each other, increasing each other's life quality by mutual support and collaboration. Only such a society is able to establish a system that offers livelihood options. This might be the most critical and debatable part of the proposed necessity pyramid. It implies that, as soon as a society reaches a certain level of integral peace and co-operation, people will feel the desire to act as parts of this society, bringing in their skills and abilities. They do that, as is assumed here, out of self-motivation and not because the social system forces them to. This fourth level in the pyramid is rather referring to livelihood options as a multitude of ways to unleash one's productivity potentials because that is what we naturally fill our lives with when the lower three levels are secured.

When survival is certain and the personal integrity secured, we start being concerned about our identity. We define ourselves through our social ties with family, friends and peers, but also – and maybe predominantly – through our social roles as competent experts in a particular field of skills or knowledge. Ultimately, when there is sufficient capacity and time for it, we form habits of thought or action that agglomerate to what we call culture. People use their creativity and intellect to engage with art, philosophy and spirituality. They choose hobbies (*spare time activities*) and fill leisure time with joyful and pleasurable endeavours. Some of those are part of the identity formation mechanisms, others are simply a *luxury* in the sense of *they are not really necessary for our life*. However, in any case, it is usually those aspects of life that give us the feeling that it is worth living.

As for the needs pyramid, the necessity pyramid can also be understood as a development description, an analogue to the one given above. More interesting – and the main reason why this way of putting it produces further insights – are the political and ethical dimensions in it. In both fields (politics and ethics) we ask *"What shall we do?"*. When taking this pyramid as a decision guideline, the answer is: *"Start at the bottom, fix the problems, and work your way up!"*. In reality, however, we observe trends that proceed in the opposite direction. Governments are eagerly promoting industrial aims for the sake of job creation and material wealth while resources and energy demands ruin the environment and the eco-system. The climate changes in an accelerated fashion under the influence of human activity, but important decision-makers and consumers seem not to care owing to the conveniences they desire on the fifth level (self-fulfilment needs and cultural necessities). Religious and societal conflicts dominate the News (for example terrorism, racism or homophobia, unemployment rate) while the serious global problems arising from atmosphere warming, pollution and species extinction are marginalised and only peripherally brought to people's awareness, at least not as an urgent issue, not to speak of one that is wholeheartedly worked on.

This brings us back to reflections on the sustainability of chemical activity and progress. Basically, the majority of philosophers and sociologists of technology regard the creation and usage of technology as the result of needs and desires. People invent and apply artefacts in order to make their life easier. The oldest known tools (if understood as technology) helped their users to ensure a sufficient supply of food, clothes, housing and warmth. Still today, many branches of technology serve the purposes of survival, be it for food production, medical technology, housing, protection from natural forces, and so on. Other items serve social purposes, for example transportation systems or mass media. Relationship needs are addressed in various forms of communication technology, but also indirectly in the form of making work processes less time-consuming, thus enabling more time with loved ones and for socialising. Technical artefacts enable many new forms of jobs and ways to be a productive member of a community, for example scientists and engineers. Moreover, technological solutions are strongly interwoven into cultural practices, arts, entertainment, and alike. However, at the same time, technology also has a negative impact on all levels of human needs and necessities: technology-caused environmental destruction and pollution, social imbalances owing to unjust distribution of access to technology-induced wealth, interpersonal and individual conflicts arising from misuse of technology, limitations of livelihood options owing to replacement of the human workforce by technological solutions, and personal numbness and blunting as a consequence of mindless consumption and application of *cold* technology.

Technologies and their social and environmental impact could be evaluated on the basis of this pyramid of necessities: in the first instance, technology must be *environmentally friendly*, that means its design, production, implementation and application must not interfere with the environmental integrity and balance. If it does, no matter how useful it is in serving the needs of the upper levels, refrain from it! In the second instance, it should be ensured that it serves social stability by promoting justice and fairness through its general availability and non-discriminatory effects. Then we can start asking in which way it affects people's life habits (interaction within families, among friends, with colleagues) and people's options to choose doing anything meaningful in their life. Then – and only then – may we take into account all those intended purposes and anticipated effects that the technology in focus has on the amenities of human daily life. There is a lot of technology (in the widest sense) currently firmly implemented in our daily life that would fail this assessment: individual auto-mobility (cars and motorcycles), cosmetics, agricultural techniques (especially meat production), energy production from fossil fuels, just to name a few examples.

We revisit Case 11.3 one more time: how can the various stakeholders' expected value gain be mediated and, in case of conflicts, be weighed? The industrial demand for rare earth metals conflicts with the interest in biodiversity and a healthy ecosystem. Let's assume for the moment that the mining of the ore and its processing, as well as the manufacturing and dissemination of the consumer goods containing the rare earth metal, is crucial for maintaining jobs and economic competitiveness of an entire region or even country. This, as part of societal integrity and life quality, conflicts with environmental integrity, too. Yet, in view of the above considerations, environmental integrity must be regarded as the highest priority, especially if the envisioned activity (the mining) endangers drinking water or clean air supplies. Thus, chemists' most urgent task would be to identify alternatives to the destructive mining, either in terms of milder acquisition and processing methods of the rare earth metal, or in terms of new materials that substitute the rare earth metal and make the mining of the ore obsolete.

Exercise Questions

1. Which of the following describes the concept of sustainability?

   A: It is of historical significance (past).
   B: It is future-oriented.
   C: It is rooted only in "the present moment".
   D: It is independent of time and space.

2. Which of the following correctly describes the origin of the basic idea of sustainability (impact of current choices on the options for future choice making)?

   A: It originated in ancient cultures.
   B: It is a brand-new concept developed since the 1990s.
   C: A civil servant in a German forestry regulation office invented it.
   D: It is an economic model that was promoted during the global finance crisis.

3. Which, of the following, is NOT one of the spheres that constitute the "triple-bottom-line of sustainability"?

   A: Environment
   B: Economy
   C: Science
   D: Society

4. How many "Sustainable Development Goals" were proposed by the United Nations in the Agenda 2030, presented at the 2015 summit in New York?

   A: 3
   B: 8
   C: 17
   D: 359

5. In principle, the sustainability concept provides a framework of _____ for the integration of various interests. (Choose a word/term below to fill in the blank.)

    A: Justice
    B: Morality
    C: Risk assessment
    D: Global unity

6. The viewpoint that the integrity of the ecosystem should always be given priority over human needs is promoted by which of the following concepts?

    A: Weak (anthropocentric) concept of sustainable development.
    B: Shallow (anthropocentric) environmental sustainability.
    C: Deep (ecocentric) economy.
    D: Strong (ecocentric) concept of sustainable development.

7. Can the environment have "interests" that are of concern in the sustainability discourse?

    A: Yes, there are features like biodiversity, recoverability, ecosystem health, and so on.
    B: No, all that matters is whether the environment can provide all the resources mankind needs.
    C: No, only humans have interests. Projecting our human thinking onto natural entities is called "naturalistic fallacy".
    D: Yes, the interest of the environment is to be free from negative human impact.

8. Which of the following represents the "three pillars of sustainability"?

    A: Health, ecosystem and human beings.
    B: Society, environment and economy.
    C: Ecosystem, industry and economy.
    D: Anthropocentrism, ecocentrism and cosmocentrism.

9. What is the major asset of chemical leasing?

    A: It forces companies to apply sustainable practices.
    B: It makes processes transparent and allows for proper punishment and sanctioning of misconduct.
    C: It gives academic chemists a stronger position in competition with industry.
    D: It sets incentives for sustainable practices by making different actors "pull in the same direction".

10. Which of the following describes the views of an ecocentrist?

    A: He/she grants value "for its own sake" to all environmental entities (animals, plants, landscapes, the ecosystem as such), not only instrumental value (like anthropocentrists).
    B: He/she believes that economy is the most important social sphere. Its interests should always be given priority.
    C: He/she only buys organic or biodegradable eco-friendly products.
    D: He/she believes that mankind is like a virus or cancer for this planet and would prefer if mankind would become extinct.

11. Which of the following best describes how REACH sets out to promote sustainability?

    A: By prohibiting all toxic and harmful chemicals.
    B: By permitting the public (citizen) to decide about the use of chemicals in industry.
    C: By creating a registry that mediates between economic feasibility, safety, social interests, and environmental integrity.
    D: By protecting the environment and saving scarce resources.

12. In the contemporary Western history of sustainability, which of the following have been the main drivers for its initiation?

    A: Politicians.
    B: Environmentalists.
    C: Bankers.
    D: A group of CEOs from multinational corporations.

13. Which of the following Summit meetings did not actually take place?

    A: The Earth Summit in Rio de Janeiro in 1992.
    B: The World Summit on Sustainable development in Johannesburg in 2002.
    C: The Universe Summit on Space Sustainability in Paris in 2008.
    D: The Sustainable Development Summit in New York in 2015.

14. Which of the following issues was not referred to by Mengzi in his statement introduced in this chapter?

    A: Taxes as an example of political/bureaucratic measures.
    B: Husbandry (agriculture) as an example of know-how.
    C: Fishing nets as an example of technical artefacts.
    D: Axes and bills as an example of techniques or procedures.

15. Which of the following describes the relationship between sustainability and the three relevant spheres society, environment and economy?

    A: Sustainability limits economy, balances society, and protects the environment.
    B: Sustainability is the backbone that maintains the hierarchy of environmental interests at the top (the highest priority), the economy in the middle, and society as the lowest priority.
    C: Sustainability is conceptualized in the social sphere, applied in the economic sphere, and observed/measured in the environmental sphere.
    D: Sustainability is located in the field where all three spheres overlap, that means where the interests of all three realms are met.

16. The etymology of the word *sustainability* is correctly described by which of the following statements?

    A: It comes from a Latin term, meaning "to linger on, to have an effect that reaches far into the future".
    B: It comes from a Greek term, meaning "environmentally friendly".
    C: It comes from a French term, meaning "to sound pleasant, to evoke positive emotions".
    D: It comes from a Hebrew term, meaning "the root of all evil".

17. Sustainability is an issue for...

    A: ...politicians.
    B: ...industrialists.
    C: ...scientists.
    D: ...all of A–C.

18. Which of the following can NOT plausibly be *sustainable* (in a categorical conceptual sense)?

    A: Economic practices.
    B: Technological development.
    C: Climate.
    D: A country's portfolio of energy production.

19. Which of the following conditions must be met for practices, decisions, or developments to be considered *sustainable*?

    A: They protect the environment from the negative effects of human activity.
    B: They satisfy current needs while, at the same time, not limiting the options of future generations to satisfy their needs.
    C: They are profitable now and will be long into the future.
    D: Future generations would do the same thing or, at least, agree to it in principle.

20. Which of the following statements suggests that chemical leasing can be considered to be a *sustainable* business practice?

    A: It increases the profit of the company that produces the end product.
    B: It supports a healthy competition among suppliers.
    C: It makes both supplier and buyer have an interest in reducing the amount of solvent or other chemicals used.
    D: It facilitates proper recycling of the applied chemicals.

References

1. *The Works of Mencius*, J. Legge, Dover Books, New York, 1970.
2. C. Blackmore, Sustainability, in *Encyclopedia of Applied Ethics*, ed. R. Chadwick, Elsevier, London, 2nd edn, 2012.
3. S. I. Rodriguez, M. S. Roman, S. C. Sturhahn and E. H. Terry, *Sustainability Assessment and Reporting for the University of Michigan's Ann Arbor Campus (Report No. CSS02-04)*, University of Michigan, Ann Arbor, 2002.
4. J. Elkington, *Cannibals with Forks: The Triple Bottom Line of 21st. Century Business*, New Society Publishers. Gabriola Island, 1998.
5. UN, Transforming Our World: The 2030 Agenda for Sustainable Development, Resolution adopted by the UN General Assembly on September 25 2015 (A/70/L.1).
6. http://www.un.org/sustainabledevelopment/, accessed on August 17 2020.
7. T. Welton, UN Sustainable Development Goals: how can sustainable/green chemistry contribute? There can be more than one approach, *Curr. Opin. Green Sustain. Chem.*, 2018, **13**, A7.
8. A. Albini and S. Protti, *Paradigms in Green Chemistry and Technology*, Springer, Heidelberg, 2016.

9. A. S. Cannon, J. L. Pont and J. C. Warner, Green chemistry and the pharmaceutical industry, in *Green Techniques for Organic Synthesis and Medicinal Chemistry*, ed. W. Zhang and B. W. Cuejr. John Wiley & Sons, New York, 2012.

10. J. H. Clark, Green chemistry: challenges and opportunities, *Green Chem.*, 1999, **1**, 1.

11. J. A. Linthorst, An overview: origins and development of green chemistry, *Found. Chem.*, 2009, **12**, 55.

12. P. T. Anastas and J. C. Warner, *Green Chemistry: Theory and Practice*, Oxford University Press, New York, 1998.

13. P. T. Anastas, T. C. Williamson, D. Hjeresen and J. J. Breen, Promoting green chemistry initiatives, *Environ. Sci. Technol.*, 1999, **33**, 116A.

14. ACS, *12 Design Principles of Green Chemistry*, https://www.acs.org/content/acs/en/greenchemistry/principles/12-principles-of-green-chemistry.html, accessed August 17 2020.

15. F. Moser, V. Karavezyris and C. Blum, Chemical Leasing in the context of Sustainable Chemistry, *Environ. Sci. Pollut. Res.*, 2014, **22**, 6968.

16. S. Erler, *Framework for Chemical Risk Management under REACH*, iSmithers, Shropshire, 2009.

17. L.-J. Cockcroft and T. Persich, *Insights on the Impact of REACH & CLP Implementation on Industry's Strategies in the Context of Sustainability*, Final Report prepared for ECHA, ECHA/2015/50 Lot 1, 2017.

18. https://echa.europa.eu/regulations/reach/understanding-reach, accessed August 17 2020.

19. C. U. Becker, *Sustainability Ethics and Sustainability Research*, Springer, Dordrecht, 2012.

20. C. Rösch, Ethics of Sustainability – An Analytical Approach, in *The Ethics of Technology*, ed. S. O. Hanson, Rowman & Littlefield, London, 2017.

21. VDI Richtlinie 3780, *Technikbewertung – Begriffe und Grundlagen (Technology Evaluation – Terminology and Fundamentals)*, VDI, Düsseldorf, 1991.

22. H. von Hentig, *Ach, die Werte! Über eine Erziehung für das 21. Jahrhundert*, Beltz Taschenbuch, Weinheim, 2001.

23. J. Mehlich and M. M. Tseng, Navigating in the Vastness – Making Sense of the Dynamics of Consumer Choices, in *Innovating in the Open Lab. The New Potential for Interactive Value Creation across Organizational Boundaries*, ed. A. Fritzsche, J. M. Jonas, A. Roth and K. M. Möslein, De Gruyter, Oldenbourg, 2020.

24. J. Russ, *Sustainability and Design Ethics*, 2nd edn, CRC Press, Boca Raton, 2018.

25. C. Shelley, *Design and Society: Social Issues in Technological Design*, Springer International, 2017.

26. *Handbook of Ethics, Values, and Technological Design*, ed. J. van den Hoven, P. E. Vermaas and I. van de Poel, Springer, Dordrecht, 2015.

27. A. H. Maslow, A theory of human motivation, *Psychol. Rev.*, 1943, **50**, 370.

28. A. H. Maslow, *Motivation and Personality*. Harper, New York, USA, 1954.

29. M. A. Wahba and L. T. Bridwell, Maslow reconsidered: A review of research on need hierarchy theory, *Organ. Behav. Hum. Perform.*, 1976, **15**, 212.

12 Responsibility

Overview

Summary: Chemical activity (science, research, engineering, innovation) – through its entanglement with technological development – affects and impacts normative and other value-related discourses concerning social and environmental dimensions of science and technology (S&T) progress. It is now time to introduce the concept of responsibility in order to clarify the position of chemists in this discourse.

Many responsibility attributions, especially from the public, apparently, are not justified and are mere accusations. Others are justified but chemists might not be aware of them. The difference is often one of a conceptual dimension: who can be held responsible by whom, for what exactly, and in view of what rules, competences and knowledge? Apparently, responsibility attributions are only legitimate when the agent that is held responsible has the cognitive capability to understand and act in accordance with certain expectations and obligations. Moreover, different types of responsibility need to be differentiated in order to make justified claims: legal, social, political, organisational, and moral responsibilities.

The considerations in this chapter, basically, have two goals. First, it will help chemical practitioners define their roles in progress and public discourse. This implies acceptance of some responsibility ascriptions and refutation of others. In any case, plausible arguments are required to claim or reject responsibilities. This chapter will provide such a line of argumentation. Second, it will convince chemists that their most general responsibility – contributing with their expertise and competence to the collective endeavour of sustainable S&T progress – arises from an obligation to serve the common good.

Key Themes: Four dimensions of responsibility (who is held responsible, by whom, for what, in view of what rules or knowledge?); responsibility *versus* accountability; types of responsibility (legal, social, organisational, moral).

Good Chemistry: Methodological, Ethical, and Social Dimensions
By Jan Mehlich
© Jan Mehlich 2021
Published by the Royal Society of Chemistry, www.rsc.org

> **Learning Objectives:**
>
> After studying this chapter, you will be able to:
> - Oversee, and apply, the four dimensions of responsibility attribution.
> - Respond to unjustified responsibility attributions and accusations convincingly and with proper arguments.
> - See more clearly where exactly the responsibilities of chemists as professional actors in academia, industry or governance lie and how they manifest themselves in particular calls for action and participation in public discourse on the social and environmental impact of chemistry.

12.1 Introductory Cases

Case 12.1: Complicity in war crime?

[Real case] Dutch chemist and businessman Frans van Anraat sold thousands of tons of chemicals to the Hussein regime in Iraq in the 1980s, among them thiodiglycol, a precursor for mustard gas and nerve gas. Chemical weapons that have been produced with his chemicals were used during attacks on Kurds in Northern Iraq in the late 1980s, killing thousands. Van Anraat was arrested and sentenced to 17 years in prison for complicity in war crimes. He himself pleaded innocent, claiming that a businessman is not responsible for what clients do with the deliveries, and that he believed his chemicals were for the Iraqi textile industry.

Case 12.2: I Don't Like POPs!

[Real case] A chemist working at the United Nations for the Stockholm Convention on persistent organic pollutants (POPs) gives a talk about his work as an invited speaker at a university symposium. When coming to the responsibility of chemical sciences and R&D, a chemistry professor in the audience stands up, apparently very upset, and complains about the speaker accusing chemists of being the root of all kinds of problems in the world. *"You want to blame us and make us feel guilty! Fine! Then let's stop all chemical research and go back to stone age!"*, he declares full of bitter sarcasm. He leaves the room without listening to the speaker's reply.

Case 12.3: Different View

Matthew, Peter and Conny studied chemistry together 15 years ago and meet again at an alumni meeting of their university. Remembering their joyful study years, they also come to talking about their current situations. Matthew is working in the research and development (R&D) section of a big chemical company. Peter got a job in an analytical lab of the National Environmental Protection agency. Conny is an assistant professor at a renowned university. Matthew claims: *"Sometimes I wish I had chosen an academic career! I miss the academic freedom! In industry, in recent years, we have so many restrictions, and expectations on 'responsible research and innovation' and such things! I feel a heavy burden of responsibility!"*. Peter objects: *"But it is good like that! I mean, all these regulations are in place to guide chemical progress in the right direction! Responsible research and innovation is a useful tool to help chemists act in responsible ways!"*. Conny adds: *"And don't think that academic chemistry is free from responsibilities! All the public scrutiny and the trend towards applied science with its dual use risks don't stop at faculty doors!"*.

12.2   Four Dimensions of Responsibility

The concept of responsibility plays a crucial rule in many fields of applied ethics and has received notable attention from scholars in the field and also in popular literature (see, for example, Hans Jonas' *The responsibility principle*).[1-5] Indeed, there is a lot to say about it and it is not as simple as it might seem at first. A first hint for its complexity is the obvious ambivalent character: we can associate responsibility with praise and with dispraise. On the one hand, the label *a responsible person* is usually intended as a compliment or admiration for someone who fulfils his or her duties very well. On the other hand, the statement that *someone is responsible* (for this mess, for example) puts a burden or pressure on that someone. It can't be easily decided whether responsibility is a blessing, a virtue, a duty, or an onerous burden.

Upon closer examination, we find that responsibility is never just one-dimensional as in *someone is responsible*. There must, at least, be a second dimension, that which someone is responsible for (Figure 12.1). Moreover, it can be analysed what *is* means in this case. Where does responsibility come from? Usually it is attributed by someone to someone, or in some way expected by someone from someone, or delegated by someone to someone. These two *someones* should be in any way related to each other so that responsibility claims are justified. Usually, the relevant relationship has to do with what the addressed someone is held responsible for: it should be in one way or another the responsibility-attributing someone's business. Last but not least, there is a fourth dimension: someone may legitimately

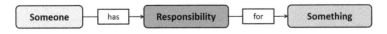

Figure 12.1   Two-dimensional concept of responsibility.

attribute responsibility to someone for something only *in view of a certain body of rules or a level of knowledge* that is related to what that someone is held responsible for. A necessary precondition for being a carrier of responsibility is the ability to fulfil the duties and obligations that go along with it. Above all, the person claimed responsible must be in a position of knowing the rules or of having relevant knowledge (Figure 12.2).

To be clear about what constitutes a legitimate responsibility attribution to chemists, we need to look at a few further characteristics of responsibility. The first is a time dimension: we can be responsible for our actions and decisions in the past that have an effect on now. In that case, we usually refer to it as accountability, or retrospective responsibility. We can also be responsible for future effects of our current actions and decisions (or those to come in the nearer future). In both cases, we are accountable or responsible NOW, which is important for practical and legal reasons. The words express the idea: the accountability for past actions and decisions is evaluated like an *account* of positive and negative positions (motivations, success, failure, conduct, *etc.*) in order to determine the overall contribution (causal or correlative) to a present state. The responsibility for future actions and decisions is formulated as the expectation on someone to be able to respond to certain questions and inquiries related to that case, for example in terms of knowledge and expertise, or leadership and other social roles.[6]

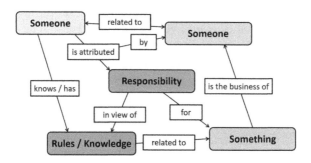

Figure 12.2 Four-dimensional concept of responsibility.

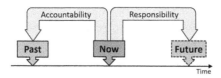

Figure 12.3 Difference between accountability and responsibility.

It is also important to distinguish moral, legal, organisational, and political responsibility, all of which are categories of social responsibility. Many responsibilities in the context of science and technology (professionalism, output such as knowledge, artefacts, processes) are governed in legal terms, for example warranty regulations for malfunctioning devices. When there are no laws or regulations, we still feel morally responsible, for example for professional integrity, avoidance of misconduct, or the reduction of misuse and dual use possibilities. Organisational responsibility refers to role expectations within organisational hierarchies in corporations, institutes, universities and other organisations. For example, engineers and researchers have certain obligations towards their bosses or the company they are employed at. This can, of course, conflict with their personal legal or moral duties and responsibilities, for example when it comes to the design and safety of technical items or innovations.[7,8] Finally, there is also political responsibility of elected social representatives and other decision-makers from economy, sports, and so forth, whose decisions and actions may have political relevance (for a chemical example, see the van Anraat Case 12.1, discussed below in detail).

12.2.1 For What?

First of all, chemists as scientists, researchers, innovators, regulators, or academic scholars have an *internal responsibility*, which means that they are expected to do their job well and with high scientific integrity, expressed in the term *responsible conduct of research*. Their professions have certain codes of conduct to follow, an ethos. These aspects of professional ethics and research ethics are the subject of Part 2 of this book (Chapters 5 to 9). Then, they have – together with other stakeholders like industrialists, entrepreneurs, regulators, politicians or consumers of technology – an *external responsibility*, which covers those aspects that go beyond their professional realm but touch society at large and sometimes the environment. It is important to note that we seldom talk about responsibility for things as such, or manifestations of scientific and technological development. For example, nobody can be held responsible for marine pollution or climate change. Responsibility is usually attributed to values and normative aspects related to such things or manifestations: Aspects of safety, justice, integrity, obligation fulfilment, and so forth. In this respect, scientists are not held responsible for the manifestation of their main product *knowledge* (for example in various applications or as the basis for other research and development), but for the truthfulness of their claims, including statements of certainty. For example, chemists are not responsible for plastic microparticles in the oceans, but rather for providing reliable knowledge that can be exploited for tackling the problem, such as scientifically sound models, life cycle assessments, alternative chemical processes, and so on.

The social and environmental impact of S&T development is expressed in terms of risk, benefits and values like safety and efficiency. Here, as far as scientific knowledge and insight are concerned, scientists may be held responsible for delivering viable and insightful information that helps protect these values and avoid undesirable outcomes. Here, it may be discussed whether scientists are responsible for making use of public funding for purposes with a clear societal benefit, like engaging in *sustainable chemistry* or *green chemistry*.[9]

12.2.2 Who?

In order to clarify responsibilities, there are two strategies: we can start from cases and ask who is responsible for what in which way; or we can start from roles and ask what is the particular person's situation in terms of responsibility. The former is often perceived of as accusation and blame, as in Case 12.2.[10] Even though certainly not intended as an accusation of guilt (the author was present at this incident), the communication was unlucky. When talking about a massive environmental problem arising from chemicals in the food chain and material fluxes around the globe, and then leading over to the responsibility of chemists whose scientific findings enabled the large-scale synthesis of these compounds, academic scientists' alarm bells go off and trigger their overreaction. It is useful to distinguish clearly between accountability and responsibility, here. Starting from a past case (the pollution with POPs), two different questions are possible: how did it come to this (accountability), and who is in the position, now, to do anything about it (responsibility)? The UN delegate intended to talk about the latter while the professor expected that he would explicate the former with all the causal elements from chemical synthesis to global pollution.

The latter strategy – starting from roles rather than from cases – appears to be more useful for our purposes: what can chemists in their various roles (researchers, developers, experts, individuals and organised in collectives) be held responsible for?[11] When we discuss responsibility attributions in Section 12.3, we will start from chemists' roles and the associated competences, not from cases in which chemists play a role. Moreover, we need to be aware of the important conceptual differentiation between *individual* responsibility of chemists in their above-mentioned roles and *collective* responsibility of research institutes, academies, corporations, companies, political organs and the state, or even entire societies. Even though it seems simpler and more plausible – especially in view of possible sanctions – to think of carriers of responsibility duties as legal persons, there are also situations in which collective entities can be attributed legal, moral, or political responsibilities.

12.2.3   Attributed by Whom?

Responsibility in legal and professional contexts must always be attributed and is never simply given. This attribution is only legitimate when the attributor is, in one way or another, authorised to claim it. When your boss or supervisor demands of you to brush your teeth for at least three minutes every morning and evening, otherwise you will be sanctioned, you are justified to claim that it is not your boss's business how and when you decide to brush your teeth. Your morning and evening routines are your personal decisions, and your boss has no authority to judge your private lifestyle, unless it affects your work performance. In a very general sense, attributors of responsibilities are those who are in any way affected by the actions and decisions of someone and who share the same social network (locally or globally) of cause-effect-pathways and role expectations. Consumers of technology may hold engineers and producers responsible because they are entrusted with this job in a society that is based on functional differentiation and expertise-based co-operation. Whether engineers may hold consumers responsible for proper application of their inventions is a much more complex question that would require the clarification of many preconditions. Let's see who is in the position to hold anyone responsible for certain implications of S&T in each of the four fields of responsibility.

Legal responsibility is maybe the most obvious and the easiest as the pathways of responsibility claims are either regulated by those particular laws and guidelines themselves, or they are justified by the implementation and justification of the law-and-order system as such. In the end, legally responsible is someone in front of the law and its enactors, like policymakers and legislators. These are – at least in well-functioning democracies – empowered by the society, so that in the end it is the wider public or the society as such that expects bearers of legal responsibility to act accordingly. Furthermore, legal responsibilities are often attributed by organisations and institutions to their staff and employees. When there are regulations and legally binding guidelines in place that have been communicated to the employee, the organisation legitimately expects the employees to act in accordance with these rules. Note, here, the difference between obligations and responsibilities: rules and laws describe obligations that people have in certain roles. Responsibility describes the expectation to always act in a way that these obligations are met. Thus, those who formulate and assign legal obligations may be different from those who attribute legal responsibility.

Political responsibility may be expected from elected representatives of a society (politicians, legislators) by voters. Yet, in our globalised world, many more actors from various spheres make decisions that have an impact on the political integrity of countries and unions among them. When, for example, an arms and ammunition manufacturer makes deals with regimes that are under international scrutiny and surveillance, this company may also be attributed, among other forms, political responsibility. In one way, the public and their representatives may attribute political responsibility to companies that act in ways that affect political relations. In another way, economic stakeholders attribute political responsibility to political decision-makers in resorts that affect economic wealth and profitability (market conditions, job opportunities, taxes, technology regulations, *etc.*), in the sense of *"You are responsible for setting the margin for a successful and profitable economy!"*. Whether economic enactors are justified to claim this kind of responsibility from politicians is the subject of heated debate. Again, the case of a trader of chemicals (Case 12.1) is an example in which even a chemist qua his chemical expertise may be attributed political responsibility.

Organisational responsibility in corporate hierarchies are attributed "top-down" from directors, chief executive officers (CEOs) and superiors to employees (workers, engineers, designers, managers, *etc.*). Every actor has to fulfil a role that goes along with expectations and responsibilities. In the end, it is the society that attributes this form of responsibility as we as a society decided to organise ourselves in this way. This has, of course, practical reasons: nobody can be an expert on everything, and nobody can be responsible for everything. We will highlight this point further in the next section.

The more difficult case is moral responsibility. In principle, every member of a society or a culturally connected group has moral claims on all other members. Moral responsibility is often not claimed by people in their professional or social roles, but as concerned society members. For example, a chief executive officer (CEO) may hold an employed engineer legally and organisationally responsible for what he produces, but if he also has a claim on moral responsibility he does so as a *member of the society* (or as a human), but not as CEO. Claims of moral responsibility can be made by everyone who sees the social integrity at stake by certain forms of conduct or practice. The boss cannot hold you morally responsible for brushing your teeth twice a day when he is not affected by your decision. However, when you have kids that learn their habits from you, and they also feel no urge to brush their teeth properly, risking bad teeth conditions, your action or habit has an effect on other members of the society (your kids, the health care system), so it might be justified to hold you morally responsible.

These considerations show that two important social values are firmly connected with responsibility: trust and justice. The attribution of responsibility is usually grounded upon the trust that the respective holder of responsibility is able to succeed in that. In the private sphere, this trust results from personal connections, emotions, or judgments based on knowledge or intuition concerning the person that is held responsible. In the professional sphere, the trust is based on education (proven by certificates or experience), contracts or legal regulations. The latter are a matter of justice as the fundamental principle of the law-and-order system that a society gives itself. Applied principles of justice also determine the way a violation of responsibility duties is sanctioned. The lightest form of punishment is declaratory measures: someone simply speaks out about the fact that a lack of responsibility or accountability concerning a certain incident has been shown. This can range from a calm remark up to loud scolding, but, in any case, it has the effect that the person who showed a lack of responsibility will feel guilty and understand what went wrong. Stronger measures are various forms of compensation. The damage caused by the lack of responsibility must be compensated by the holder of responsibility, for example in monetary forms or in terms of investing time for reparation. Hard treatment would include physical punishment. Frequently applied are exclusionary measures: in the case of legal violations, the accused person may be imprisoned or lose his or her job position. Informal exclusionary methods (in the private interpersonal sphere) are applied in the form of exclusion from group activities or relationships. All of these are enacted in various ways in the context of chemical science, research, development, application, and so forth.

12.2.4 Related Rules and Knowledge

The four-dimensional conception of responsibility is practically extremely relevant because it puts a strong focus on the missing link between the responsibility-attributed agent and the object of its responsibility (the *'for what?'*). *Ought* implies *can*. For past-related accountability (see Figure 12.3), it means that the agent must have made a deliberate choice among alternatives under awareness of the consequences of these alternatives. Sanctionable choices are only those in which either a deliberate intended violation of moral, legal or professional norms or a gross misjudgement or negligence can be proven. For future-related responsibility, it means that the agent can legitimately be expected to possess and apply knowledge, skills and competences to (re-)act appropriately. For example, a young adult gets the permission to drive after passing the driving test and, with this, is attributed the responsibility to drive safely, because the successful test is taken as a proof that he knows how to drive a car and has comprehended and internalised the traffic regulations. Accordingly, no caught adult and sane thief could claim at the police station that he was not aware that stealing is illegal, because educated healthy people know the legal and moral frameworks of the society they grew up in. In a similar fashion, certified chemists with university degrees and proven competences may be expected to have the ability to foresee chemical implications of a chemistry-related process (for example, material properties, environmental effects of compounds, potential usage of chemicals), especially when these chemists occupy offices (university chairs, lab leadership, public service authorities, *etc.*) after passing a competence and capability test (for example, a job interview).

Entities that carry collective responsibility do so in view of certain rules and capacities. Companies and corporations are aware of (and therefore asked to comply with) codes of business conduct and the possible political relevance that their activities bear. This may be the same for research institutes like those at universities and in other academies. Governments and parliaments may be held responsible in view of their social justification and the principles of social justice that legitimate democratic organisations are committed to. A more difficult case might be the question of whether the society at large can be attributed collective responsibility in view of anything. Nowadays, thinking of knowledge society and the widespread availability of information and education, we may assume that societies (not its individual members, but as a collective that is formed by constant communicative interaction and media) develop a problem awareness and the internal pressure to tackle present and prevent expected problems. This point is especially important in the context of responsibilities for global problems like marine pollution with plastics and climate change.

12.3 Responsibility Attributions to Chemists

The goal of this section cannot be to cover all possible responsibility attributions to chemists. Given the diversity of jobs and roles that chemists occupy, that would be an impossible task. Here, the most common roles are listed and explained. Chemists in other domains will be able to extract the essence of these elaborations and translate that to their particular situation. As highlighted earlier, we locate chemical professions in academia, the private sector, and in public service other than universities. Here, we focus only on responsibilities that are attributed in view of the particular professional role, and not those that are attributed to all members of a society. All chemists have in common that their professional obligations, that they have the responsibility to fulfil, arise from their chemical knowledge and competence. Moreover, all attributions are, in the end, legitimised by society and its representative authorities. Regardless of the job description, what all chemists are held responsible for in one way or another is the truthfulness of epistemic testimony as chemists' main contribution to society. This unwieldy expression describes all those situations in which chemists communicate their specific knowledge. Academics report their research results and discuss them with peers and sometimes with non-chemists; innovators contribute with research results to product development and economic competitiveness; regulators and agency staff make decisions after carefully considering chemical judgments. All these are, in the widest sense, forms of testimony, the sharing of knowledge and insight with others. This sharing must be truthful and reliable. Most of the attributed responsibilities hinge on this point.

In addition to these similarities, the various roles of chemists have their specific conditions for responsibility ascriptions. Table 12.1 provides a non-exhaustive overview of ten exemplary roles, three of which are academic chemists' roles, four are private-sector chemists' roles, and three are service-sector chemists' roles. Academic chemists work as scientists, collaborators, and at universities as educators, sometimes all three in one. As scientific researchers, they are attributed legal and moral responsibility to maintain scientific integrity by

Table 12.1 Responsibility attributions of chemists in different fields and roles.

Who?	Role	Relevant competence and knowledge	Attributor	Responsible for what?
Academic chemist	Scientist	Methodology codes of good scientific practice	Scientific community, home institute	Scientific integrity
	Educator	Teaching methods	Students, employers of chemists	Teaching quality, mentorship,
	Collaborator	Communication	Collaborator	Scientific integrity
Chemist in the private sector	Innovator (R&D)	Company's rules and guidelines	Executive board, customers, regulators	Risk & EHS assessment
	Negotiator	Economic codes of conduct	Executive board, stock- and shareholders	Compliance with rules
	Trader	Context of the trade	Dependent on the scale	Implications of the trade
	Company representative	CSR, company policy	Colleagues and superiors	Compliance with codes
Chemist in public service	Service provider	Client's contexts	Clients, agency, state	Service quality
	Regulatory authority	Laws and regulations	Legislative bodies	Accordance and compliance with laws
	Civil servant	Duties and obligations relevant for the job	Employers	Service quality
		Chemical knowledge, general and specific	Society and its representatives	Truthful epistemic testimony as contribution to society

the scientific community and their home institutes in view of their expected competence to apply appropriate methodologies and to comply with professional codes of good scientific practice. In collaboration with others, when agreements and rules have been communicated, cooperation partners for whom a lot is at stake may legitimately expect professional integrity (see Chapter 8). Those who contribute to higher education in the form of lectures and lab courses are attributed a quasi-moral responsibility by students and future employers of chemists to spend efforts on teaching quality and mentorship. There are no legal and often no organisational frameworks for how to teach, and the responsibility for that is not moral in the strict sense. Yet, as teaching is part of the job description of professors, it is their professional obligation to try their best to offer good teaching quality. Therefore, it is called *quasi-moral*, here.

Chemists in the private sector, usually, are attributed organisational responsibility by the executive boards, stock- and shareholders of their companies owing to stricter contracts and regulations. In application-oriented research and development (R&D), for example, they are held responsible for truthful testimony in risk and environmental, health and safety (EHS) assessments. As negotiators in business interactions, being familiar with business ethics and economic codes of conduct, they are held responsible for compliance with business rules. As representatives of a corporation in general, in view of corporate social responsibility profiles and company policies, colleagues and superiors surely demand compliance with these codes. A special context is trade with chemicals (see the van Anraat case below) in which, dependent on the scale, even political responsibility can be attributed to the decision-making chemist. Note, again, that this paragraph lists only those attributions that add to the more general societal attributions of responsibility to chemical experts for contributing appropriately to society.

Chemists in public service jobs like patent offices or environmental protection agencies are expected to be familiar with public policies, laws and regulations, and, if applicable, with clients' demands and contexts. On this basis, they are held responsible for the provided service quality, compliance with laws and regulations, and decision-making that is in accordance with the goals and objectives of the organisation they are working for. Legitimate attributors of this responsibility are the clients, the employer (for example, the office, agency or, in the case of civil servants, the state), and legislative bodies that set the legal frameworks within which offices and agencies operate.

Besides these specific roles, chemists are always also members of the chemical community and of the society. As the former, they know codes of conduct (good scientific practice) and the rules and guidelines of their institutions. As the latter, we may expect that they have education, an idea of common-sense morality, and social and cultural competence. In other words, chemists have a *moral conscience* without being experts in philosophical ethics. In many situations in which chemists have influence as epistemic authorities, it may be of greater importance to exhibit a solid integrity in character, attitude and moral conscience.

To reiterate why we elaborate on responsibility attributions in such detail and, occasionally, close to triviality. We want to be able to distinguish legitimate attributions from accusations and too demanding expectations. The four dimensions are important because responsibility ascriptions must be analysed, accepted or refuted in all four separately. First, responsibilities vary from role to role. An academic chemist doing basic research at university has obligations that are very different from chemists in R&D or in analytical labs of environmental protection agencies. Second, chemists' expected knowledge, competence and expertise varies in different professional contexts and is not always limited to chemical expertise alone. Third, the attributors matter! Accusations of responsibility violations expressed in mass media, social media, or other channels of public communication, for example, may be refuted on the basis of the attributors not having the authority to make that judgment. Fourth, not every chemist is responsible for all sorts of 'chemical' impacts. A differentiation of role-dependent outputs and their evaluation is crucial. Remember that the *'for what?'* dimension in this 4D-concept of responsibility is concerned about values, norms, and intents rather than things and events.

12.4 Responding to Responsibility Attributions

With these abstract considerations in mind, we are ready to respond to common responsibility claims on chemists or chemistry as such. The following are exemplary claims that are more or less frequently expressed:

- *"Chemistry produces substances that cause global problems (POPs, PCB, plastic in the ocean, etc.)!"* This claim is too general! It lacks a clear separation into *who* and *for what* in view of competences and decision-making power! It is true, obviously, that chemical research enables the synthesis, large scale production and dissemination of such products, so chemists' activities are a direct cause for their existence. Yet, for example, polymer chemists are clearly not responsible for industrial production, manufacturing, consumption and disposal of plastics, as that is outside of their decision-making and their knowledge. If the statement is intended as an accountability attribution, it lacks legal justification as there are no inventor pays or producer pays regulations in place. It would be an entirely different thing to claim that chemists are responsible for developing chemical methods to tackle the mentioned problems. Yet, again, in that case it must be clarified who exactly is addressed: academic science, R&D, regulatory measures with chemical competence?

- *"Chemists (researchers) should spend their resources on research for sustainability!"* This may be partly true. Many of the environmental and societal problems caused by adverse effects of chemicals can be solved by exploiting the profound chemical knowledge that we have and still advance for the development of alternative processes and substances. Keywords, here, are green chemistry or sustainable chemistry (see Chapter 11). However, it would be fatal to claim this for all chemical activity! Academic freedom in basic research must be maintained (see Chapter 8). Not all research can be purpose driven, even though it may be discussed whether in emergency situations like global threats all available resources should be channelled into an immediate solution development.

- *"Chemists have a social responsibility and should be concerned about what happens with their findings and achievements!"* In principle, this claim can be agreed to. To the best of their knowledge, chemists – especially those in applied and special-interest science – should reflect on the implications of their work. Here, we need to be aware of the differentiation of legal and moral responsibility, both sometimes interfered with by organisational responsibilities (duties arising from one's company's hierarchies and work contracts). More importantly, caution is needed as the claim can only be made from the *now*-perspective! Long-term future implications, unpredictable risks and dual use potentials cannot be anticipated, even with the best available models and scenario tools. Again, it is a question of competence and expertise: the claim may be translated into a call for being up-to-date on scientific state-of-the-art and current trends, and for communicating one's expert judgments with other relevant stakeholders, especially decision-makers in industry and in the public sector (regulation, S&T governance).

We will now revisit the three example cases from Chapter 10 (Cases 10.1–10.3) and this chapter (Cases 12.1–12.3) and apply the four-dimensional conception of responsibility to inform an evaluation of the cases:

Case 10.1: The Manhattan Project[12] The controversial discussion of this case revolves around two views concerning the involved scientists' responsibilities. On the one hand, it is acknowledged that the scientists – archetypes of basic researchers like Einstein and Bohr – upon foreseeing the detrimental implications of near-future scientific progress (here: the exploitation of nuclear fission for atomic bombs by the Nazis) raised their voice to urge societal leaders (here: the US government) to initiate countermeasures. On the other hand, some regard it as an inexcusable act of scientific irresponsibility to contribute to a project of which it was known that its goal was the most disastrous weapon ever built. Oppenheimer's statement that until the end he believed that the main goal was deterrence and that such a weapon would never really be used sounds like a naïve excuse. On the other hand, his public activism against nuclear weapons and the development of the hydrogen bomb is an example of a concerned scientist taking responsibility for what his contributions have unleashed, just a little bit too late, perhaps. The experience from this case is that whenever scientific research is goal oriented and has a high dual use potential, the contributing scientists cannot simply hope that someone else will make sure that the negative effects are prevented or reduced. Not only do the scientists know the connection between phenomenon and application, but they are also authorities in the public discourse whose voices are heard and respected. Dilemmas as in the Manhattan project – political duty *versus* risk of immoral application of the outcome – can be solved in the public discourse and must not be a burden for the scientist. Yet, the scientist delivers important insights and must not shy away from raising concerns. This responsibility is attributed by the general public that legitimates the scientific profession as *for the common good*.

Case 10.2: Agent Orange Arthur Galston is the master example of a chemist who finds his research output taking a direction that was not intended and that is clearly unethical. He regarded it as his responsibility (a self-attribution out of moral conscience and a fundamentally *social* interpretation of professionalism) to *actively* prevent the *misuse* of *his* compounds.[13] All of the terms in the previous sentence that are written in Italics may be discussed. It is important to note that we do not discuss his responsibility as causal contributor to the US using

Agent Orange in the Vietnam war. He was not able to predict that a fertilizer that has defoliation effects in high concentrations would be exploited deliberately as a warfare product. The responsibility attributions of him to himself and, as he hopes, all scientists to themselves, refers to being aware of what happens with research output and, if necessary, to actively seek participation in its discourse.[14]

Case 10.3: Explosives research Scientific research with inherent and obvious dual use potential like the development of explosives for mining, construction and missiles requires a careful assessment of conditions and communication channels to support the beneficial application and to control the detrimental or unethical application. The scientist, here, has a chance to critically question the intentions of funding sources and may insist on decision-making authority in the use of the produced compounds, for example *via* patents, written down in project proposals and agreements and, thus, legally binding. Where this is practically limited, the considerations of the previous two examples apply: where it is not realistic that the scientific researcher can prevent unintended use of the research output, it is of utmost importance to follow developments, usages and policies concerning these outputs with awareness and public voice. Similar cases are research with or on warfare agents,[15] some nanomaterials,[16] or pathogens (see Case 7.3).[17]

Case 12.1: Frans van Anraat His jail sentence for complicity in war crimes is a clear responsibility attribution.[18,19] As a chemist, he could not plausibly claim that he did not know that his chemicals can be used to fabricate chemical weapons. As an act of negligence, he illegitimately put his personal monetary interest over the humanitarian interest. Moreover, besides a moral responsibility, political responsibility can also be attributed to him. Knowing the situation in Iraq, trade deals like this need to be indicated to the Dutch authorities, which van Anraat did not do.[20] Miles Jackson writes:[21]

> "Van Anraat is a paradigmatic example of what it is to be accomplice. There is no doubt that complex questions of principal responsibility arise in respect of the chemical attacks themselves; among the Iraqi pilots who dropped the bombs, military officials who ordered them to do so, and civilian superiors who decided the course of action. Moreover, there is little doubt that these chemical attacks involved a number of actors and were, in a sense, collective. But that collectivity does not mean that we should not hold van Anraat responsible for what he did: contribute as an accomplice to the wrongdoing of others."

Case 12.2: Guilt Accusation? The upset professor is right: blaming chemists for all kinds of problems in the world is an illegitimate attribution of responsibility in most of the cases. Yet, the UN delegate is right, too: in order to tackle the problem, we need to get all stakeholders at one table to discuss the options for how to proceed, including chemists and those in the chemical industry. The friction point, here is a misunderstanding, as indicated above (Section 12.2.2). Both accountability and responsibility attributions should have a pragmatic goal: how can the issue be solved and, if possible be prevented in the future? Accusations of guilt and shame are not goal-oriented. In this respect, it would have been more responsible of the academic chemist to wait for the reply and engage in the important discourse concerning this matter, rather than just leaving the room.

Case 12.3: Responsibilities in Different Fields of Chemical Activity The three figures represent the main three fields of occupation for chemists with university degrees. Apparently, private-sector chemists feel (and, perhaps, really have) more restrictions and control mechanisms to ensure that employees do their work within the given frameworks of responsibility (corporate social responsibility (CSR), responsible research and innovation (RRI), EHS standards, *etc.*). Responsibility, here, is perceived as a burden and a constraint of freedom. The alternative is to regard responsibility as a virtue to which chemists should be motivated to live up to. The public service chemist in this example represents this view: a chemist should be committed to working for the common good. Being responsible, thus, is part of the commitment and fills the chemist with a sense of professional achievement when successfully maintained. The remark of the academic chemist in this example summarises the main message of this chapter: No professional activity is free from responsibility attributions. Differentiating in the right way (here: along the lines of the four-dimensional responsibility conception) allows the formulation of these attributions in plausible and practical ways that are helpful for the discourse and for chemists to see orientations for their participation in that discourse.

Exercise Questions

1. From which of the following do responsibilities arise?

   A: Duties and obligations.
   B: External forces.
   C: Human nature.
   D: Cosmic harmony.

2. Which of the following is not one of the "four dimensions of responsibility" introduced in this chapter?

   A: Attributed by whom?
   B: For what?
   C: Who?
   D: In view of what kind of sanctions and punishments?

3. As an employee in a company, you are attributed responsibility...

   A: ...by your boss.
   B: ...by legal rules and regulations.
   C: ...by common ethics and moral guidelines.
   D: ...all of A–C.

4. Which of the following correctly describes the difference between accountability and responsibility?

   A: *Accountability* is only relevant to business and finance, whereas *responsibility* is a broader concept.
   B: Someone may be held *accountable* for a past incident but *responsible* for a future development.
   C: *Accountability* (a negative term) is about punishments and sanctions, whereas *responsibility* (a positive term) is about reward and achievement.
   D: The two terms describe the same thing, but *accountability* was coined in the field of sociology, whereas *responsibility* is a term mostly used in philosophy.

5. Which of the following attributions of responsibility to scientists/researchers is not tenable?

   A: Responsibility for methodological competence and expertise.
   B: Responsibility for the acquisition of knowledge and keeping it up to date.
   C: Responsibility for the long-term social and environmental impact of research output.
   D: Responsibility for good scientific practice and commitment to institutional codes and guidelines.

6. Which of the following would not represent a legitimate attribution of responsibility?

   A: A researcher is held responsible by other members of the scientific community for the proper representation, dissemination, and interpretation of data.
   B: A scientist is held responsible by religious leaders of his/her society for the proper representation of religious teachings and worldviews in his scientific claims.
   C: A researcher in the private sector is held responsible by senior management at his/her company for fulfilling the obligations and duties as stated in the work contract.
   D: A scientist is held responsible by society or society's representatives for using his/her expertise and competence to point out the potential risks and harm associated with contemporary scientific and technological developments.

7. Which of the following statements describes the *Social Responsibility of Scientists*?

A: Society attributes specific responsibility to scientists in their professional roles.
B: Scientists are held responsible for the well-being of society and its members.
C: Scientists hold the general public responsible for acquiring at least a basic level of scientific knowledge.
D: Scientists have a duty to spend some of their working time in communicating their research results to the public.

8. Which of the following consequences would never be the result of a scientist behaving irresponsibly in his professional role?

A: Loss of job.
B: Exclusion from scientific community.
C: Jail sentence.
D: Loss (or forfeit) of human rights.

9. Science as such, as well as the position/job of an individual scientist, depends on...

A: ...social scrutiny.
B: ...social justice.
C: ...social justification.
D: ...social networking.

10. It is sometimes stated that a scientist *lives in an ivory tower.* Which of the following statements explains what this means?

A: The scientist lives in an ivory tower.
B: The scientist lives in a big mansion in the countryside.
C: The scientist is applying improper methods that are better characterized as alchemy or quackery than as good science.
D: The scientist is – in his thinking and his worldview – detached from the real world and the social reality around him/her.

11. What is meant by the claim that *a scientist should also have common sense?*

A: Be able to make sound judgment of a situation that is not based on specific (scientific) knowledge.
B: Be accepted by everybody.
C: Have a high level of general knowledge.
D: Have a feeling for contemporary and near-future trends in technological development.

12. What is the most important form of responsibility? (Select the most appropriate answer from the choices below.)

A: Moral responsibility.
B: Legal responsibility.
C: Social responsibility.
D: This question is implausible. Different responsibilities matter to varying degrees in different situations, and may even conflict with each other, but none in principle has priority over the others.

13. You have recently been hired to carry out an innovative R&D project, but quickly decide that a (time-consuming and cost-intensive) risk assessment that exceeds the original plan will be needed to meet required safety standards. However, your boss insists that you stick to the work as described in your work contract, leaving everything else to senior management (who certainly won't initiate the necessary assessment). Which of the following conflict of responsibility is occurring here?

   A: Legal *versus* Social.
   B: Organizational *versus* Moral.
   C: Social *versus* economic.
   D: Moral *versus* Political.

14. Which of the following is the crucial point that makes it legitimate to hold Frans van Anraat responsible for his actions in selling precursors for chemical weapons to the Iraqi regime?

   A: His chemical expertise and knowledge (having studied chemistry at university).
   B: His personal friendship with Saddam Hussein.
   C: Doing a job (trade) that is not related to what he is educated in (science).
   D: His naivety concerning the political dimension of his activities.

15. According to the definition of the four dimensions of responsibility used in the chapter, which of the following describes what legitimizes what someone may be held responsible for?

   A: Someone's social status including ethnic origin, gender, and religious commitment.
   B: An attributor's wise judgment.
   C: Someone's knowledge, position and capability.
   D: Zeitgeist ("spirit of the age", that is, political, cultural, historical, societal background).

16. In terms of their responsibilities, which of the following may chemists be required to be concerned about?

   A: Long-term future effects of chemicals.
   B: Unpredictable risks and side effects of S&T development.
   C: Dual use of chemicals.
   D: Communication of critical knowledge with other stakeholders.

17. Was Robert Oppenheimer, scientific leader of the Manhattan project, responsible for the casualties from the nuclear bombs on Hiroshima and Nagasaki?

   A: No.
   B: Yes, because his role is a direct causal factor for the existence of these bombs.
   C: Yes, because he had a chance to refuse to work on this project.
   D: Yes, because after realizing the weapon's destructive capacity, he did not attempt to impede its usage.

18. Which of the following statements is the only tenable one among the four? Chemists are held responsible...

A: ...for none of their scientific outputs, because science is neutral by definition.

B: ...for the well-being of society, given the tremendous impact that chemistry has on society.

C: ...for communicating their expert judgments with other relevant stakeholders in the attempt to reduce risks and increase the chance of benefits.

D: ...only for good scientific practice such as not committing any fraud, proper publishing, avoiding conflicts of interest, and so forth.

19. Which of the following is not among the common (and acceptable) measures/sanctions to be applied in cases of violation of responsibilities?

A: Declaratory

B: Torture

C: Compensatory

D: Exclusionary

20. Who may make legitimate claims on a scientist's moral responsibility?

A: A fellow scientist.

B: A priest.

C: The general public (represented by an elected politician, for example).

D: All of A–C, as long as the claim is substantiated by reasonable arguments and can withstand critical scrutiny.

References

1. G. Williams, Responsibility, in *Encyclopedia of Applied Ethics*, ed. R. Chadwick (Chief), Elsevier, London, 2nd edn, 2012.
2. J. Nihlén Fahlquist, Responsibility Analysis, in *The Ethics of Technology*, ed. S. O. Hanson, Rowman & Littlefield, London, 2017.
3. C. Mitcham, Responsibility. Overview, in *Encyclopedia of Science, Technology, and Ethics*, ed. C. Mitcham, Thomson Gale, Famington Hills, 2005, vol. 3.
4. H. Lenk, Responsibility. German Perspective, in *Encyclopedia of Science, Technology, and Ethics*, ed. C. Mitcham, Thomson Gale, Famington Hills, 2005, vol. 3.
5. H. Jonas, *The Imperative of Responsibility: In Search of an Ethics for the Technological Age*, University of Chicago Press, Chicago, 1984.
6. I. van de Poel and M. Sand, Varieties of responsibility: two problems of responsible innovation, *Synthese*, 2018, DOI: 10.1007/s11229-018-01951-7.
7. S. Arnaldi and L. Bianchi, *Responsibility in Science and Technology*, Springer, Wiesbaden, 2016.
8. C. Coenen, A. Dijkstra, C. Fautz, J. Guivant, K. Konrad, C. Milburn and H. van Lente, *Innovation and Responsibility. Engaging with New and Emerging Technologies*, IOS Press, Netherlands, 2013.
9. N. Saenko, O. Voronkova, M. Volk and O. Voroshilova, The Social Responsibility of a Scientist: the Philosophical Aspect of Contemporary Discussions, *J. Soc. Stud. Educ. Res*, 2020, **10**, 332.

10. C. Glerup, S. R. Davies and M. Horst, 'Nothing really responsible goes on here': scientists' experience and practice of responsibility, *J. Responsible Innov.*, 2017, **4**, 319.

11. C. Glerup and M. Horst, Mapping 'social responsibility' in science, *J. Responsible Innov.*, 2014, **1**, 31.

12. A. Wellerstein, Manhattan Project, *Encyclopedia of the History of Science*, October 2019, accessed August 26th 2020, , DOI: 10.34758/9aaa-ne35.

13. A. Galston, *Science and Social Responsibility: A Case History*, Annals New York Academy of Sciences, 1972.

14. C. Jacob and A. Walters, Risk and responsibility in chemical research: The case of agent orange, *HYLE*, 2005, **11**, 147.

15. H. Frank, J. E. Forman and D. Cole-Hamilton, Chemical weapons: what is the purpose? The Hague Ethical Guidelines, *Toxicol. Environ. Chem.*, 2018, **100**, 1.

16. R. McGinn, Ethical Responsibilities of Nanotechnology Researchers: A Short Guide, *Nanoethics*, 2010, **4**, 1.

17. E. Ristanovic, Ethical Aspects of Bioterrorism and Biodefence, in *Defence against Bioterrorism. NATO Science for Peace and Security Series A: Chemistry and Biology*, ed. V. Radosavljevic, I. Banjari and G. Belojevic, Springer, Dordrecht, 2018.

18. *Public Prosecutor v. van Anraat*, District Court, Case No. 09/751003-04, December 23rd 2005 *Prosecutor v, van Anraat*, Dutch Supreme Court, Case No. 07/10742, June 30th 2009.

19. *Public Prosecutor v. van Anraat*, District Court, Case No. 09/751003-04, December 23rd 2005 *Prosecutor v, van Anraat*, Dutch Supreme Court, Case No. 07/10742, June 30th 2009.

20. H. van der Wilt, Genocide, Complicity in Genocide and International v. Domestic Jurisdiction: Reflections on the van Anraat Case, *J. Int. Crim. Justice*, 2006, **4**, 239.

21. M. Jackson, The Attribution of Responsibility and Modes of Liability in International Criminal Law, *Leiden Journal of International Law*, 2016, vol. 29, p. 879.

13 Risk, Uncertainty, and Precaution

Overview

Summary: Almost all aspects of the discourse on societal and environmental impacts of scientific and technological development can be framed in terms of risk and uncertainty. It is an unavoidable component of progress and innovation that some effects are unpredictable and unknown. Therefore, this topic deserves its own chapter in the context of chemical progress in science, research and development (R&D) and innovation. Here, we will shed light on the conceptual and practical definitions of risk and uncertainty, approaches to risk assessment and risk management, the role of chemists in different risk discourse contexts, and contemporary institutional implementation of handling uncertainties in the form of the precautionary principle.

Risk is one of those terms that different people associate with very different things. Chemists – that is, people with an educational background in a natural science, often working in environments in which technical problem-solving is achieved by using expertise, knowledge, skills and competences – often understand risk as something empirically comprehensible (for example, the likelihood of a malfunction or contamination) or a result of ignorance that can be tackled by doing more research (that means, a cognitive challenge). We will learn that parts of the risk discourse revolve around normative and evaluative aspects. In accordance with the claims in the previous chapters, decision-makers and actors in chemistry contexts benefit from an awareness of these discourses as important contributors to an interdisciplinary endeavour: mitigating risks on a solid evidence-based factual foundation (delivered by science and empirical research) under consideration of a well-informed plausible normative framework.

Key Themes: Definitions of risk and uncertainty; risk assessment; risk management; risk discourse types; precautionary principles.

Good Chemistry: Methodological, Ethical, and Social Dimensions
By Jan Mehlich
© Jan Mehlich 2021
Published by the Royal Society of Chemistry, www.rsc.org

Learning Objectives:

With the content of this chapter in mind, you will have:

- A sharpened awareness of various levels of risk types and the demands on their respective discourses.
- An idea of the role of chemical scientists and researchers in such discourses.
- The motivation to participate actively in multi-stakeholder discourses in ways that your professional position provides, so that the goal of reducing risks and increasing benefits can be reached.

13.1 Introductory Cases

Case 13.1: Big Issues with Small Particles

Sunscreens with titanium dioxide nanoparticles are commercially available. Toxicologists could not identify any adverse health effects of the NPs. Based on these studies, consumer products have been approved by regulators. Yet, the success of such innovative sunscreens is below that expected. Cosmetics companies classify the market potential as risky. Consumerists, too, employ the parlance of risk trade-offs in the discourse on risk and benefits of nano-sunscreens. Was there a misunderstanding?

Case 13.2: Chemometrics

A good analytic chemist knows that every toxicological study goes along with false positives (for example, a substance showing a certain toxicity in a test while, actually, it is not toxic) and false negatives (a substance showing no effects while, actually, it is toxic). According to textbook knowledge, a threshold should be set where the rates of both types of error are equal, because such a scientific approach should be neutral. Yet, the good analytic chemist wonders if this is plausible. In a real-world context, wouldn't it be safer to overestimate the toxicity of a compound rather than to risk toxic substances slipping through the test? That means, shouldn't the threshold be set in a way that false positives are taken more seriously (with a bigger impact) than false negatives?

Case 13.3 Better Safe than Sorry!

Dr Philipps works in the R&D department of a chemical company, synthesizing and testing organophosphates and carbonates for application as pesticides. Several of his products are on the market. Yet, he is puzzled about a letter from the management that says that the current procedure of sending compounds for new marketable products to the EPA for approval has now been changed because of the enactment of a

REACH regulation. From now on, instead of the regulators demonstrating that a product is not safe before removing it from the market, companies have to prove their products are safe before they are introduced into commerce. Dr Philipps sighs. This requires a big load of extra work and maybe external contractors for the toxicity studies. Probably, this is what *responsible research and innovation* is about: better safe than sorry!

13.2 Perspectives and Definitions

When asking different technological stakeholders about their definition of risk, we will get different answers (Figure 13.1).[1] Colloquially, risk is often used synonymously with harm, danger, hazard or uncertainty. When risks are communicated in and by the wider public, for example in political discourses, roundtables, public information events, and so on, it is often expressed in terms of fears, worries, and other emotional concerns, sometimes irrationally and unreasonably. While the justification and legitimacy of the public perception of risk may be questioned, it is important for other actors in the context of technology and risk assessment not to ignore these concerns as they express the *real* atmosphere or mood in the society.[2] On the other hand, we have a scientific approach to risk: natural and technical sciences (including engineering) study risk factors associated with technology empirically and numerically, often with a focus on the technology.[3] Semi-empirical sciences like the social sciences and humanities study risk rather with a focus on the affected people. In both cases, dealing with risk is a very rational and pragmatic endeavour rather than an emotional or intuitive one.[4] In economic perspectives, the potential or actual risks are of a different

Figure 13.1 Perspectives on risk.

nature: for companies, enterprises, entrepreneurs and other business people, risks are associated with product quality (potential functional failures, but also market failures) and with business performance (competitiveness, market share, return of investment rates, *etc.*).[5] In the macroeconomic view, risk means uncertainty of market developments, labour (for example, unemployment), growth, and the wealth of the society (expressed, for example, in the gross domestic product (GDP)).[6] A special case of private-sector businesses is insurances: their business model is based on statistical likelihoods of damages or undesirable events that the client overestimates and is willing to buy a certificate of insurance for.[7] In governance, ideally, the various actors (legislators, policy-makers, deputies, regulators, *etc.*) intend to enact policies and regulations that reduce the various risks that the society that is governed is exposed to. These range from very broad (for example, international relations, or the reliability of critical infrastructure like energy supply, mobility, or the healthcare system) to very specific (for example, regulations concerning concentration thresholds of substances in food packaging).[8] Often, these decisions are informed by risk and life cycle assessments, so that the regulatory guidance as a response to risk levels have a higher chance of proving successful.[9] Last but not least, questions of risk are always also philosophical questions,[10] especially in ethics: what is a risk and for whom, what kind of value is at risk in a particular situation, what does it mean to act under uncertainty, and how can precautionary measures be ethically and legally justified? Ethicists define the normative framework in which a risk debate is held.[11,12]

Case 13.1 gives us an idea of how different concepts of risk can collide in science and technology (S&T)-related contexts. The scientists, as they believed, have tackled the risk by providing empirical evidence that nanoparticles in sunscreens have no adverse health effects. Yet, the social acceptance of commercial products containing nanoparticles is determined by factors other than mere statistical risk assessments. The perceived health risks are weighed against interests, utility, economic considerations, and other not strictly scientific elements of the risk discourse.[13] This will be the focus of this chapter.

In addition to all these different nuances in the perspectives on risk, we may try to find a common ground in the form of a definition of risk.[14] The first way of saying it is this:

Risk is an unwanted event which may or may not occur.

There are two things to pay attention to. The first is the *may or may not* formulation. In a case in which we are sure that something will occur, we call it a harm or a threat or a hazard. In a case in which we are not so sure – it means there is a degree of uncertainty – we call it a risk. The second is the focus on *unwanted*. Obviously, risk is always associated with something negative, undesirable, displeasing. In this simplest definition the risk is the event itself. An example could be the statement "*Lung cancer is one of the major risks that affect smokers.*". However, sometimes the word risk is used differently:

Risk is the cause of an unwanted event which may or may not occur.

The exemplary statement would then be "*Smoking is by far the most important health risk in industrialized countries*". Both of these understandings of risk – an unwanted event or its cause – are qualitative. In technical contexts, we need a more quantitative definition, like this one:

Risk is the probability of an unwanted event which may or may not occur.

Here, we could state the exemplary statement "*The risk that a smoker's life is shortened by a smoking-related disease is about 50%.*" In many contexts, the mere probability is not sufficient. It must be combined with the severity of an event:

Risk is the statistical expectation value of an unwanted event which may or may not occur.

The expectation value of a possible negative event is the product of its probability and some measure of its severity. It is common to use the number of killed persons as a measure of the severity of an accident. With this measure of severity, the risk associated with a potential accident is equal to the statistically expected number of deaths. Other measures of severity give rise to other measures of risk. Although expectation values have been calculated since the 17th century, the use of the term risk in this sense is relatively new. It was introduced into risk analysis in the influential Reactor Safety Study, WASH-1400, in 1975. Today, it is the standard technical meaning of the term risk in many disciplines. It is regarded by some risk analysts as the only correct usage of the term. It is important to note that risk is differentiated from uncertainty as described by this definition:

Risk is the fact that a decision is made under conditions of known probabilities (decision under risk as opposed to decision under uncertainty).

When there is a risk, there must be something that is unknown or has an unknown outcome. Therefore, knowledge about risk is knowledge about lack of knowledge. A scientific approach to risk, consequently, puts a methodological focus on generating more knowledge in order to reduce risk levels. This has two strains of argumentation: first, the more we know the more problems we shift from (unmanageable) uncertainty-related to (manageable) risk-related issues. Second, the clearer we are about probabilities of occurrences and events the easier it is for us to intervene or prepare for it. In other words, the first endeavour is to reduce uncertainty, and the second is to manage risks and react on them properly. This is a task for *risk governance*.

13.3   Risk Governance

Figure 13.2 summarises the cycle of risk governance that proved useful and practicable for S&T governance as compiled by the International Risk Governance Council, communicated by risk researcher Ortwin Renn.[15,16]

Every assessment starts with the awareness of a particular problem or conflict. Someone must point out that there is or might be a

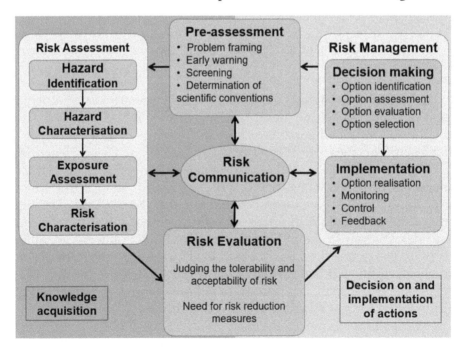

Figure 13.2   Elements of risk governance.[15] Adapted from ref. 27 with permission from Macmillan Publishers Ltd, Copyright 2008.

risk. In this pre-assessment phase, problem-framing takes place and early warnings are expressed. A superficial screening under consideration of scientific conventions and viewpoints reveals whether or not an articulated problem is manageable as a risk and, thus, makes it to the next stage, the assessment phase. Here, the risk is analysed thoroughly. First, the particular hazard must be identified, for example the contamination of a river, the chance of injury from misuse of a technical artefact, or social injustice as the result of misgovernance of new technologies. This hazard must be characterised in numbers, for example pollutant concentrations, their source, their effect, and so on. Then, it must be determined who or what is exposed to the risk and to what extent (how many, how much) – in the river example: fish, citizen, and so on. With this knowledge, the actual risk can be characterised: *"There is a risk of losing up to 20% of the fish in that river due to the release of pollutant from the upstream lacquer manufacturing plant."* When the risk is thus determined, it must be evaluated whether or not the risk is tolerable and acceptable, and whether or not there is the need for risk reduction measures. As we will see later, this might be the most difficult part of the chain. Before discussing this point in more detail, let's see what happens when it is decided to do something about the risk: the risk needs to be managed, which usually means an attempt to reduce it. Based on the available knowledge, options for action must be identified, assessed and evaluated. Finally, the best options are selected. After this decision-making procedure, the options are implemented and realised, including monitoring and control of the process. With this feedback it can be decided whether the measures are successful or not, whether risks remain or not, or whether new risks arise instead or not. Here, the cycle is complete by subjecting the risk analysis to another pre-assessment stage to eventually start the cycle again. All parts are necessarily connected by the important aspect of risk communication. In order to ensure the efficacy and usefulness of risk governance, all involved parties need to establish channels of clear, efficient and fast communication so that important information finds its way into the decision-making process.

We can see clearly in this scheme that there are, basically, two main parts of work to do: first – represented by the left half – the acquisition of knowledge related to the particular case of risk, second – with the right half being in charge of it – the decision-making on actions and their appropriate implementation. In most of the cases, the people who deal with the tasks involved in this process are different for the assessment and the management phase.

Although scientists, researchers, and other kinds of (technical) experts elaborate and compile knowledge, it is often legislators, regulators, managers, executive board members, agencies, governmental offices, and so on who debate and decide on strategies and actions. It is, however, quite unclear who is in charge of the evaluation phase (the Risk Evaluation box). Is it a third group of people (for example ethicists or social scientists)? Is it the experts from the left side or the decision-makers of the right side, or both? When technical experts are equipped with the task and responsibility to make evaluative and normative judgments, there is a danger of a technocratic decision-making system. When political or economic stakeholders in leading positions have that power, there might be the danger of biased decisions and severe conflicts of interest. Today, different levels of risk governance (for example, within companies, on the national governance level, internationally, globally), different countries and different political organs handle the organisation of the evaluation phase in different ways for different types of risk. Some parliaments established independent institutions (for example the Office of Technology Assessment in Germany), others delegate the evaluation of risks to those who are also in charge of managing them. For high-impact social risks of sociotechnical systems like bio- or nanotechnologies, even the wider public is participating in various channels in the risk evaluation. We will return to this topic in greater detail in Chapter 14.

As the examples (river pollution, injuries from technical artefacts, social injustice from the impact of a new technology) indicate, many different kinds of risk can be subjected to this procedure. Classically, this cycle was applied mostly to mere technical risks like contaminant concentrations or malfunction of devices or their parts. However, it is certainly possible to apply the same strategy to ethical and social risks. Remember the *larger picture* overview in Chapter 10 (Figure 10.3). The classical risk assessment is, according to this tiered structure, only the first step of a more complex assessment.[17] However, in S&T governance, not only risk perceptions need to be assessed and addressed. A major goal is the public acceptance of a technology and its development, and ultimately this can be reached by the societal embedding of the development process. This requires the inclusion of social and ethical implications into the assessment. Risk governance as an assessment tool, therefore, should not be limited to technical risks, but may be expanded to those kinds of conflicts that bear social tensions and ethical (or broader: normative) ambiguities.

13.4   Risk and Ethics

As these considerations illustrate, risk discourse is strongly interwoven with ethics.[18] Even the definition of *what* counts as a risk and what counts as a benefit is an evaluation that requires normative premises derived from ethical reasoning strategies. Moreover, what is one stakeholder's benefit is another stakeholder's risk. The question *for whom* something is a risk requires careful analysis and argumentation, too. Some say risks can only exist for individual persons as only those have expressible interests and a personal integrity that can be *at risk*. However, it was formulated that societies or even mankind as such may face risks. Moreover, certain risks are certainly also threatening non-human stakeholders, the biosphere, the environment, or the world as such. Here, again, we see the necessity to apply the centrisms (see Chapters 9 and 11) in order to support our arguments. Anthropocentrists and bio- or ecocentrists will argue differently, or – as we called it in the context of justice – will add different entities to the equation.

Ethics becomes especially significant when arguments have to be compared, weighed and prioritised. Ethical cases in this context, with different interests and colliding perspectives, are usually not a matter of compliance with morals (plural *ethics*, see Chapter 1), but of ethical argumentation and reasoning (singular *ethics*). Even though it is often impossible to identify one (or some) viewpoint(s) as *correct* and others as *wrong*, it is often possible to give good arguments why one viewpoint is *more convincing* or *plausible* than another. The resulting risk trade-off – a decision on a proper distribution of risks – is based on principles of justice and fairness. As usual in our science and technology discourses, the most common arguments are either consequentialistic or deontological.

Furthermore, the risk debate has clear connections to that of responsibility. As seen in Chapter 12, S&T-related responsibilities are often attributed in view of the possession of competences or knowledge related to risks and their prevention or reduction. Moreover, technological risks are usually related to one or more of the values we identified in Chapter 11 (Section 11.5), for example functionality and safety, or health, environmental integrity, economic wealth, and so forth. Clarifications on these relationships help identify conflicting interests and mediate options to proceed further. The normative framework of sustainability with its applications of justice and fairness principles provides a helpful reference for risk evaluation and management.

Case 13.2 provides an insightful example of the inherently normative character of risk assessment, even those that are considered fundamentally *scientific*. Chemometrics is a statistical method to identify potentially harmful substances. Classically, it follows a frequentist approach as explicated in Chapter 4. It sets a threshold level as a trade-off between acceptable levels of false positives (type I error, an observed hit (or effect) that actually is none), and false negatives (type II error, no observation even though the event (experiment) is a match, or: missing a hit). The conventional approach that assumes value-neutrality in scientific reasoning sets the error rates as equal ($\alpha = \beta$), that means detection limits are neutral (Figure 13.3). This is done because only physicochemical evidence shall be established as *scientifically grounded*.[19]

It has been argued that in many applications, this approach is not useful, or even unethical.[20] It is more plausible to favour type I errors over type II errors for health-impacting applications like drugs, food (*e.g.* packaging), construction materials, or other fields of chemicals that need to be regulated (see Figure 13.4). In simple terms: we would rather over-regulate a substance that is actually harmless (but has been found as potentially harmful, a false positive error) than risk letting a harmful substance slip through the testing (a false negative).

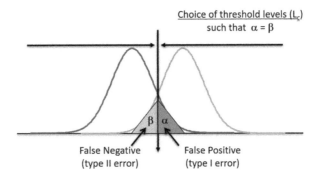

Figure 13.3 Conventional understanding of errors in chemometrics.

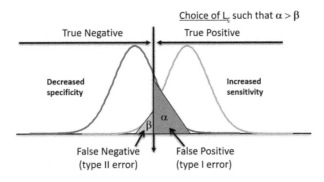

Figure 13.4 Intended threshold setting based on normative categories.

Interestingly, in jurisdiction we would apply the opposite approach, favouring type II errors over type I errors: we would rather let a guilty person run (false negative) than imprison an innocent one (false positive). In both cases, however, the rationale for this decision is not a purely scientific one, like the underlying premise $(\alpha = \beta)$ of the frequentist approach to statistics. Even such a down-to-earth risk consideration is made with normative aspects in mind: what situation do we favour, what is at stake, and how can we adapt our scientific approach in order to meet these interests and values (like safety, health, and integrity)?

13.5 Risk Discourse Types

Of course, not all cases of risk and their respective conflicts have the same character and impact. It would be waste of resources and energy to treat a simple risk with a commission on ethical and social implications, same as it would be dangerous to mandate a classical risk management group with the evaluation of ethical and social implications of, for example, nanotechnology. The abovementioned International Risk Governance Council (IRGC) report suggested a classification of risks and their management (Table 13.1).[15] As there is an obvious increase in the complexity of actors, conflicts, remedies, and discourse types from the simplest to the most sophisticated risk type, this overview is communicated as the *risk escalator*.

According to this classification, we distinguish four types of risk: simple risks (cases of known probabilities of a clearly defined hazard), risks induced by the complexity of the case, risks caused by a high degree of uncertainty, and risks arising from ambiguity and

Table 13.1 The four risk types and their management according to the IRGC's risk escalator.

Risk type	Discourse type	Actors	Conflict type	Remedy
Simple	Instrumental	Agency staff	—	Statistical risk analysis
Complexity-induced	Epistemic	Agency staff External experts	Cognitive	Probabilistic risk analysis
Uncertainty-induced	Reflective	Agency staff External experts Stakeholders	Cognitive Evaluative	Risk balancing Risk modelling
Ambiguity-induced	Participative	Agency staff External experts Stakeholders General public	Cognitive Evaluative Normative	Risk trade-off Deliberation

strong disagreement among stakeholders on priorities and normative values. In the first case, an instrumental discourse is sufficient: an agency (*e.g.* an environmental office) discusses the case on the basis of the known facts and performs the risk assessment as a statistical risk analysis in order to conclude what is the best strategy to deal with this risk. There is, most likely, no conflict arising from this kind of risk. 'Chemical' examples include the toxicity of nanoparticles in confined spaces such as food packaging, the life cycle assessment of a hydrogen fuel cell and its materials, or the effect of an agent in a cosmetic ointment on skin.

When the case is too complex, there is usually a lack of sufficient knowledge about the relevant factors. Then, the discourse should be epistemic, which means it should be focused on the generation of more profound knowledge that can help to solve the problem. In order to do so, the input from external experts in fields that have to do with the case is required. Arising conflicts are usually of a cognitive nature: two experts disagree on a key aspect and try to convince each other by presenting knowledge and facts that they regard as significant. As statistical data is often not available for cases like this, the best remedy is a probabilistic risk analysis. In chemical contexts, examples of complexity-induced risks are the toxicity of nanoparticles in open spaces like human organisms or the environment, persistent organic pollutants (POPs) in the food chain, or the effect of sequestering carbon dioxide with phytoplankton on the ocean ecology.

In case of simple and complex risks, it is clear what is at stake. That means, the cost of inactivity can be compared to the cost and effect of taking action to tackle the risk. The discourses revolve completely around factual questions (see Figure 1.2 in Chapter 1). The complexity increases drastically when the risk arises from uncertainty, that means when the near-, mid-, or long-term effect of a development on a particular value cannot be predicted because probabilities of events cannot be estimated. All involved parties agree on the value that is at stake, but nobody knows which direction a development concerning this value will take and with what degree of intensity. A related discourse can only be reflective, which means that it is conducted in small steps of action-feedback loops. Agency staff, external experts and various additional affected stakeholders try to figure out the best options, implement them, re-evaluate them and proceed step by step, like walking in a dark unknown room. Conflicts, here, are not only cognitive ("*What are the most reasonable options?*"), but also evaluative ("*What would we do, if...?*"). Risks like this often can't be reduced or dissolved. Therefore, solutions are found not merely in probabilistic risk modelling, but require a balancing of risks to an acceptable level, for example by exploiting principles of distributive justice. Here, chemistry-related examples would be the impact of progress in the field of nano-scaled compounds on global equity and justice, the relationship between progress in the capacity improvement of energy storage materials and the national environmental policy strategy, or the economic and environmental impact of commercialising fusion reactors for the production of electricity.

Ambiguity-induced risks arise when the interests and integrities of affected parties collide and conflict in intractable disagreement. Either one side's interests are neglected or the other's. Only a participative discourse that brings all stakeholders – including the general public and other third parties – together has a chance to solve the issue. The conflicts are not merely of a cognitive or evaluative character, but of normative impact (*"What do we value and in which order of priority? How do we want to live?"*). Therefore, an objective risk assessment based on data and probabilities is insufficient. The solution can only be found in a trade-off of risks and a careful participative deliberation on how to proceed – in other words: in making compromises. Here is an example from the nano-field again: nano-enabled human enhancement tools include eye implants that enable vision beyond the natural human range of up to 700 nm. It is envisioned that it can help truck drivers to drive more safely at night or in the twilight owing to an increased ability to see in shady conditions. Others worry that such an enhancement could create social injustices, especially in view of people having personal objections against cyborg-like implants. Yet, the pressure created by the availability of such a technology would limit those people's freedom of choice (when the alternative is being jobless). Here, interests have to be protected and values (freedom, autonomy, justice) preserved. Another chemical example is the assessment of the application of rare earth metals in microelectronic parts. Although they are crucial for devices that, apparently, find high acceptance in the population thanks to their convenience and perceived improvement of life quality, available resources are depleting, and recycling methods are still in their infancy. Consumer interests, economic interests, environmental interests, and even political interests (in view of large supplies of rare earth metal ores in the North Korean mountains, or the rare earths trade dispute with China) collide. The chemical assessment of alternative materials, recycling methods, or the environmental effect of electronic trash disposal is but a small part of the overall discourse of stakeholders and their interests. Yet, the normative conflicts and their possible solution depend strongly on the chemical assessment and near-future technological developments in this field.

Certainly, many chemistry-related risks fall into the category of simple risk discourse. The toxicology of compounds and materials in controlled and confined environments, for example in food packaging, construction materials or in consumer products is standardized and internationally communicated and applied. Here, no discourse on disagreements is necessary. However, the case of toxicology easily gets very complex, for example when the materials enter *open spaces*, like an organism or the environment. Here, it needs the help of experts and researchers to elaborate suitable toxicological measurement methods and assessment strategies. Owing to the complexity a disagreement on the proper methodology and interpretation of the results might arise. Therefore, a cognitive discourse is needed, but not so much a normative one, as the affected values (*e.g.* safety and health) are endorsed by all involved discourse participants.

Chemists and other scientific and technical experts feel competent, familiar and comfortable in these risk discourses. Value- and norm-related discourses are left to other epistemic authorities. Yet, as the overview and the examples hopefully show, reflective and participative discourses as a response to uncertainty- and ambiguity-induced risks can only be reasonably held under participation of the scientific and technical experts. Recall two previous insights: normative arguments always have two types of premises, factual *is*-statements and normative *ought*-statements (Figure 1.1, Chapter 1); and epistemic dependence makes interdisciplinary explorations of the interspace between domains of knowledge the most effective form of collaboration (Figure 8.3, Chapter 8). For evaluative and normative risk discourses it means that epistemic authorities of the fields of facts and norms need to collaborate and mutually inform their positions to avoid the danger of speculation and science-fiction scenarios. If discourses are held on matters that have no factual foundation, a lack of credibility of the risk approach is created. There is a strong call for pragmatism by those who deal with S&T-related risk issues. It is very difficult to predict the future and perform plausible technology foresight. Instead of debating *future presents* (situations that might or might not occur in the future) it is more reasonable to debate *present futures* (possible pathways or trajectories of development that can be followed from the present situation onwards). Thus, chemical experts are an inevitable part of norm-related risk discourses.[21]

Case 13.1 describes this common misunderstanding of the role of scientific expertise in risk assessments. The toxicologists have done their job, and most likely they have done it well in view of the methodologies, models, interpretation of data, and communication. Yet, their part is not done with that. They may have solved the cognitive conflict of the epistemic discourse concerning a complexity-induced risk (effect of titanium dioxide nanoparticles in the human organism), but the risk discourse proceeds on another level. Are the consumers satisfied with the regulations regarding consumer protection in products containing nano-scaled compounds? Is their expectation reasonable? Are special regulations required to handle nano-scaled products or are the existing ones sufficient? These are uncertainty-induced risk discourses that were held in the early 2000s and resulted in a re-conceptualisation of the European chemical registry REACH. As discussed in the context of interdisciplinarity, the contribution of chemical expertise is crucial for the elaboration of fruitful guidelines, because it informs all other contributions.

Box 13.1 Exercise: Risk Assessment, Small and Large

Prepare a risk assessment for either a research project that you are working on, or a practical programme that you are in charge of. The assessment should be structured as indicated in the following.

1. *Introduction.* Briefly outline the scope of the project/programme that you are assessing.
2. *Hazard Identification.* Identify all associated hazards under the four headings, Chemical Hazards, Biological Hazards, Physical Hazards and Ergonomic Hazards.
3. *Hazard Evaluation.* Identify the potential seriousness of the identified hazards?
4. *Exposure Assessment.* Consider potential levels of exposure (*who? how much?*) to the various hazards, taking account of any control measures that are in use.
5. *Risk Characterisation.* Use the above considerations either to justify that risks are being adequately controlled or to identify the need for additional controls.
6. *Concluding Remarks.* This should include a brief reflective paragraph on any ethical considerations (*e.g.* why further assessment would be advised).

Now, after learning about risk discourse types and the normative components of risk, what additional factors that were not part of your "classical" risk assessment would you have to consider? Why can't you perform a "social risk assessment" for uncertainty- and ambiguity-related risks of your research output by yourself alone? Yet, what could be your contribution?

13.6   Precautionary Principle

If we are forced to make decisions in the face of risks or uncertainties, we might need to apply guidelines or principles that give an orientation on what is a proper way to proceed. The most established tool in the field of S&T governance is the precautionary principle, or better: *precautionary principles* as there are many variations of it.[22] The simplest way of saying it is the proverb *"Better safe than sorry!"*. We may express it this way: when there is an obvious risk that might cause significant or even catastrophic damage, we need to be active and do something about it, even if our knowledge about it is incomplete and we are not sure about every detail concerning this risk. Politically, the precautionary principle (PP) was enacted first in the Rio declaration that was written at a UN conference in Rio de Janeiro in 1992 (Earth Summit, see Chapter 11):

> *"Where there are threats of serious or irreversible damage, lack of full scientific certainty shall not be used as a reason for postponing cost-effective measures to prevent environmental degradation." (Rio Declaration, 1992, Principle 15)*

This formulation is difficult to understand and on first reading might sound confusing. In order to understand it correctly it is important to know that at that time many legislations refused to fight environmental degradation and global phenomena such as climate change, claiming that there is no scientific evidence that the problems are really caused by anthropogenic (human) activities. The idea is: even though it is not scientifically certain in which way environmental degradation happens, we should do everything we can to prevent serious or irreversible damage, because if we wait too long, it might be too late to react! This has been, since then, the major theme of precaution: act as long as it is still possible. Precaution almost never means *"Stop doing research!"* or *"Ban a technology!"* as some proponents of moratoria against genomics, nanosciences or synthetic biology have suggested. It was conceptualised as a tool to preserve human control over technological progress and to empower humans to protect the values they hold important.

The PP has been applied in various ways. Put negatively, it is simply a rhetoric tool exploited by politicians in order to silence critical voices, as in *"No worries, we have everything under control!"*. In a few actual cases, the PP was employed as a decision-making aid, for example in prioritising EU research funds for critical techno-scientific fields.[23] It has been attempted (mostly in academic essays) to use the PP as a moral principle in the same way we use sustainability, autonomy or justice as principles. Moreover, some legislations made use of the PP as a legal principle in law- and policy-making, for example in several regulations of the EU on nanotechnological applications such as nano-scaled drugs.

Precaution was debated heatedly among philosophers and ethicists.[24] Strictly speaking, it is not a principle that helps in situations of risk (a probabilistic estimate of the possibility that a (negative) event might occur), but only in cases of uncertainty (a situation where it is not possible to estimate risk probabilistically). Knowledge is created in order to elevate an issue of uncertainty into an issue of risk which is easier to handle. When this is not possible, the PP applies until knowledge is available. This causes a problem that the promoters of the PP intended to avoid: it supports scientistic and technocratic approaches of risk governance rather than holistic solutions under inclusion of ethical and social implications. With the intention to reduce uncertainties and deal with risks, technical and scientific strategies are promoted. Ultimately, this approach of avoiding facing moral and social risks blocks sustainable scientific and technological development, according to the critics of the PP.

Chemists have practical experience with the PP when they are affected by the enactment of REACH (see Case 13.3). The risk governance approach of the chemical registry is dedicated to a PP that manifests itself in the substitution of regulatory proof of unsafety by producer evidence of safety.[25] The former is inefficient and cost-intensive because thousands of substances used in industrial products need to be tested with a large array of methods to identify and specify hazards in order to meet precautionary measures. Delegating the analysis and risk assessment to the producers and appliers of industrially relevant substances following a pragmatic and practice-oriented protocol of requirements solves two issues at the same time: the process is more efficient in view of decentralized, less transportation-intensive and, thus, safer procedures; and the precautionary measures are easier to be met because the appliers apply thresholds in their risk assessments that are feasible and plausible for the intended context of the tested substance. This regulation establishes a balance between the workload of industrial R&D and an acceptable required degree of product safety. Critical substances require more testing, while the bar is lower for uncritical chemicals (in terms of scale or exposure). Informed by a sophisticated catalogue of criteria for the risk evaluations, thresholds and precautionary measures, it is believed that REACH meets high scientific standards while being firmly embedded in socially justified normative frameworks. A final remark on Case 13.2: the mentioned normative judgment concerning error rates in chemometrics can be perfectly justified by precautionary principles. When 100% certainty cannot be reached, a plausible strategy for dealing with uncertainty informed by normative (ethical, legal) priorities may be the best response to tackle risks.[26]

Exercise Questions

1. Which of the following constitutes the final step in a risk assessment?

   A: Hazard characterisation.
   B: Risk characterisation.
   C: Exposure assessment.
   D: Hazard identification.

2. A hazard may be defined as...

   A: ...any deviation from standard operating procedures.
   B: ...anything with a potential to cause harm.
   C: ...the probability of a harmful effect occurring.
   D: ...a universal natural phenomenon.

3. How is risk related to a hazard?

   A: Risk and hazard are synonyms.
   B: Risk is an intrinsic property of any chemical and is independent of hazard.
   C: Risk depends on the degree of exposure to a hazard.
   D: The relationship varies and depends on the nature of the hazard.

4. Which of the following defines when it would it be acceptable not to carry out a risk assessment before introducing a new chemical into an on-going process?

   A: If a risk assessment for a similar chemical has already been carried out.
   B: If the urgency of continuing the process justifies a delay to a later date.
   C: If the workplace is limited to five or less employees.
   D: Never.

5. Which of the following would not be considered as an ethical dimension of a risk assessment?

   A: Profit margins.
   B: Society.
   C: The environment.
   D: Individuals.

6. Deontological considerations deal with...

   A: ...cost-benefit analyses.
   B: ...benefits to society.

C: ...individual rights (and duties).

D: ...temporary legislation.

7. Which of the following is the appropriate type of discourse when dealing with a complexity induced type of risk?

A: Reflective

B: Instrumental

C: Epistemic

D: Participative

8. For which two of the following would quantitative risk analysis be unlikely to prove adequate?

A: Global equality in the harvest of economic potential from nanotechnology development.

B: Exposure levels in a chemical factory.

C: Controlling effluent discharges from a factory into a river.

D: Regulating consumer behaviour (for example purchase, use and disposal of plastic bags).

9. The Precautionary Principle suggests that...

A: ...nothing should be done until all the facts are known.

B: ...steps should be taken to reduce potential harm without waiting for absolute proof of the cause.

C: ...all stakeholders should be consulted before controls are introduced.

D: ...economic considerations should be balanced against the rights of individuals.

10. Which of the following steps should you take before starting to work with a new compound that you have synthesized, analytically identified, but not yet fully characterized?

A: Await the results of animal testing before beginning to work with it.

B: Assume that it may be harmful and adopt respective safety measures.

C: Assume that it is safe and carry on using it.

D: Assume that risks are the same as for similar compounds.

11. Which of the following steps should an approach to dealing with an uncertainty induced risk involve?

A: Probabilistic risk analysis.

B: Risk balancing.

C: Risk trade off.

D: Statistical risk analysis.

12. Following any risk assessment, which of the following should be the next step?

A: Risk evaluation.

B: Risk reduction.

C: Exposure assessment.

D: Risk communication.

13. On which of the following does the level of risk associated with a particular chemical primarily depend?

    A: The hazards posed by the chemical.
    B: An individual's experience of working with the chemical.
    C: The risk assessment.
    D: The probability of harm occurring and the potential severity of this harm.

14. Which of the following should be considered in an ethical risk-benefit analysis?

    A: Society.
    B: The environment.
    C: Individuals.
    D: All of A–C.

15. What of the following would NOT be of concern in a risk analysis from the perspective of a consequentialist agenda?

    A: The consequences of inaction.
    B: Desirable and adverse impacts on populations.
    C: The rights of individuals.
    D: Cost-benefit analysis.

16. Which of the following would be considered to be an unethical use of resources?

    A: Providing safety training.
    B: Paying 'danger money' (extra payment for working in risky environments).
    C: Installing local ventilation systems.
    D: Providing PPE (personal protective equipment).

17. How many deaths would be expected if accidental release of a toxic gas results in 150 workers being exposed to a level associated with a 2% risk of fatality?

    A: None.
    B: 3.
    C: 15.
    D: 30.

18. When undertaking safety testing for chemicals to be used as food preservatives, it is considered beneficial to have higher levels of false positives, in which safe chemicals may be discarded rather than have false negatives in which potentially harmful chemicals are allowed into food. Why?

    A: This complies with the precautionary principle.
    B: It is easier to conduct the test this way.
    C: False positives are easy to identify and discount.
    D: This is the most cost effective approach to testing.

19. Which of the following describes the characteristics of a risk discourse concerning an ambiguity induced risk?

A: Two experts have different opinions on what would be the best strategy to deal with the risk.

B: There are insurmountable disagreements between stakeholders so that no solution can ever be achieved.

C: Two or more different and possibly conflicting values are affected by the decision-making concerning the management of the risk.

D: In order to avoid emotional and irrational discussions, risk discourses in this risk type domain exclude public stakeholders so that the experts can efficiently determine the best solution.

20. Why do uncertainty and ambiguity induced risk discourse types require the contribution of scientific expertise?

A: Evaluative and normative arguments should be firmly grounded in facts and knowledge. This knowledge is, ideally, provided by scientists.

B: Owing to their high level of education, academic scientific experts are considered the wisest among the discourse participants. Therefore, they are in the position to make the best ethically and socially relevant decisions.

C: Most Western and many Asian societies chose technocratic forms of governance and policy-making. Thus, in respective countries, scientists and technologists are in charge of making policy-relevant decisions.

D: It is legally required by risk assessment directives, that committees and commissions dealing with risks in the EU should include scientists.

References

1. N. Möller, The Concepts of Risk and Safety, in *Handbook of Risk Theory*, ed. S. Roeser, R. Hillerbrand, P. Sandin and M. Peterson, Springer, Dordrecht, 2012.
2. S. Chilton, J. Covey, L. Hopkins, M. Jones-Lee, G. Loomes, N. Pidgeon and A. Spencer, Public Perceptions of Risk and Preference-Based Values of Safety, *J. Risk Uncertain.*, 2002, **25**, 211.
3. National Research Council, Committee on Improving Risk Analysis Approaches, *Science and Decisions: Advancing Risk Assessment*, National Academies Press, Washington, 2009.
4. J. O. Zinn, *Social Theories of Risk and Uncertainty: An Introduction*, Blackwell, Oxford, 2008.
5. W. D. A. Bryant, The Microeconomics of Choice under Risk and Uncertainty: Where Are We?, *Vikalpa*, 2014, **39**, 21.
6. S. Makarova, *Risk and Uncertainty: Macroeconomic Perspective*, UCL SSEES Economics and Business Working Paper No. 129, 2014.
7. S. Asmussen and M. Steffensen, *Risk and Insurance*, Springer, Cham, 2020.
8. H. Matthee, Political risk analysis, in *International Encyclopedia of Political Science*, ed. B. Badie, D. Berg-Schlosser and L. Morlino, SAGE Publications, Thousand Oaks, 2011.
9. O. Renn, *Risk Governance – Coping with Uncertainty in a Complex World*, Earthscan, London, 2008.
10. T. Lewens, *Risk: Philosophical Perspectives*, Routledge, Abingdon, 2007.
11. S. O. Hansson, *The Ethics of Risk: Ethical Analysis in an Uncertain World*, Palgrave Macmillan, New York, 2013.
12. *The Ethics of Technological Risk*, ed. L. Asveld and S. Roeser, Earthscan, London, 2009.

13. *Nanotechnology Environmental Health and Safety: Risks, Regulation, and Management*, ed. M. Hull and D. Bowman, Elsevier, Amsterdam, 3rd edn, 2018.
14. S. O. Hansson, Risk, in *Stanford Encyclopedia of Philosophy*, ed. E. N. Zalta, Fall Edition, 2018, https://plato.stanford.edu/entries/risk/, accessed August 26 2020.
15. O. Renn and P. Graham, *Risk Governance – Towards an Integrative Approach (White Paper No. 1)*, International Risk Governance Council, Geneva, 2005.
16. O. Renn, Risk Governance in a Complex World, in *Encyclopedia of Applied Ethics*, ed. R. Chadwick, 2nd edn, Elsevier, London, 2012.
17. *Shaping Emerging Technologies: Governance, Innovation, Discourse*, ed. K. Konrad, C. Coenen, A. Dijkstra, C. Milburn and H. van Lente, IOS Press, Netherlands, 2013.
18. S. O. Hansson, Ethical Risk Analysis, in *The Ethics of Technology*, ed. S. O. Hanson, Rowman & Littlefield, London, 2017.
19. P. Gemperline, *Practical Guide To Chemometrics*, CRC Press, Boca Raton, 2nd edn, 2006.
20. M. C. Ortiz, L. A. Sarabia and M. S. Sánchez, Tutorial on evaluation of type I and type II errors in chemical analyses: From the analytical detection to authentication of products and process control, *Anal. Chim. Acta*, 2010, **674**, 123.
21. J. Mehlich, Is, Ought, Should - The Role of Scientists in the Discourse on Ethical and Social Implications of Science and Technology, *Palgrave Commun.*, 2017, **3**, 17006.
22. S. Holm and E. Stokes, Precautionary Principle, in *Encyclopedia of Applied Ethics*, ed. R. Chadwick, 2nd edn, Elsevier, London, 2012.
23. Science for Environment Policy, The Precautionary Principle: decision making under uncertainty, *Future Brief 18. Produced for the European Commission DG Environment by the Science Communication Unit*, UWE, Bristol, 2017, http://ec.europa.eu/science-environment-policy, accessed on August 26 2020.
24. D. Steel, *Philosophy and the Precautionary Principle. Science, Evidence, and Environmental Policy*, Cambridge University Press, Cambridge, 2015.
25. J.-P. Llored, Ethics and Chemical Regulation: The Case of REACH, *HYLE*, 2017, **23**, 81.
26. I. T. Cousins, R. Vestergren, Z. Wang, M. Scheringer and M. S. McLachlan, The precautionary principle and chemicals management: The example of perfluoroalkyl acids in groundwater, *Environ. Int.*, 2016, **94**, 331.
27. O. Renn, White Paper on Risk Governance: Toward an Integrative Framework, in *Global Risk Governance*, ed. O. Renn and K.D. Walker, International Risk Governance Council Bookseries, vol.1, 2008, pp. 3–73.

14 Science Governance, Technology Assessment

Overview

Summary: After introducing concepts such as sustainability, responsibility, risk, and the connection between scientific activity and ethical values, we still miss a crucial link: why would this matter to chemists, and what can they do about the impact of chemical research and development (R&D) on society and environment, that is within their power? This chapter will introduce channels and established procedures for chemical professionals in science, research and innovation to contribute their competence and expertise to discourses in the context of science and technology (S&T) governance and policy, in public stakeholder panels, or in any form of S&T assessment.

In Chapter 10, we discussed the Manhattan project as an example for scientists taking social responsibility. Einstein wrote a letter to President Roosevelt to warn him of a threat. Today, scientists don't need to write letters to political leaders. Instead, a variety of communication and exchange platforms have been created. In the European Union and its member states, offices of technology assessment are associated with parliaments or governments in order to inform S&T governance and policy with state-of-the-art scientific knowledge and a competent estimation of the expectable trends in the nearer future. Decision-making in the context of societally important topics like health care, energy supply, mobility, infrastructure, food supply, and so on, requires the input from experts who, ideally, are skilled in interdisciplinary discourse and communication with non-experts. After an overview of the role of scientific expertise in policy-making and the implemented approaches for a fruitful contribution, a guide for successful policy-relevant knowledge reporting is presented. The considerations presented in Chapter 1 – the role of ethics as a discourse methodology for the clarification of facts and norms through an ethical prism – become most effective in this chapter.

Good Chemistry: Methodological, Ethical, and Social Dimensions
By Jan Mehlich
© Jan Mehlich 2021
Published by the Royal Society of Chemistry, www.rsc.org

Key Themes: The role of scientific expertise in S&T policy and governance; technology assessment; EU research programs' accompanying work packages (ELSI, RRI, 3O), knowledge reporting.

Learning Objectives:

This chapter shall help you to:

- Set the insights from the previous chapters (sustainability, responsibility, risk discourses) into perspective and understand their meaningfulness and relevance for chemical professions.
- Determine the possibilities of chemists to engage in S&T-related discourses on desirable and undesirable implications and effects of chemical progress and its role in innovation.
- Avoid common fallacies and misunderstandings concerning the role of scientists in such discourses and bring in your competences in the most credible and fruitful way.

14.1   Introductory Cases

Case 14.1: We Need You!

Prof. Stone is part of a larger multi-national research consortium that develops novel materials for infrastructure construction. The complex grant application with its references to several of the sustainable development goals has been approved by the European Commission, so that the 15 partners from academia and industry receive funding for five years. He can finance three PhD students and a postdoc for research on physicochemical surface coating methods that shall increase the durability of concrete-steel structures in combination with plant cover for green architecture. Unexpectedly, he receives an invitation for a meeting of the *work package on ethical, legal and social implications* of the project from the leader of that work package (WP), a social scientist. What does he have to do with that?! As his schedule is extremely full, Prof. Stone rejects the invitation and decides that he will let those sociologists do their thing while he does his.

Case 14.2: Speculation

At the end of the first decade of this millennium (2007–2009), a large number of publications and research projects addressed ethical and social issues in the field of nanoscience and nanotechnology. Many essays were written by authors who themselves were not nanoscientists, and scientific-technical introduction chapters in multi-author books were only loosely connected to the other chapters. This led to the accusation that *nanoethics leaps ahead* and discusses science fiction rather than the real and more urgent developments in state-of-the-art nanoscience. At the same time, national and international technology assessment offices were asked by parliaments and governments to support efficient policy-making with scientific input on ethical, legal and social implications of nanotechnology. How can this dilemma be solved?

Case 14.3: Don't Let Me Be Misunderstood

Prof. Ramirez, leading expert on industrial scale polymerisation reactions for the manufacture of plastics, is asked to speak at a UN panel about the chemical possibilities of fabricating environmentally friendly plastics (for example, biodegradable or recyclable), alongside other experts who address topics like trapping plastic microparticles in water cycles, policies for limiting consumption, or alternative materials from plants like cellulose. Prof. Ramirez is planning his speech which is limited to 10 minutes plus a 10 minutes question and answer session. How can he explain all the science that is necessary to understand the complexity of polymerisation reactions in large throughput reactors? His students learn that in 8 hours of lecture time. What else would he have to report so that the panel will be convinced of the high economic value of these reactions?

14.2   Scientific Expertise in S&T Governance

In Chapter 10, in an overview of assessments for greater public acceptance and societal embedding of S&T progress (Figure 10.3), we learned that this growing degree of complexity is met by extending the procedure of risk assessment towards a broader and interdisciplinary assessment that includes many stakeholders.[1] We also learned that this is based on the constructivist view of S&T progress being influenceable, controllable, designable and at all stages debatable. In the past decades, manifold strategies and methods have been elaborated under the broad term *technology assessment* (TA) in order to provide a helpful tool for decision-makers in economy and politics to facilitate a socially sound and sustainable development of science and technology through supportive and efficient governance and regulation.[2] TA has been institutionalized in the European Union (EU) and the United States of America in the form of *Parliamentary TA*.[3,4] Every EU-funded research project since the *framework program 6* agenda, for example, must include a work package on *Ethical, Legal and Social Implications* (ELSI) in order to implement a more mature analysis of these issues as a basis for EU-wide regulatory governance and policy-making on the one hand, and to facilitate a more democratic governance procedure

and incorporate public participation to examine the risk issues arising from S&T policy innovation on the other hand (see Case 14.1).[5]
The ELSI concept has been further elaborated in the more recent
approach labelled *Responsible Research and Innovation*[6-8] and is taken
up by the latest vision *Open Science, Open Innovation, Open to the
World*.[9] In principle, they all share the same idea: ethically and socially
sound progress is achieved by a high degree of interdisciplinarity,
integrating ethics and sociology into R&D activities and establishing
an ongoing accompanying normative discourse on S&T issues. Moreover, the degree to which public participation is integrated into S&T
related policy-making and governance has increased significantly. In
particular, scientists face situations like the following, sometimes voluntarily, sometimes obligatorily:

- Description of possible application fields, as well as the social
  and ethical dimensions of research projects in grant proposals,
  even for basic research.
- Communication of research results and scientific expertise with
  non-experts and scientific laymen, for example in the form of
  press releases, interviews with journalists, public hearings and
  information events.
- Expert testimony in court cases or for government committees.
- Participation in ELSI work groups of research consortia, as for
  example in EU-funded research projects, or in other platforms of
  policy debate.

It must be acknowledged that the openness of political authorities
and holders of power for evidence-based science-informed policy-
making is a blessing for the political landscape in democratic Europe.
This is far from self-understanding and is the result of efforts that
have been made since the 1980s by academic authorities that had
a voice in the parliamentary and governmental institutions in their
countries and on the EU level. With the political systems of different
countries being quite diverse, it is inappropriate to generalise the procedures of science consultation. Yet, there are a few common themes
that scientific actors should know.

Figure 14.1 illustrates the pathways of mandates and inputs in an
idealized fashion.[10] Political organs appoint special portfolios (for
example, ministries, preserves, or departments of a party) to elaborate insights into a specific topic that requires governance, but that
is too complex to be comprehended by the legislators. Examples are
technological solutions in fields such as energy supply, transportation

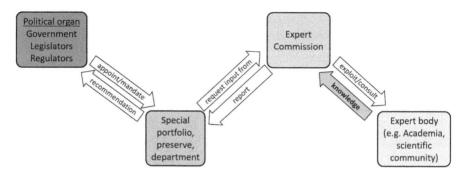

Figure 14.1 Role of scientific knowledge in policymaking.

systems, healthcare, infrastructure, communication, manufacturing, or agriculture and food supply. The members of such portfolios are usually politicians or state secretaries with a professional background that is not sufficient for solving the complex matters and questions of the respective topic. Therefore, they request input from an interdisciplinary expert commission that consists of advisors and professionals with competences and expertise in specific fields. This commission, knowing what has to be found out and where and how to get the necessary information, consults the larger *expert body*, that is the huge pool of scientists, researchers, academic scholars, and other epistemic authorities and their knowledge base. It is important to note that in recent years, owing to a paradigm shift away from positivistic scientism towards a more integrated view on participative deliberative democracy, the role of social sciences, psychology and humanities (for example, philosophy and ethics) increased significantly, so that for obviously *technological* topics this field of expertise is also consulted.[11] The expert commission collects all the relevant information in a report which can easily exceed 1000 pages. It is the task of the special portfolio that requested the input from the expert commission to condense this report into understandable recommendations for the political organisation that appointed it.[12]

For example, the European Commission (EC) recognised the importance of regulating recent developments in *nanomedicine*. Apparently, with new diagnostic and therapeutic tools, much more critical patient data is generated than with established common methods. In order to be able to make the regulatory and governance process sustainable (for example in terms of social, but also economic or environmental aspects), a special portfolio was mandated in the *Science and Technology* section of the EC with the task to elaborate recommendations on how to solve current and possible future issues in nanomedicine. This portfolio formed a commission from the pool of experts that are connected (for example, through EU-funded research consortia, distinct research institutions, non-governmental organisations (NGOs), *etc.*). This group on nanomedicine consisted of a nanotechnologist (a scientist who knows what nanomedical methods are capable of and how they work), a toxicologist (an expert on the effects of nanoparticles in the human body), a medical practitioner (who knows how nanomedical treatments are applied in hospitals and/or doctors' offices), an industrialist from a pharma company (expert on availability and market aspects of nanomedicine), a representative of a patient organisation (for example the *Rheuma league*), a regulator (expert on current legal frameworks and regulatory issues of nanomedical agents and drugs), a social scientist (expert on health systems and the impact of changes in it on society), and an ethicist (expert on reasoning strategies for applying certain legal and normative principles, for example, precautionary principles, autonomy, privacy, *etc.*). Each of these experts is able to extract relevant knowledge from his or her field of expertise, generated and documented by the worldwide scientific and academic community. From this perspective, the expert commission is just a representation of all available knowledge, the basis of pragmatic and empiric decision-making. The extensive report that this commission compiled contains sections on technical aspects of nanoparticles and their application for medical purposes, economic aspects, patient and consumer safety, legal aspects of responsibility, privacy and data safety, the impact of certain regulatory decisions on social fabric and on ethical values, and so on. It is evaluated and summarised by the portfolio. The recommendations that result from the extensive report are short, clear in language and content, and skip all the background information, so that a member of the government or council (here: the EC) can base his or her decision on simple and easily accessible information. In the case of the nanomedicine commission, it looked like this: the toxicologist explained on approximately 20 pages (including graphs and study results) that a lot of data on the toxicity of nanoparticles at their active sites are available, but that nothing is known about where they remain afterwards and what kind of effects they might have on metabolic processes and organs (liver, pancreas, *etc.*). The ethicist explains on 10 pages that in the case of a lack of knowledge the common European approach of dealing with this situation is precaution (confirmed by the regulator informing current EU directives on precaution) which can be justified and argumentatively supported by Kantian deontology and modern concepts of contractualism as the normative foundations of the European society. In the final recommendation, this is condensed into a statement such as *"Given the lack of knowledge on long-term effects of nanoparticles in the human body, a precautionary regulation approach as in EU directive XYZ is advised."*

As mentioned in earlier chapters, the key to successful guidance of S&T developments into sustainable trajectories is the interdisciplinary collaboration of epistemic authorities.[13] In simple English: for orientational knowledge to be effective and insightful for decision-makers, different scientists and experts have to work together.[14,15] Case 14.1 describes the reaction of a chemist who refuses to collaborate with other experts with the argument that he has nothing meaningful to contribute to the ethicists' and sociologists' research. In this chapter, we are going to refute this argument.[16] First, we will highlight the scientific approach to technology assessment (Section 14.3). Then, we need to understand the roles of ethics (Section 14.4) and what the chemist has to do with all that (Section 14.5). To summarise the point, we will finally make it clear how and what chemists of all kinds can contribute as competent and responsible experts to the governance of S&T progress (Section 14.6).

14.3 Technology Assessment

The endeavour of scientific policy-advising for a more sustainable S&T progress as described in Section 14.2 is now commonly referred to as *technology assessment*. Historically, it developed from a rather technological or economic tool (for example, analysing the components of a technological artefact concerning their probability to malfunction and risking the loss of the object or harm for someone or something) into a political and sociological tool that aims to take the *larger picture* of technological development into account. It was highly promoted and extended in the 1980s by the academic field of *Science, Technology and Society* (STS, sometimes also *Science and Technology Studies*), a sociological discipline.[17] An important precondition is the acceptance of social constructivism as a paradigm of technological development instead of technological determinism (see Chapter 10). The social constructivist approach allows intervention into any development process and guiding it in the *right* direction. If progress was deterministic it would be a meaningless endeavour.

In the *Encyclopedia of Applied Ethics* we find the following definition. Almost every word in this definition is important and requires further explanation.

"Technology assessment (TA) is a scientific, interactive, and communicative process that aims to contribute to the formation of public and political opinion on societal aspects of science and technology."[18]

Science and technology: This is the object that TA is studying. Understanding the term '*science and technology*' as a grammatical singular is a deliberate choice to demonstrate the tight connection between the two in *techno-scientific systems*. The considerations of Chapter 10, especially the overview in Figure 10.1, apply here. S&T in the sense that is relevant for TA covers all parts of the development chain, from design and planning *via* research, fabrication, product innovation, marketing and sales to consumption, application, and finally disposal and/or recycling. TA looks at all relevant stakeholders, from scientists, engineers and designers to entrepreneurs, manufacturers, workers and providers, to consumers, appliers and disposers, to regulators, policy-makers and legislators, to the environment, and to third parties like future generations and animals.

Societal aspects: TA understands itself as accompanying research on societal aspects of S&T, which include ethical and legal implications of S&T progress, represented by the acronym ELSI. This modern conception of TA stands in contrast to the classical TA approach as an engineering tool. The latter studied technical artefacts with the goal to identify improvement potentials in either functional terms (utility, resource managements, *etc.*) or economic aspects (efficiency, market performance, *etc.*).

Scientific: TA understands itself as an academic discipline that is devoted to scientific methodologies and procedures.[19] Institutional TA can be found at universities or in independent research facilities, as well as in the form of offices or agencies that perform professional TA for governments (*e.g.* the Office for Technology Assessment at the German Bundestag, Büro für Technikfolgenbewertung beim Deutschen Bundestag) or corporations, as a kind of science consulting. As the former, TA is close to disciplines such as technology management or the sociology of technology. Its methodologies include foresighting, scenario analysis, conceptual analysis, interdisciplinary research spanning a multitude of disciplines and their specific methodologies (for example, the scientific rigour of life cycle assessments with the logic coherence of normative sciences like ethics). Yet, also the latter (public sector TA) fulfils strict scientific standards in order to meet its goal of providing effective orientational knowledge with credible epistemic authority.

Interactive: Bringing together expertise from various fields such as natural and technical sciences, social and normative sciences, industry, politics, jurisprudence, and public interest groups, the nature of TA is highly interdisciplinary and interactive.[20-22] Reaching out to actors and stakeholders outside of 'disciplines' (non-experts, laymen, the general public and its representatives, NGOs, *etc.*) is a rather novel approach to realising participative discourses that are so important for S&T assessments, as seen in the previous chapter.

Communicative: It is often highlighted that TA is communicative because the generation of orientational knowledge necessarily requires the exchange of information among experts, a solution-oriented debate and the communication of conclusions, strategies and/or recommendations to the relevant stakeholders, decision-makers and – in some cases – the general public. In terms of Figure 14.1, TA operates between the expert body and expert commission element. The more or less direct communication of insights with decision-making authorities is the most important objective of TA.

Public: The public is both the stakeholder and target group: representatives of public interest groups participate in TA processes and debates (for example, patient groups in medical topics or environmental activists in projects with a potential environmental impact); and at the same time the whole effort is undertaken in order to facilitate a socially sound and healthy development.

Political opinion: As explicated in Section 14.2, the decisions of political authorities shape and guide S&T development to a large extent. This decision-making that aims at tackling risks, reducing adverse implications, harvesting chances and facilitating opportunities requires solid evidence-based factual input from as many epistemic agents as possible. Organising and operating this communication for efficient policy deliberation and subsequent decision-making is the main goal of TA.

Contribute: The sustainable development of society and its lifeworld environment as a societal, political, and economic goal requires many players to participate in the discourse on how to achieve it. The mission of TA is not and cannot be to solve all the problems and overcome all the obstacles just by itself. It is but one contribution to a plethora of approaches in the wider context of S&T development, yet a very important one. It attempts to bundle together the synergies of cross-disciplinary and cross-sectoral collective efforts to increase the chances of S&T progress taking a direction towards the kind of future that we envision by how we wish to live.

In the past two decades, the methodological and practical foundations of TA have improved constantly and overcome many obstacles. Interdisciplinarity in scholarly exchange is extraordinarily difficult on the procedural level. The chemist in Case 14.1 is shying away from discussing issues that are outside of his familiar epistemic microworld. His understanding of the content and methodologies of different scientific disciplines is limited and he is worried about spending time inefficiently. Unfortunately, the ELSI assessment of innovations in the field of construction materials can only yield useful results when it is informed by a material scientist who provides the factual basis that is assessed.

In some conceptions, TA is more than a governance tool. While *parliamentary TA* and *participative TA* aim at policy-making influence, *constructive TA* intends to impact S&T processes from the very start.[23] This is especially relevant for chemical actors in the development chain. The ELSI work package in the research consortium described in Case 14.1 is operating in this context because it aims to guide the researchers in aligning their decisions concerning research directions and assessment of alternative options with the values manifested in the later outcome, meaning that the scientific progress represents values that are regarded as important and deserving protection.

The following real example describes the procedures and effects of accompanying a multi-disciplinary research project with the tools of constructive TA.[24] The *Nanopil* is a project by a Dutch team of nanoscientists and researchers in the framework of the Nanotechnology Program in the Netherlands. Based on the idea of an oncologist, they are attempting to develop a lab-on-a-chip for the detection of colon cancer. This form of cancer is one of the most frequent ones in Europe (or maybe worldwide). Therefore, the government is interested in simple and cost-effective screening methods. The current method requires the patient to send a stool sample to the doctor's office and has a relatively high error margin. The nanoscientists conceptualised a pill-sized lab-on-a-chip – a tiny sensor array with sample filters and detectors – that is swallowed and that travels through the digestive tract, collecting bowel fluid, detecting cancer-specific molecules, and responding in the case of positive detection (see an animation of the process on youtube.com, ref. 25). There are two possible options for revealing the results: either the lab-on-a-chip carries a radio-frequency identification (RFID) sender that transmits the detected concentration of the target molecule to a receiver, for example a smartphone, or the chip has a small capsule of blue dye which is released when the target molecule (the indicator for cancer) is found. The scientist's idea is the method's simplicity: patients don't have to collect and send those humiliating stool samples anymore. Moreover, the sensitivity of the screening is greatly improved, making it much more efficient. The promoted values, therefore, are health, dignity, efficiency, and personal integrity.

The project was accompanied by a team of technology assessors who examined the ethical and social implications of the Nanopil. There are not many toxicological issues as the lab-on-a-chip does not release any substances into the intestines. Therefore, this project can serve as a perfect example for a nanotechnology-related development which impacts values that go beyond mere toxicological risks. We may assess this project with a three-tiered plausibility analysis. First, we may ask whether the expected values are plausible and whether they are indeed plausibly affected by the new technology. Second, we ask whether the expected and promoted values are really desirable for those who are affected. Third, we point out the options that the Nanopil designers and developers have in order to make the Nanopil support desirable values, that means how they can inscribe the desirable means-ends-relationships into the new product.

The developers claim that the Nanopil is a much simpler, less humiliating, and more elegant detection method. Is this really true? It turned out that during the research phase the biggest obstacle for the Nanopil to work was clogging and jamming with solid substances in the digestive tract. Therefore, in order to work, the patient has to drink 4 litres of laxative, a disgusting liquid used to clean the intestines. This part of the detection was not mentioned in the animation but adds an important procedure that has a strong significance for its applicability. The scientists also debated which indication method, RFID signal or blue dye, was more suitable. This has a technical dimension (the RFID chip is more difficult to implement than a blue dye capsule), but also presents a significant user dimension: in the case of the RFID signal questions regarding data protection and privacy arise, for example who may receive the result on his smartphone, the patient only or also the doctor? The blue dye capsule implies that it is conceptualised as a self-test: the patient might feel insecure in interpreting the results, so that there is no added value of the Nanopil as an efficient screening method. Another important value, the screening efficiency, will strongly depend on the availability and distribution of the pill. If it is only given out by doctors to patients, the demand of produced Nanopils might be low, meaning that it will be more expensive. If it is offered freely in pharmacies, the price will be low, but patients might show different patterns of usage, so that the detection method must be taken into consideration again.

The case has to be debated outside the confinements of the researchers' lab, for example with doctors, patients, pharmacists, health insurances, and so on. Doctors report that most patients don't have any problem with the stool sample, actually. In contrast, it will be very difficult to convince the people to drink the laxative before using the Nanopil. If this problem cannot be solved, the Nanopil will not be accepted, according to their point of view. Patients are mostly concerned about the affordability of the screening. Here, coordination with health insurance companies and distributors is required. Again, this has an impact on the actual pill design, so that insights on the actual value preferences in society has a direct impact on the research and development process. Last but not least, the design of the pill directly influences the doctor–patient-relationship: patients feel insecure with the interpretation of the results, especially in the case of the blue dye option, and prefer a data exchange with the doctor. This, in turn, raises issues in data handling and its safety, especially in case of the RFID solution. Here, in terms of Section 13.5, the case is an ambiguity-induced risk discourse that cannot be held by the scientists alone, not even together with the doctors. It requires the inclusion of the public and a decision by legislators (elected representatives), which directly affects the final design of the Nanopil. In any case, these considerations show that it is possible to assess a technology during its R&D phase and implement the acquired insights into value relationships in the design and conceptualisation of the item. This is the idea of constructive TA.

14.4   The Role of Ethics

A few clarifications concerning the role of ethics in S&T assessments are due. The conceptual and procedural connection of factual and normative accounts of knowledge that both play a role in S&T progress requires approaches to discourse practice that bridge the gap between empirical and normative sciences, their parlances, and their methodological differences. Practical experiences have shown that discourse participants from social and normative sciences are more or less familiar with the methodological self-understanding of natural sciences and engineering fields, whereas natural scientists have little experience with ethics as a normative science.[26] In Chapter 1, we learned a basic definition of ethics, either understood as morals (and compliance with them) or as ethical reasoning and argumentation.

TA deals with cases in which ethics in the latter sense is required. In practice, how are these discourses organised and conducted? See Figure 14.2.

Ethics as a philosophical discipline (moral philosophy) has developed a few well-established ethical theories as reasoning strategies for the ethical or moral judgment of cases. Moral philosophers spend significant efforts arguing which of the theories is the most viable and convincing one. Once a theory is chosen as the preferred one, it is exploited to solve real and hypothetical cases with the means and rationality that it provides. This is called the *top-down approach* of ethics. It proved unhelpful in many *applied ethics* contexts because neither the experts nor the general public can agree on which ethical theory is preferable and why. Yet, the opposite approach is not any more practicable: the bottom–up approach is followed if particular cases are solved one by one with case-selective suitable theories. In practice, this casuistry is often performed unconsciously by philosophical laymen in *applied ethics* contexts, like doctors in medical ethics, animal rights activists in bioethics, businesspeople in business ethics, or the general public in common-sense morality. This approach can be extremely tedious and time-consuming because the reason for the case-specific judgments must be provided and explained over and over again. As an alternative, a middle way between top–down

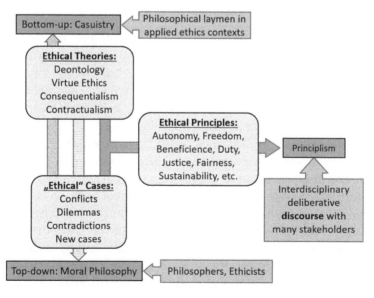

Figure 14.2 The practical-pragmatic role of ethics in deliberative discourse on science and technology.

and bottom–up is the strategy to apply ethical principles to particular cases. This is known as *principlism*. The principles are informed by and derived from ethical expert knowledge, the cases are brought in by experts and laymen in the affected fields.

The idea is very pragmatic: a discourse is only worth the effort when it comes to practicable, viable, plausible, down-to-earth solutions. While top-down strategies are too theoretical and intellectual, and bottom–up procedures are tedious, repetitive, time-consuming and inefficient, principlism has the advantage of narrowing down the theoretical content to handy principles while allowing many stakeholders with different backgrounds (therefore, interdisciplinary) to engage in the discourse. Stakeholders are entities for which something is *at stake*, which means something they value and wish to be preserved. As illustrated in Section 14.2, there are established arenas of such ethical discourses, for example science and technology policy and governance, technology assessment institutions, or academic research on the ELSI of S&T. Hence, we are not discussing something theoretical or academic here, but very practical approaches that are implemented to certain extents in all routines of our daily life.

This pragmatism, though, cannot guarantee that the procedure always works successfully. As Case 14.2 shows, the discourse on the ethical and social implications of emerging S&T fields like bio- or nanotechnology in the early 2000s and, currently, of digitalization, AI, big data, internet of things, robotics, and cloud computing often starts with little connection between the normative and the factual side of the argument.[27] This may lead to ethical discussions of science-fiction-like scenarios that are detached from the reality of laboratory work or innovation teams in R&D. In the worst case, this has such a negative impact on the credibility of the ethical discourse that it is entirely dismissed by the S&T experts as useless.[28] The *Collingridge dilemma* describes the difficulty of finding the right time for governance intervention in innovation and S&T progress.[29] Starting ethical assessment and subsequent regulation too early can kill potentials, while doing it too late can lead to loss of control over adverse impacts. The S&T-accompanying discourse on social and ethical implications, with its governance relevance, can constructively support the development when all involved experts and stakeholders communicate efficiently with each other and inform each other of factual and normative aspects of what is at stake and what the options for proceeding are.

14.5 The Role of Chemists

So far, in this chapter, reasons for the interdisciplinary discourse on S&T involving chemical expertise have been provided. What is missing, yet, is practical guidelines for how expert-based scientific testimony can be efficient and helpful. Robert Pielke described four stereotypical scientific knowledge contributors in S&T-related discourses that serve as an insightful orientation, here.[30] His reflections apply to basic and applied researchers, academic and private-sector scientists, and to engineers. They are plausible in contexts in which technical means are discussed to solve important issues, but also in governance-oriented assessments of S&T developments as such. Moreover, it will be easy to transfer the insights of this section to all kinds of interdisciplinary discourse situations, even on a rather small scale. The competences that shall be sharpened with what follows are discourse performance and a broader vision in science communication.

Imagine a political or citizen panel seeking counsel regarding geo-engineering responses to climate change. A scientific expert is consulted to inform about feasible options that tackle the problem. Pielke's first character, the *pure knowledge exponent*, responds like a detached bystander, spelling out in detail the various chemical and/or mechanical engineering processes that can sequester carbon. Politicians and citizens might well feel like they have inadvertently walked into a technical engineering class. Presenting all the scientific background and the advantages and disadvantages of possible options, he himself remains neutral on any decision-making.

The *advocate*, by contrast, acts like a salesperson and immediately argues for a bioengineering-related seeding of the ocean with iron to stimulate phytoplankton growth that would consume carbon dioxide. But the argument is made with a peculiarly technical rhetoric that deploys information about the chemical composition of the iron, transport mechanisms, relationship to phytoplankton blooms, and more. Politicians and citizens might well think they were standing at a booth demonstrating a proprietary innovation at an engineering trade show. The advocate seems to have made a choice on what he believes is *the best solution* and promotes it regardless of the normative judgments made by others (politicians, public).

The *arbiter* acts more like a hotel concierge. Steering a course between that of neutral bystander and advocate, such an expert starts by asking what the politician or citizen wants from a geoengineering response: simplicity, low cost, safety, dramatic results, public acceptability, or what? Once informed that the aim is safety, the arbiter engineer would identify a matrix of options with associated low risk factors. The arbiter engineer engages with the public and communicates knowledge guided strongly by publicly expressed needs or interests. The concierge might on another occasion work as a medical doctor or psychologist counselling a patient.

Pielke suggested a 4th type of scientist/engineer that in his view would be the most helpful one in a normative discourse: the *honest broker* reaffirms some modest distance from the immediate needs or interests of any inquirer in order to offer an expanded matrix of information about multiple geoengineering options and associated assessments in terms of simplicity, cost, safety, predictable outcomes, and more. The effect will often be to stimulate re-thinking on the part of the inquirers, maybe a re-consideration of the needs or interests with which they may have been operating, even when they did not originally take the time to express them. The experience might be more analogous to a career fair than a single booth at a trade show.

Pielke and associates argue that the most appropriate path is to recognize the limits of science and engineering, and to distance advice from interest-group politics while more robustly connecting it to specific policy alternatives.[31] Research will not settle political and ethical disputes about the kind of world in which we wish to live. But chemists can connect their research with specific policies once citizens or politicians have decided which outcomes to pursue. In this way, chemists (in addition to other scientists and engineers) provide an array of options that are clearly related to diverse policy goals. Rather than advocate a particular course of action, either openly or in disguise, chemists should work to help policymakers and the public understand which courses of action are consistent with the current – always fallible – technical knowledge about the world and the current – always revisable – visions of the good.

The protagonist in Case 14.3, Prof. Ramirez, finds the answer to his question about how and what to talk about to the panel in the 10 minutes of speech time with the help of a reflection on Pielke's archetypes. Familiar with the scientific view, he surely tends to report like a *pure knowledge exponent*, describing in great length what he deems important and what the panel most likely won't understand and, more importantly, won't even need to know for a policy-relevant decision. According to his thoughts as expressed in the case description, there is also a danger that he would choose to present in the style of an *advocate*: He seems to have decided for himself what the most important value is that every rational person must endorse, namely the economic value of polymerization reactions in large-scale industrial production. He would be well advised to react more flexibly on the panel's expressed value preferences and relate his scientific judgment to that. If the main goal is to find a solution quickly, regardless of the costs, then approach A will help best, but if the priority is on economic sustainability, approach B has more advantageous effects. As an arbiter, he would accept the panel's value preference and deliver, somewhat neutrally, what science is able to provide. As an honest broker, however, he would use his socially responsible and value-aware scientific mind to critically assess the expressed value preferences and attempt to illustrate the more complex relationships between the choice of a scientific-technical remedy for the problem and the various implications on different stakeholder's interests and the underlying norms.

Box 14.1   Exercise: Advise your Government

As an expert in the field, the government invites you to a hearing to provide expert opinion on how to deal with persistent organic pollutants (POPs). More binding regulations concerning the fulfilment of the sustainable development goals require a more rigorous policymaking approach to banning POPs. Yet, in order to meet industrial demands, alternatives should be in place. The legislators want to understand what is at stake. Prepare a 5–10 minutes speech.

   How would Pielke's four different characters reply (in principle, not in scientific detail)? Try to reflect on the risk types detailed in Chapter 13 as well: are there uncertainty- and ambiguity-induced risks that require the input of chemical expertise? What concept of sustainability is underlying your replies?

14.6   A Trivial Conclusion?

As mentioned above, the interdisciplinary character of the institutionalised (that is: politically desired and procedurally implemented) ELSI debate bears methodological difficulties that result in a reluctance among S&T enactors to participate in roundtables or ELSI work package group meetings for the research programs they are involved in as, for example, Prof. Stone in Case 14.1. Here is an example from a nanomedical research consortium that the author was involved in:[32] One work package has been assigned for the analysis and assessment of ethical, legal and social aspects of nanomedicine. The leader of the work package was a Germany-based technology assessment institution. According to the grant agreement, one representative of each of the 15 collaborating institutions – if possible, the principle investigator (PI) – was requested to participate in meetings and roundtable discussions that were scheduled biannually. Some of these roundtables took place during general project meetings, so that participation required as little effort as possible. However, the resonance was markedly small. In most of the meetings, no more than three to five consortium members participated. When invited in person – in the most direct possible way – a scientist refusing to join the roundtable replied: *"You do your Ethics. I have nothing to say about it!"*. This could be due to a lack of knowledge about ELSI questions. It could also be scientific humbleness. Without the input from the scientific and clinical experts, however, the work group was not able to fulfil its objectives satisfyingly. A final report was evaluated by the European Commission as "too theoretical, too abstract, too general", thus wasting valuable monetary public resources.

Science and technology experts and R&D enactors play an important role in the ELSI debate. This will be illustrated by the common form of an ethical argument that was introduced in Chapter 1, here in shorter form:

$$P_i + P_o \rightarrow C_S$$

In the ethical assessment of S&T issues, the *is*-premise necessarily needs the input from those who are experts in the respective debated fields, here referred to as *S&T enactors*: scientists, engineers, product developers, industrialists, entrepreneurs, regulators (policy-makers), and others. Debating ethical issues that are based on incorrect assumptions is a waste of time and effort. For example: in the early phase of the nanoethical debate, people discussed implications of *nanobots* circulating in human blood vessels or the risks of *grey goo* from autonomous, self-organizing, self-replicating nanoparticles, even though there was and is no hint that nanoscientists will be able to fabricate that kind of nanoparticle in the nearer future. These approaches damaged the reputation of nanoethics as it was criticised for being speculative and mixing up scientific progress with science fiction (see Case 14.2). ELSI assessment requires a constant dialogue with those who *know what is going on* concerning both the current state (for example in research and industry) and the future prospect (for example risk assessment).

The normative premise (the *ought*-premise) is often brought in by *ethical laymen*. In principle, ethics (*i.e.* ethical reasoning, evaluating, debating) can be performed by everyone as all members of a society usually have an intuitive understanding of basic values and virtues, be it only by common sense. However, it is very likely (and various experiences support this claim) that ethical and social issues arising in the context of S&T progress are beyond this common-sense morality. As soon as viewpoints conflict with each other, as soon as ethical dilemmas occur or sophisticated problems are identified, the help from a professional ethicist appears helpful or even obligatory as the large majority of scientists and engineers to date have little or no expertise in the field of professional ethics. It is the role of the ethicist to moderate the debate among stakeholders, to identify argumentation lines and their errors, to sort and correct arguments, to bring in well-reasoned ethical principles and argumentatively elaborated normative premises, and to guide the discussion to an output that supports those decision-makers that shape and regulate technological development. As Cotton puts it:[33]

> *"Scientists and engineers certainly possess expertise, but expertise and familiarity with a research topic and its consequences should not be confused with expertise in the application of normative ethical theory, nor in providing robust moral judgements. [...] Scientists are not ethics experts, and if technology policy is significantly shaped by the proscribed moral viewpoints of scientific authorities, then this is, in essence another form of technocracy, one that would likely exacerbate further public conflict."*

The ELSI approaches, responsible research and innovation (RRI), and certainly also the newer "3O" campaign, all highlight the significant importance of interdisciplinary collaboration. Synergies are created by teaming up experts with different backgrounds so that none of them needs to act beyond the realms of their professional knowledge (like in the past, with ethicists writing about biotechnology without understanding its scientific background, or scientists claiming an "ethically sound outcome" without overseeing the whole area of potential conflicts and their solutions). The ambitious goal of sustainability through constructive S&T assessment can only be reached by collaboration between the sources of empirical scientific and technological (and sociological, political, *etc.*) knowledge and those who have the competence to constitute the normative framework of the discourse. Scientists clearly belong to the former and, therefore, are not expected to *do ethics*. However, without their willingness to feed the arguments with their input, the ethicists' attempts to do ethics in S&T-related fields won't be fruitful either.

To engage the example of the nanomedical research project one more time: the ELSI analysis report written by ethicists and technology assessors, naturally, contained a lot of platitudes and commonplaces on precautionary principles, good scientific practice, responsibility, autonomy and freedom, distributive justice, and so on. Recommendations for policy-makers and regulators, then, often had the form *"If it happens that case A occurs, affecting social value p in this or that way, and if value p is seen as important or worth protection, then follow strategy X.".* This, indeed, is very abstract and not very helpful for regulators. An alternative report written by the scientists, however, enthusiastically promoting the advancements and benefits enabled by the newly developed technique, method, device, or compound (and their safety), was not helpful either. The idea behind the implementation of constructive ELSI, RRI or "3O" approaches is not to *generate acceptance* by cheerfully concluding that no (new) ethical pitfalls could be identified. The strong point of these synergetic and fruitful collaborations between experts of completely different fields is the interdisciplinarity in which each participant brings in his or her core competence to come to meaningful insights that wouldn't be possible when they all work alone by themselves. Moreover, the collaboration with social scientists and ethicists can sharpen their awareness of the social and ethical dimensions of S&T.

The progress of S&T has become a multidisciplinary constructive endeavour in which the role of the scientist must be redefined. In addition to core competences as a *researcher* or *developer*, the functional traits of a *communicator* and *evaluator* gain more significance. Experiences have shown that many scientists could not familiarise themselves, yet, with their shifted roles and the expectations going along with that, for example, participation in ELSI workgroups, explaining ethical and social dimensions in grant proposals, public roundtables, and so on. Here is a simple plea for a higher awareness of the ethical and social aspects of scientific activity, hopefully understood as a motivator for bringing in scientific expertise as an essential contribution to S&T discourse without being worried about a lack of ethical competence. On the one hand, this protects chemists from optimistically promoting the benefits and prospects of their work or its application while ignoring public concerns. On the other hand, this improves the efficiency and usefulness of ELSI/RRI analysis and technology assessment in general. Then, the goal of building bridges between *the empiric* and *the normative*, between *the professional* and *the responsible*, and between *the innovative* and *the sustainable* can be reached.

Exercise Questions

1. In which of the following ways do scientists contribute to S&T governance?

   A: As elected members of a parliament or government.
   B: Only very indirectly through their publications and research reports.
   C: In no way because governance and politics are detached from academic expertise.
   D: As sources of knowledge and expertise in policy-advising commissions.

2. Which of the following is NOT a concept employed by the European Commission in the research and development projects that it funds?

   A: Ethical, Legal, and Social Implications (ELSI).
   B: Responsible Research and Innovation (RRI).
   C: Sustainable Millennium Strategy (SMS).
   D: Open Science, Open Innovation, Open to the World (3O).

3. Which of the following best describes the current understanding of technology assessment?

   A: In a technology assessment, elements and compartments of a technical device are analysed concerning the likelihood of malfunctioning.
   B: Technology assessment is a scientific, interactive, and communicative process that aims to contribute to the formation of public and political opinion on societal aspects of science and technology.
   C: Technology assessment is an economic tool to calculate profit margins, value gain and investment return in R&D projects.
   D: Technology assessment aims at identifying technology-related risk potentials and reducing them to zero.

4. Technology assessment as an academic discipline is... (choose the one tenable statement)

   A: ...highly interdisciplinary.
   B: ...a purely sociological inquiry.
   C: ...rather irrelevant for the general public.
   D: ...a matter only for engineers and technology designers.

5. Technology assessment addresses... (choose the most plausible answer)

   A: ...technology assessors.
   B: ...decision-makers in the public and private sector.
   C: ...academics in a wide variety of disciplines.
   D: ...consumers of technologies.

6. The knowledge elaborated by technology assessment is considered to be...

   A: ...tacit.
   B: ...politically delicate.
   C: ...post-factual.
   D: ...orientational.

7. Which of the following best describes the role of ethics in technology assessment?

   A: It is brought in "top down" by moral philosophers.
   B: The most common and most efficient strategy is to apply well-defined principles (like justice, responsibility, sustainability, *etc.*) when discussing contemporary and possible future cases arising in the context of technological development.
   C: Ethical aspects are discussed case by case ("bottom up") whenever such questions arise.
   D: Ethics is irrelevant for technology assessment because it deals with intangible aspects of humanity.

8. Which of the following is NOT one of the policy-advising types of scientist suggested by Robert Pielke?

   A: Pure knowledge exponent.
   B: Advocate.
   C: Honest broker.
   D: Goalkeeper.

9. Robert Pielke suggested that the most valuable contribution to S&T-related governance is made by "honest broker" type scientists and engineers. Which of the following best describes the characteristics of such an expert?

   A: He/she can relate the implications of making choices (*e.g.* of technical methods or engineering approaches) on goals and priorities (*e.g.* profitability, global justice, social harmony, *etc.*) and can trigger a constructive debate on the usefulness and feasibility of such goals and priorities.
   B: He/she is able to convince political and economic decision-makers by applying rhetorical skills and by appearing charming and trustworthy.
   C: He/she knows the stock market very well and, thus, is able to figure out the most economically feasible option.
   D: Throughout his/her career, this expert was able to build and maintain a wide network of colleagues, peers and collaborators from academia, industry and politics, so that his/her judgments will find wide acceptance among the stakeholders in the respective discourse arena.

10. Among the types of scientific expert described by Robert Pielke, the *arbiter* was compared to the communication style of which of the following?

    A: Postman.
    B: Salesperson.
    C: Hotel concierge.
    D: Coach of a football team.

11. Which of the following best describes the dangers of scientific expertise in political decision-making?

    A: Dogmatism and theocracy.
    B: Elitism and oligarchy.

C: Idealism and communism.
D: Scientism and technocracy.

12. Which of the following phenomena is commonly referred to as *post-factualism*?

   A: The common practice in scientific publishing to always write interpretations and conclusions *after* presenting the facts (experimental data, observations, *etc.*).
   B: The turn away from a *knowledge society* towards a *spiritual society*.
   C: The trend that beliefs and opinions have the same epistemic value in political debates as scientific and other empirically acquired knowledge.
   D: The paradigmatic shift from *representative democracy* (in which politicians present the results of their debates like facts to the public) towards *deliberative democracy* (in which politicians merely make vague suggestions that are open for public debate).

13. Which of the following statements best describes the relationship between post-factualism and science/scientists?

   A: Scientists as sources of empirically sound and justified knowledge should fight trends towards post-factualism whenever and as best as they can!
   B: Science as a social institution will benefit from trends towards post-factualism because more and more academic fields will achieve the status of a real scientific discipline, even those that don't conduct experiments or perform empirical studies.
   C: The role and status of sciences won't be affected by trends towards post-factualism because the latter is only of concern for politics and public debate, in which scientific knowledge is insignificant.
   D: Scientists should embrace post-factualism. At least, it will give them visibility on mass media (see, for example, the protests of scientists against President Trump in USA) and a stage for expressing their claims on funding and the meaning of their profession.

14. Which of the following does the acronym ELSI stand for (in the context of this chapter)?

   A: European Life Science Institute.
   B: Ethical, Legal, and Social Implications (of S&T).
   C: Excellent and Liberal Sciences Initiative.
   D: Extra Labile Synthetic Isotope.

15. Which of the following does the acronym RRI stand for (in the context of this chapter)?

   A: Responsible Research and Innovation (a European concept for sustainable R&D).
   B: Russian Research Initiative (the largest political science funding program in the world).
   C: Red Ribbon International (a social movement for raising solidarity of people living with HIV/AIDS).
   D: Research on Risk Impacts (of S&T) (a new transdisciplinary academic field).

16. Which of the following represents the best strategy for technology assessment?

    A: *Multidisciplinary*: experts from various disciplines compile their respective views in reports and edited books on critical topics.

    B: *Transdisciplinary*: technology assessment is a new discipline synthesized from certain sub-branches of other disciplines (like political science, applied ethics, sociology, green chemistry, and others).

    C: *Interdisciplinary*: experts from various disciplines elaborate knowledge from the "inter-space" between disciplinary margins and exploit the synergies thus created.

    D: *Anti-disciplinary*: the public character of technology assessment requires that disciplinary jargons and lifeworld-detached expert knowledge should be avoided at all costs.

17. Which of the following constitutes the main target group for technology assessment projects?

    A: Industrialists and businessmen.

    B: Academic scholars.

    C: The public and its political representatives (decision-makers).

    D: Engineers and product designers.

18. Which of the following expert commissions are likely to require contributions from a chemist?

    A: Future mobility solutions.

    B: Sustainable energy supply.

    C: Personalized medicine and its impact on the health care system.

    D: All of A-C require chemical expertise, for example material science, biomedical chemistry, and so forth.

19. In which way was the turn from technological determinism towards social constructivism in the 1960s/70s a pre-condition for the establishment of TA as a S&T governance tool?

    A: Only the latter promotes the view that technological progress is debatable, controllable and designable. Only under this paradigm is the endeavour of TA in any way meaningful.

    B: The older generation of academic scholars that opposed this kind of intellectual innovation had to be replaced by young visionary scientists before a new discipline such as TA could arise.

    C: Political parties, in the 20 years after WWII, followed the deterministic paradigm in their programmes. The move towards constructivism changed the political landscape so that the acceptance of TA among parliamentarians could grow.

    D: Technological determinists channelled all the available tax money into technical innovation and R&D projects. Only the constructivists were willing to spend a fraction of the budget for research on the ethical and social implications of S&T.

20. The goal of science and technology governance, in general, is...

    A: ...to make S&T development sustainable.
    B: ...to maintain a development that supports social justice and harmony.
    C: ...to harvest economic potential and, at the same time, manage risks efficiently.
    D: ...all of A–C.

References

1. *Shaping Emerging Technologies: Governance, Innovation, Discourse*, ed. K. Konrad, C. Coenen, A. Dijkstra, C. Milburn and H. van Lente, IOS Press, Netherlands, 2013.
2. A. Ely, P. van Zwanenberg and A. Stirling, Broadening out and opening up technology assessment: Approaches to enhance international development, coordination and democratisation, *Res. Policy*, 2014, **43**, 505.
3. *TA Practices in Europe*, ed. J. Ganzevles and R. van Est, EPTA-PACITA, EU, 2012.
4. *Policy-Oriented Technology Assessment across Europe*, ed. L. Klüver, R. O. Nielsen and M. L. Jörgensen, Palgrave Macmillan, Basingstoke, 2016.
5. A. Hullmann, *European Activities in the Field of Ethical, Legal and Social Aspects (ELSA) and Governance of Nanotechnology*, European Commission, DG Research, EU, 2008.
6. European Commission, *Options for Strengthening Responsible Research and Innovation*, 2013, DOI: 10.2777/46253.
7. *Responsible Innovation 1. Innovative Solutions for Global Issues*, ed. J. van den Hoven, N. Doorn, T. Swierstra, B. J. Koops and H. Romjin, Springer, Dordrecht, 2014.
8. *Responsible Innovation 2. Concepts, Approaches, and Applications*, ed. B. J. Koops, I. Oosterlaken, H. Romjin, T. Swierstra and J. van den Hoven, Springer, Dordrecht, 2015.
9. European Commission, *Open Innovation, Open Science, Open to the World – a Vision for Europe*, 2016, DOI: 10.2777/552370.
10. D. B. Pedersen, The Political Epistemology of Science-Based Policy-Making, *Society*, 2014, **51**, 547.
11. H. Zwart, Ethical Expertise in Policy, in *Encyclopedia of Applied Ethics*, ed. R. Chadwick, Elsevier, London, 2nd edn, 2012.
12. *Experts in Science and Society*, ed. E. Kurz-Milcke and G. Gigerenzer, Kluwer, New York, 2004.
13. *Interdisciplinarity – Reconfigurations of the Social and Natural Sciences*, ed. A. Barry and G. Born, Routledge, Abingdon, 2013.
14. A. S. Balmer, J. Calvert, C. Marris, S. Molyneux-Hodgson, E. Frow, M. Kearnes, K. Bulpin, P. Schyfter, A. Mackenzie and P. Martin, Taking Roles in Interdisciplinary Collaborations: Reflections on Working in Post-ELSI Spaces in the UK Synthetic Biology Community, *Sci. Technol. Stud.*, 2015, **28**, 3.
15. A. S. Balmer, J. Calvert, C. Marris, S. Molyneux-Hodgson, E. Frow, M. Kearnes, K. Bulpin, P. Schyfter, A. Mackenzie and P. Martin, Five rules of Thumb for Post-ELSI Interdisciplinary Collaborations, *J. Responsible Innov.*, 2016, **3**, 73.
16. *Evaluating New Technologies – Methodological Problems for the Ethical Assessment of Technology Developments*, ed. P. Sollie and M. Düwell, Springer, Dordrecht, 2009.

17. R. van Est and F. Brom, Technology Assessment, Analytic and Democratic Practice, in *Encyclopedia of Applied Ethics*, ed. R. Chadwick, Elsevier, London, 2nd edn, 2012.
18. D. Bütschi, R. Carius, M. Decker, S. Gram, A. Grunwald, P. Machleidt, S. Steyaert and R. van Est, The practice of TA; Science, interaction and communication, in *Bridges between Science, Society and Policy: Technology Assessment – Methods and Impacts*, ed. M. Decker and M. Ladikas, Springer, Berlin, 2004.
19. A. Grunwald, *Technikfolgenabschätzung – eine Einführung*, edition sigma, Berlin, 2nd edn, 2010.
20. *Interdisciplinarity in Technology Assessment. Implementation and its Chances and Limits*, ed. M. Decker, Springer, Berlin, 2001.
21. *Practising Interdisciplinarity*, ed. P. Weingart and N. Stehr, University of Toronto Press, Toronto, 2000.
22. C. F. Gethmann, M. Carrier, G. Hanekamp, M. Kaiser, G. Kamp, S. Lingner, M. Quante and F. Thiele, *Interdisciplinarity Research and Trans-disciplinary Validity Claims*, Springer, Cham, 2015.
23. A. Rip, J. W. Schot and T. J. Misa, *Managing Technology in Society: The Approach of Constructive Technology Assessment*, Pinter Publishers, New York, 1995.
24. F. Lucivero, *Ethical Assessment of Emerging Technologies*, Springer, Heidelberg, 2016.
25. https://www.youtube.com/watch?v=5v5quH0BAp0, accessed August 26 2020.
26. *Scientific Imperialism. Exploring the Boundaries of Interdisciplinarity*, ed. U. Mäki, A. Walsh and M. F. Pinto, Routledge, Abingdon, 2018.
27. A. Nordmann and A. Rip, Mind the gap revisited, *Nat. Nanotechnol.*, 2009, **4**, 273.
28. A. Nordmann, If and Then: A Critique of Speculative NanoEthics, *Nanoethics*, 2007, **1**, 31.
29. W. Liebert and J. C. Schmidt, Collingridge's dilemma and technoscience, *Poiesis & Praxis*, 2010, **7**, 55.
30. R. A. Pielke, *The Honest Broker: Making Sense of Science in Policy and Politics*, Cambridge University Press, Cambridge, 2007.
31. C. Mitcham, Rationality in Technology and Ethics, in *New Perspectives in Technology, Values, and Ethics*, ed. W. J. Gonzalez, Springer, Heidelberg, 2015.
32. J. Mehlich, Is, Ought, Should – The Role of Scientists in the Discourse on Ethical and Social Implications of Science and Technology, *Palgrave Commun.*, 2017, **3**, 17006.
33. M. Cotton, *Ethics and Technology Assessment: A Participatory Approach*, Springer, Berlin, 2014.

15   Science Communication

Overview

Summary: Former chapters have highlighted the importance of communication and discourse as an element of the scientific method itself (Chapters 2 and 3), communication with peers and members of your scientific community (publications, conference talks) (Chapter 7), with collaboration partners and practitioners from outside your own field (Chapter 8), and with regulators, non-expert decision-makers and other stakeholders (Chapter 14). This chapter elaborates further on the communication of chemical issues in informal environments or with the general public, either through channels of mass media or face-to-face in public panels or public education (museums, science campaigns, *etc.*).

Communication between experts and non-experts is always asymmetric: specific knowledge may be misunderstood or not understood at all, and once it is refined for easier comprehension it may be misinterpreted or applied in inappropriate contexts. The dialectic (that means, *in both directions*) requirements for successful science communication in chemistry start with chemical experts' awareness of these obstacles. Chemists who want to reach out of their chemical community, for example giving an interview to a science journalist for a newspaper, providing scientific advice to a science museum, writing a chemistry book for children, or participating in a public roundtable discussion on climate science, need to practice this form of communication in the same was as they train for everything else.

In this chapter, we will discuss why competences and skills in public communication of chemical matters are important and necessary, how this competence can be acquired, and how a chemist should listen and respond to non-expert communication partners in the general public. Again, we will discuss the important differentiation between fact-premises and norm-premises as introduced in Chapter 1. Here, it will help us to understand the conflict potentials that arise in public communication of an expert field such as chemical science, research and innovation in academia, industry, and public service.

Key Themes: Science and mass media; chemists and science journalism; practicing public communication; public participation in S&T discourse; three types of risk communication (knowledge and expertise, experience and competence, values and worldviews).

Good Chemistry: Methodological, Ethical, and Social Dimensions
By Jan Mehlich
© Jan Mehlich 2021
Published by the Royal Society of Chemistry, www.rsc.org

Learning Objectives:

In this chapter, you will learn:

- That communication with scientific laymen requires training and should be practiced in order to avoid pitfalls and common mistakes.
- How to respond to public concerns and questions properly, to distinguish scientific knowledge-directed questions from those concerning worldviews and values, and to increase your credibility as an important public figure with competence and influence.
- That the ethical call for engaging in public communication of chemistry arises from the fact that chemistry is an institution based on societal acceptance and justification.
- That scientists have the authority to deliver evidence-based factual knowledge that would be filled by others if not actively provided by scientists.
- That public communication makes your scientific research better.

15.1 Introductory Cases

Case 15.1: Fruity galaxy

[Real case] The Max Planck Institute Bonn performed a spectral analysis of Sagittarius B2, a dust cloud in the centre of our galaxy. One among more than 4000 substances found is ethyl formate. Believing that space science and results like these are of interest to a larger public readership, the institute published a press release. A few days later, a News magazine sent a journalist for an interview. Shortly after the interview, the magazine published an article under the headline *"The center of the Milky Way smells like raspberries and tastes like rum"*.

Case 15.2: Crosstalk

[Real case] The following is an excerpt of a conversation at a Q&A section of a public information event about nanoparticles for the early diagnosis of arthritis at the Charité hospital in Berlin.

Patient 1:	"You say that your particles can visualise the joint inflammation 15 years before the rheumatic symptoms appears?"
Researcher:	"Yes, that is correct. The particles have protein receptors that can find special molecules in the joint that indicate an inflammation. That means, they accumulate, and..."
Patient 1:	"OK, I got that! But why would I want to know that 15 years in advance when there is no treatment method in place, anyway?"

Researcher: "Aehm... I... I am sure treatment methods are under development, too! But as a first step, it is necessary to identify the inflammation, and with our method..."

Patient 1: "But do I have a right of not-knowing? I mean, what if my employer or my health insurance find out? Can I make sure that I am not in trouble with that diagnose, especially when I show no symptoms?"

Researcher: "I don't know... I am not an expert on patient rights. But our method is safe, there is no indication that the nanoparticles have any adverse effects, so you don't need to worry about the test."

Patient 2: "Well, but in your presentation 5 minutes ago, you said that you don't know for sure where the particles end up! You said, 'probably' they are transported to the kidney and then egested. It means, you don't know, right?"

Researcher: "Right, we can't find the particles after the diagnostic tests, but we also can't observe any negative effects in any of the toxicity tests and clinical trials, which means that..."

Patient 2: "To me, that doesn't sound convincing. I don't think I would have that kind of test when it is not safe!"

Case 15.3: Not My Business?

A new faculty member has recently begun her research into the development of catalytic anti-cancer metallodrugs. Her focus has been on the role of stereochemistry in the complexes she and her students synthesise to enantio-selectively reduce metabolic intermediates generated in certain classes of cancer cells. Her research is lab-based and she currently does not have any contact with cancer patients nor have her proposed complexes been used in any clinical trials. She would characterize her research program as fundamental science with a focus on the biochemistry of cancers.

Early in her first year as a faculty member, she is invited to a dinner hosted by a local group looking to privately invest 50k € in innovative research from young scientists. The group comprises local entrepreneurs with limited scientific training. The group proposes that she should prepare a short 10 to 15-minute presentation for the committee. This video will be recorded and later posted on the organization's website to highlight the innovative research it financially supports.

In anticipation of the dinner, the young faculty member reaches out to the chair of the committee that will be making the decision on the award. During their one-on-one meeting, they plan to discuss her research and what it is that she plans to speak about at the dinner.

During the meeting, however, the funder begins to tell the professor about his personal connection to cancer within his family. *"Cancer, now that's something I know about. It took the lives of both of my parents. It is really a horrible monster that just runs through a person's body, destroying whatever it finds in its path. It is great that there are scientists like you who are finding the cure—an antidote—to such a vicious disease. I think it would be really convincing to the committee to talk about how you're helping people win the battle and to beat this foe. One day we'll eliminate cancer!"*

After her meeting, she sits down to outline her presentation and is presented with the challenge ahead of her. She knows that her choice to embrace the chair's mischaracterization of cancer will likely give her an advantage in persuading the committee to help advance her research programme, but likely at the cost of misrepresenting the context and motivation for her research. It's their funding, after all, that will help advance her research, not their understanding of the complexities of cancers and metal complexes. If she speaks too technically, however, she worries that she will lose the attention of her potential funders and miss out on an important opportunity.

15.2   Science Communication as Responsible Conduct of Chemistry

Public outreach is more and more regarded as an element of the social responsibility of scientists.[1,2] This claim is not self-understanding and far from trivial. Managers, chief executive officers (CEOs), bankers, policemen or real estate agents, usually, are not requested to explain the content and output of their professional activities to the general public, even though they sometimes do. What constitutes this special demand on scientists?

A common line of argumentation goes like this: the society affords an expensive science endeavour for two main reasons. One is the expected positive impact of scientific knowledge on innovation, products, and life quality. The other is the public demand for and interest in education. Not only scientists, but everybody, wants to understand the world we are living in, and scientists deliver the answers to important questions of the age. This demand cannot be satisfied by science teachers at schools alone. Various channels in mass media, to an increasing extent *via* the world wide web, deliver trends in science and report breakthrough findings and future prospects. Another reason was addressed in Chapter 14: the public must be able to understand the basics of science and technology (S&T) progress to make informed decisions on critical issues that arise in the context of this progress.

Recent public disputes over climate change or marine pollution with plastics show how important understandable scientific input is for the credibility of the discourse if drifting off into populism and post-factualism is to be avoided.[3]

Public communication of chemistry is a special case. Although news and stories in physics, biology and mathematics enjoy a high popularity, the public image of chemistry is rather poor. Chemistry is often associated with *artificial* as a contrast to the popular *natural*, with poison and pollution, warfare and destruction, or with industrial accidents and explosions. This fact has to be accepted, no matter how irrational and unsubstantiated these associations are. For example, when Pulitzer Prize-winning science journalist Deborah Blum intended to add the subtitle "*Chemistry, Murder and Jazz Age New York*" to her book "*The Poisoner's Handbook*" about the evolution of forensic science in 1920s America, the publisher Penguin changed it to "*Murder and the Birth of Forensic Medicine in Jazz Age New York*" because '*the word chemistry on the book's cover would tank sales*'. Pierre Laszlo coined the term *chemophobia* in this context.[4] Among the attempts to explain this low public image are the observation that there are no big historical figures associated with chemistry (like physics has Einstein and biology has Darwin), and the lack of a grand theme or narrative with interesting philosophical underpinnings such as biology or cosmology.[5] Even the most prestigious science award in chemistry, the Nobel Prize for Chemistry, is often awarded for achievements at the very edge of the discipline (for example, biomedical chemistry, physics of molecules, methodological advancements) because the greater relevance of the findings can be explained to the public more easily.

Chemists who want to communicate chemical issues to the public face several obstacles. Many non-chemists perceive it as a fundamentally difficult subject. Chemists' reliance on molecular structures for communication makes it hard to understand for those who are not familiar with these information-rich representations. While mechanics, optics or electricity in physics, or animals and plants in biology, are familiar phenomena or objects for laymen, the relationship between molecular composition details and material properties, for example, the different rigidity of a PE bottle and a PVC tube as the result of just one chlorine atom in the monomer of PVC, is difficult or impossible to explain in simple terms to laymen. Technical aspects of training public communication will be discussed in Section 15.4. In addition to these skill-related difficulties, there exist obstacles in terms of attitude. Chemistry is a rather insular field in which absolute truths are rare, field-reshaping paradigm shifts are not happening, and scientists feel no direct urge to work on communicating their insights to non-specialists. Richard Zare (Stanford University) observed that *"scientists who do speak to the public are considered by other scientists to have lost their way as to what is really important"*.[3]

It is difficult to argue for an ethical duty of chemists to engage in public communication. Rather, if there is no institutional obligation, say, to give an interview to a science journalist after a publicly and widely noticed press release, public communication is *supererogatory*. This term describes normative calls for action that cannot be morally or legally expected from someone, but express a good will and good character, like helping an elderly person cross the street. Commitment to supererogatory acts may be selfish (expecting a personal benefit in return for the good deed) or altruistic (motivated by someone else's benefit). In view of these considerations, why would chemists spend some of their valuable time on public communication? Roughly said, it pays off for both the chemist (selfish motivation) and the public (altruistic motivation)! First, an occasional reflection on the larger picture of one's research, triggered by the challenge to explain it to laymen, helps on to stay grounded and not lose track of the reality context of one's activities. Then, it supports a positive public image of chemistry and of science in general, something that should not be underestimated (remember Chapter 2).[6] Moreover, it strengthens the role of scientific stakeholders in public discourses. This is especially important in times of a threat from post-factualism and populism. Chemists as *epistemic authorities* are able to contribute knowledge-based arguments and should insist on this role. Public engagement increases our credibility in this role.[7,8] Another aspect may be the educational mission that public communication of science always also follows: the public is generally (but not each and every member, of course) interested in learning what scientists know (or find out), and they can get first-hand information from chemists.[9]

The tensions that chemists can find themselves in when communicating their science in informal environments are illustrated vividly in Case 15.3. Even though not the general public, the academic's audience consists of non-chemists. The personal interest in successful communication, here, is the vision of a research grant. Codes of scientific integrity demand that all her communications are truthful, including those outside the scientific community. A correct information of stakeholders on the goals and chances of her research is her altruistic motivation. Yet, knowing that her audience approaches her presentation with particular expectations makes her choice of presentation content a balancing act between scientific truthfulness and instrumental communication strategy. Sticking too closely to her actual research would do her no favours, here. But instilling hope for near-future availability of cancer cures is not only unscientific but also irresponsible and unethical. On the other hand, if that strategy gets her the grant, she can then focus on her core competence again and do high-quality research that may, in one way or another, benefit the society. This example shows that public outreach and science communication with laymen are important topics in the contexts of both scientific integrity and the socio-ethical implications of science.[10–14]

There are several channels of public outreach for chemical experts. Occasionally, chemists are consulted by science journalists or other mass media actors for print, audio, or video content in mass media outlets.[15] Many chemists write blogs or science columns.[16] Writing popular science books for all sorts of target groups such as children, teenagers, dummies (here as a reference to a popular book series), or science-interested intellectuals, or maintaining a video-blog platform are activities that some scientists became very famous with (for example, Bill Nye and Neil deGrasse Tyson in USA, Brian Cox in the UK, or Rangar Yogeshwar in Germany). In recent years, events with more or less entertaining characteristics such as science slams, TED talks, or other public information events have emerged around the globe. Here, researchers speak in front of layman audiences and try to refine their research topics into easily understandable and sometimes entertaining presentations. Similar occasions are engagement activities in science museums and exhibitions (like the *Nanotruck* in Germany),[17,18] or in projects with schools and even Kindergartens (like the EU-funded *irresistible* project).[19]

15.3 Chemistry and Science Journalism

The general public is widely interested in scientific findings. This interest is served by science journalism in print media, TV, and online media. Very often, though, we find funny, astonishing, or downright terrifying examples of failed science reporting. Case 15.1 describes an example that is not a very big deal and has no real ethical implications, but it shows what can go wrong.[20] It is easy to imagine what happened here: probably, the institute published a press release, a media platform sent a reporter for an interview, the researcher/scientist reported that one of the outstanding discoveries was the existence of ethyl formate in dust clouds, and the journalist asks: *"What is special about ethyl formate? Can you give an example of where on Earth we find that?"* The scientist mentions the examples of raspberries and rum. Certainly not intended by the scientist, this becomes the focus of the entire article.

There are more critical issues in which correct and non-falsified or non-exaggerated research findings are of greater importance, for example climate change related research (in view of the heated public and political debate on it), medical research (curing cancer, or other promising drug developments), consumer product safety, and so on. Here, misleading populistic headlines and catchy hooks (*click baiting* in the age of internet) may confuse and irritate the public so that rational and reasonable opinion formation on such critical topics is hindered or made more difficult.

The major source for such communication problems is a difference in interests. Scientists are often enthusiastic about their achievements, for example after time- and resource-consuming projects and studies, or when milestones have been achieved. Journalists are more interested in purposes and benefits of findings: *what is it good for?* Often, this correlates with what (they believe) readers or watchers are interested in.

We may, of course, blame the incompetence of science journalists for the problems that arise in public reporting of science. This would be too short, though. Chemists who are untrained in public communication commit mistakes, too. First of all, we are so used to speaking in a scientific jargon that we often don't realise that we do, and that non-chemists simply don't understand what we mean with particular terms. This can go so far that people imagine something radioactive when we just say the word *atom*. Then, we are not aware of journalistic methods, for example the way they extract and refine our information for writing the article, the focus they place, the narrative they create. As in the example above, one keyword in a sub-sentence can become the hook of the main storyline. Moreover, we are also unaware of the public perception of science and its importance. What matters to us the most is not relevant to the public, while their main interest is served by extracting information (often invalidly) from our statements. Using metaphors to explain complex issues can be very dangerous in that this becomes the focus of the report, or it can be misunderstood, especially if not chosen wisely. Another problem – a more ethical one – is the mixing up of scientific and normative judgments. In view of the virtues of science (Chapter 5), intellectual honesty and objective truth-seeking require from us that we stick to empirical evidences and data-based suggestions rather than making claims that have nothing to do with what our research revealed. Normativity – including the judgment of what something is good for – is not part of our expertise, and sometimes this should be communicated clearly to the journalists. Otherwise, the journalist's trained ear will filter exactly those claims, turning our competence upside down.

15.4 Public Participation in S&T

Whereas in the previous example the science communication failed because of a misrepresentation of the chemist by a mediator (the journalist), there are communication situations in which the chemist faces public stakeholders directly. To understand the pitfalls of public expert consultations it is important to point out that scientific activity has an undeniable reflexive connection with the social sphere and its worldviews, value systems and cultural, historical and political constitutions.[21] On the one hand, scientists as members of a society – coloured by its culture and Zeitgeist – conduct their profession in the paradigmatic frameworks and on the normative foundations that this society is embedded in. On the other hand, science and its consecutive translation into technological achievements has the power and the potential to challenge, refine and change those paradigms and worldviews. Occasionally, this leads to public concerns and even fears, heated debates in politics and public, and even social backlash against entire scientific fields – be it justified or not.[22] Science, however, depends on the public trust and support in its institutional justification and societal implementation. Therefore, it is also (but not only) the scientists' responsibility to create trust through a high degree of credibility and reliability as *experts* when it comes to (public) discourses on the risks and benefits of S&T or the ethical and social implications of scientific and technological progress.[23]

How is credibility created? Case 15.2 describes a scene from a public information event in the scope of the research project on nanoparticles for medical purposes that was mentioned in Chapter 14 (Section 14.6).[24] Patients of the *Rheuma-Liga*, the German equivalent of the *European League against Rheumatism* (EULAR), were informed about the usage of nanoparticles for early diagnosis of osteo-arthritis and rheumatoid arthritis by researchers and clinicians involved in the project. The scientists presented illustrations on how the nanoparticle-based imaging works, explained the benefits (early stage diagnosis) and assured the patients of the safety of the procedure. The patients, however, were much more sceptical about the promised benefits. They implied that an early diagnosis is in no way helpful as long as there is no therapy to prevent the progression of the joint inflammation.

Instead, the knowledge *in the wrong hands*, for example employer or health insurance, may have disadvantageous effects for the affected person. Patients wanted to know whether it would be possible to decide not to receive the treatment. Moreover, the scientists couldn't convince the patients of the safety of the method. Most concerns expressed by the patients were related to aspects of responsibility in the case of unexpected side-effects of the treatment, for example liver or kidney impairment caused by the nanoparticles. The scientists promised that *enough research on toxicity* will be conducted, but that could not remedy all the concerns of the patients. In particular, questions regarding responsibility had not been properly addressed. What was called a *patient forum*, intended as an *information event*, and taken for an *advertising platform* by the scientists turned out to be dissatisfying for both patients and experts.

What happened here? The expectations on how to address and approach conflicts and their solutions differ between public stakeholders and S&T enactors. Ortwin Renn pointed out three levels of risk communication according to the degree of complexity and the intensity of the conflict (Figure 15.1A).[25,26] In short, he stated that knowledge and expertise (for example provided by scientific data or professionals from a certain field) can only help in solving conflicts to a limited extend. The majority of concerns (for example those expressed by the public) cannot be answered by scientists and risk researchers alone as they are related to moral and social values and touch on or effect certain worldviews.

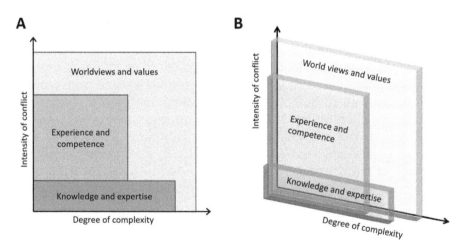

Figure 15.1 Three levels of public risk communication. (A) Two-dimensional model (B) Three-dimensional extension. Adapted from ref. 25 with permission from Elsevier, Copyright 1992.

Even though the model originally described aspects of risk communication in a debate among stakeholders, it can certainly be applied to conflict assessment in general. Concerns or problems with comparably low conflict potential can be solved by scientific and technical knowledge and expertise, even when the complexity might be high. For example, the toxicity of nanoparticles injected into a patient can be investigated in advance as long as trustworthy methods are available. This is difficult, but not impossible. As soon as clear toxicological data is available, it can convince the patient of the safety of the treatment. If the solution to a conflict requires arguments that are beyond empirical research findings and expert knowledge, or when there is no scientific data available, people trust experts that are proven to have experience and competence in a particular field. In the above-mentioned example, the inquiries on the (legal) responsibility for certain side-effects might fall into this category. The patients want to know if sufficient regulations are in place to clarify responsibilities in the case of adverse outcomes. The problem is not very complex, but bears a high conflict potential as a large number of patients might be affected as soon as the nanotechnological methods are available and approved for application in medical treatment. In a debate, a scientist or an engineer who shows and proves competence and experience in a wider range of aspects related to his research focus can make a stronger argument that is trusted by laymen, rather than, for example, a viewpoint expressed by a politician or a businessman. However, it is dissatisfying for the patient with a "what if" question - inquiring on the preparedness for unexpected cases, for example *"What if after the treatment the nanoparticles damage other organs?"* – when the reply from the S&T expert solely focuses on the empirical data and statistical extrapolation (*"So far, there is no evidence that the applied nanoparticles have undesired health effects."*). Furthermore, in many cases, a science or technology related problem is beyond any competence or expertise. When knowledge or experience is not available as the uncertainty is high, when new territory is explored or effects are unforeseeable, the strongest argument in a debate is one that refers to values and worldviews. The privacy issues of personalized (nano)medicine is such a case. This concern needs to be answered with statements about ethical guidelines, laws and regulations that preserve and protect values that a society finds important.

A small but significant detail should be pointed out here: Renn's two-dimensional scheme that is based on Funtovicz and Ravetz's knowledge classification model[27] might be interpreted in a way that for conflicts with a low intensity and low complexity, knowledge and expertise are completely sufficient to achieve a solution. However, the model can and should be regarded as three-dimensional in a way that the three domains fully overlap (see Figure 15.1B). Values and worldviews still play an important role when scientific knowledge and experts' findings are powerful arguments. Only in view of a normative framework that is constituted by value and belief systems can it be defined by what counts as *risk* and what as *benefit* (and for whom), what has the power to serve as an convincing fact or argument to solve a conflict, and what kind of incident or concern has the potential to mobilise sufficient awareness and attention so that it finds its way into the contemporary S&T discourse agenda. In this respect, ethics is a fundamental and crucial element of S&T discourse, for example in ethical, legal and social implications (ELSI) research and modern technology assessment (TA) concepts. In other words: ethics does not only come into play when science and politics are not convincing enough, but it underlies the whole debate. A scientist as a participant in this discourse who is aware of those interrelations and shows this in his arguments and viewpoints will earn more credibility and attention – and, ultimately, more influence – than a scientist whose focus is too narrowly confined to his core expertise or – as often found – who unconvincingly attempts to advertise the beneficial outcomes while neglecting the risks and potential adverse side-effects.

15.5　Practicing Public Communication

What options do chemists have to learn and improve their public communication skills? As indicated in Section 15.2, the success hinges on two different aspects: attitude, and skill. Matthew Hartings and Declan Fahy provide a useful list of strategies to support one's public outreach.[3]

- *Practice Research-driven Communication*: There is a lot of evidence-based empirical research in the field of communication studies that chemists can learn from.[10] Not only are communication techniques and strategies in place, but also the attitudes, capabilities and intents of audiences have been studied and are well understood now.
- *Understand the Audience*: When planning a presentation, speech, blog post, article, or book, the target audience should be analysed so that linguistic styles and choice of content can be adapted. Here, the available experiences mentioned in the first point of this list are very supportive.

- *Participate in the New Communication Landscape*: Chemists are communicating in a contemporary communications environment that is pluralistic, participatory, and social. Chemists can use blogging platforms and social media channels without much effort and technical know-how, but have a wide outreach and large audiences. Blogging, especially, is a great entry point for practicing non-academic writing skills.[16]
- *Tie Chemistry to Society*: The issue of detachment of chemistry from the social lifeworld (see Section 15.2) should be addressed by the better connection of chemistry with aspects of daily life that people, so far, do not primarily associate with chemistry. The effects of detergents, of the colour of wall paint, the substances in food, drugs, plastics, and so on – when chemical compounds come alive and become vivid in the minds of the audience, it is easier to convince them that chemistry is also potent in creating materials for green energy, compounds for personalised medicine, or substances that enable safer and healthier food preservation.
- *Frame Key Messages to Prompt Engagement*: The outreaching chemist must clarify, first, what the main narrative or storyline is intended to deliver. These frames help refining complex issues into clear statements concerning what a problem is, why it matters and to whom, who is responsible, and what chemistry can contribute to solve it.

The following points may be added to this list:

- *Practice in Low-stakes Environments*: In addition to the abovementioned engagement in blogging and social media, chemists have many stages, literally, for addressing non-expert audiences. Signing up for a science slam, or applying to talk at a TED event, are simple ways to test and improve one's public communication skills. Still, the best practice arena is a chemist's private environment: family members, friends, neighbours, and so on, who are willing to talk about chemical issues.
- *Make Use of Workshops, Courses, Books, and Instructional Online Material*: Science communication as a skill including linguistic, rhetoric, and technical competences can be trained with the help of experts.[28,29] Some chemistry curricula at universities include specific courses on science communication, often in combination with critical thinking and scientific literacy. Even high-quality instruction videos on YouTube can significantly improve communication skills.

Box 15.1 Exercise: Practice Public Communication

How far can you make it on this list of informal communications of your professional work?

1. Explain your work to a child. Success factor: The child's interest is sparked.
2. Talk to your (non-chemist) friends or adult family members about your latest achievements. Success factor: They anticipate your joy because they understand why that means something to you.
3. Write a blog or social media post. Success factor: Readers reply with quality comments and show interest in the topic.
4. Apply for a TED talk or science slam with a self-recorded video. Success factor: You are invited to a TED event and the audience is entertained by your presentation.
5. Volunteer at the nearest science museum. Success factor: You contribute with a popular theme or item at a museum event or program.
6. Write a popular chemistry book. Success factor: A publisher accepts your manuscript and your book sales are considerable.

Exercise Questions

1. As a journalist, which of the following topics would be the last you would choose for a popular magazine?

   A: Anticancer drug development.
   B: Toys safety.
   C: Research on enantio-selective organometallic synthesis.
   D: Greenhouse effect.

2. Which of the following best characterizes the difference between the two professions, journalists and researchers?

   A: Journalists want to address a wide cross-section of the public, whereas researchers usually write for other researchers (their peers).
   B: Journalists envy researchers for having a much higher salary (usually).
   C: Journalists think more about writing (explaining things theoretically with words), researchers think more in terms of doing (laboratory work, experiments).
   D: Journalists often work outdoors, while researchers work indoors (office, lab) all the time.

3. Which of the following is not on the list of common mistakes made by scientists in public communication?

   A: Too little concern about the public perception of science.
   B: Use of scientific jargon.
   C: Information seemingly going into too much detail.
   D: Lack of narration.

4. Which of the following is considered a communication mistake by scientists embarking on public communication?

   A: Applying simple language, understandable by scientific laymen.
   B: Putting the research progress into an interesting storyline.

C: Using improper analogies and metaphors to explain one's research.
D: Withholding experimental data.

5. Which of the following statements concerning scientists in public communication is NOT tenable?

   A: Researchers should be trained to communicate with laymen in order to avoid pitfalls.
   B: Scientists have a responsibility to create trust through a high degree of credibility and reliability as experts.
   C: Knowledge and expertise are, in most cases, not sufficient to resolve conflicts and facilitate reasonable and acceptable solutions.
   D: The judgment on 'what something is good for' is part of scientists' responsibility and should be clearly communicated.

6. You can learn and practice scientific communication focused on a wide and diverse public in... (select one *false* answer!)

   A: ...science slams.
   B: ...scientific conferences.
   C: ...science communication workshops.
   D: ...conversations with friends and family members.

7. Scientists should spend some of their time communicating with the public because it... (select one *untenable* answer!)

   A: ...leads to reflection of the context of their research.
   B: ...supports a positive public image of science, including chemistry.
   C: ...strengthens the impact of scientists in public discourse, an important counter pole to post-factualism and populism.
   D: ...looks good on a CV.

8. Scientists should spend some of their time communicating with the public because it... (select one *false* answer!)

   A: ...serves the public demand for and interest in scientific insight.
   B: ...is an important part of public education.
   C: ...is well paid.
   D: ...promotes interest in natural sciences.

9. Which of the following is NOT one of the levels of risk debate as suggested by Ortwin Renn?

   A: Worldviews and values.
   B: Experience and competence.
   C: Knowledge and expertise.
   D: Power and influence.

10. In the context of Ortwin Renn's *Three Levels of Risk Communication*, what are the two parameters that characterise a conflict?

    A: Complexity and intensity.
    B: Exposure and hazard.
    C: Factual background and normative evaluation.
    D: Emotion and rationality.

11. Which of the following statements is NOT tenable in terms of Ortwin Renn's *Three Levels of Risk Communication?*

    A: Concerns or problems with comparably low conflict potential can be solved by scientific and technical knowledge and expertise, even when the degree of complexity is high.
    B: Conflicts with a high intensity and a significant degree of complexity require argumentative strategies that address values and worldviews as expressed by the general public.
    C: Intangible public concerns and empirically acquired knowledge can't go together. That's why public communication of science almost always fails.
    D: Laymen may trust the experience and competence of non-expert discourse participants (for example politicians) when the degree of complexity of the conflict is rather low.

12. What is problematic about a researcher's claim that his new compounds "may someday contribute to curing cancer", that is subsequently published in a popular newspaper?

    A: It forces the researcher to conduct research in this direction and present data that supports this claim.
    B: It might evoke false hopes in cancer patients or their acquaintances, risking the credibility of scientific expertise.
    C: The researcher's institution may get into trouble because it is responsible when the researcher's promise can't be fulfilled in the future.
    D: There is nothing problematic here. It is an important and insightful judgment that sets the researcher's work into a larger perspective.

13. Which of the following do scientists most likely NOT need to be aware of when giving interviews to science journalists?

    A: Journalists will always try to extract catchy phrases and funny or interesting analogies from what the scientist says.
    B: Journalists will most likely not understand all the scientific details. The more such details that the scientist mentions, the more likely it is that the journalist will misunderstand aspects or even twist facts in his writing.
    C: Journalists will double check the validity of the scientific information and identify every fallacy and inconsistency in the scientist's answers.
    D: Journalists consider reader interests much more than the knowledge level of the scientist and the academic sophistication of his work.

14. Which of the following is NOT a common field of public communication for scientists?

    A: Collaborating with science museums.
    B: Inviting the public into their research facility and showing them around ("*Open Day*", German"*Tag der offenen Tür*").
    C: Performing spectacular chemical experiments (preferably with explosions and a lot of smoke) at a fairground or a folk festival.
    D: TED talks, science slams, and the like.

15. Why is there a demand from the general public for science communication in the mass media? (Select one *implausible* answer!)

A: Many people are interested in scientific achievements and contemporary science progress.

B: Many people would like to have the chance to serve as an additional level of peer-review, putting scientific projects and their results under critical scrutiny.

C: Many people beyond school age are interested in learning more about science in general and would like to improve their scientific knowledge.

D: Some people may want to see that public money (tax) is being well spent on meaningful scientific inquiries and successful research agendas that actually have a purposeful output.

16. The scientific virtues introduced in Chapter 5 also play an important role in public science communication. Which of the following pairs of terms is especially relevant in this context (and has been highlighted in the lecture)?

   A: Intellectual honesty and objective truth seeking.
   B: Fairness and collegiality.
   C: Communalism and universalism.
   D: Dedicated disinterestedness and systematized doubt.

17. Which of the following is an appropriate reaction by a scientist to public fears and irrational concerns?

   A: Ignore them.
   B: Criticize them.
   C: Attempt to allay their concerns by presenting empirical evidence.
   D: Address them by outlining the significance of state-of-the-art knowledge for the plausibility of the expressed concerns.

18. Why should chemists be concerned about the public image of chemistry? (Select one *implausible* answer.)

   A: Jobs, the profession as such, as well as the science of 'chemistry' depends on social acceptance and justification. Public support will ultimately depend on the image that the public has of chemistry.
   B: Actually, they don't need to be concerned. The chemical profession is minimally affected by what the public thinks or feels about it.
   C: The future of the chemical professions in all fields (academia, private and public sectors) depends on the young generation deciding to study chemistry. A negative image is likely to drive many talented students away from studying chemistry.
   D: A negative public image of chemistry might repel potential funders and investors, thus limiting financial support for cost-intensive chemical R&D programmes.

19. Which of the following is (or, should be) of major concern to chemists involved in public communication?

   A: Admiration.
   B: Financial reward.
   C: Power.
   D: Credibility.

20. At a public information event dealing with a new nanomedical diagnostic tool, concern was expressed over the safety of the applied method and inquires were made about insurance. Which of the following is considered the most (or only) appropriate reply?

    A: "Current data suggests that there is no evidence that the applied nanoparticles have undesired health effects."
    B: "We are aware of the risks and work closely with a commission on developing appropriate regulations that protect patient interests."
    C: "We can assure you that our toxicological team will exclude any harm with 100% certainty!"
    D: "Insurance is not my business. I am only responsible for the research and development part of the project, and I find our results pretty convincing and promising!"

References

1. S. E. Brownell, J. V. Price and L. Steinman, Science Communication to the General Public: Why We Need to Teach Undergraduate and Graduate Students this Skill as Part of Their Formal Scientific Training, *J. Undergrad Neurosci. Educ.*, 2013, **12**, E6.
2. A. Campbell, Communicating science effectively to the public, *Chem. Eng. News*, 2017, **95**, 34.
3. M. R. Hartings and D. Fahy, Communicating chemistry for public engagement, *Nat. Chem.*, 2011, **3**, 674.
4. P. Laszlo, On the Self-Image of Chemists, 1950–2000, in *The Public Image of Chemistry*, ed. J. Schummer, B. Bensaude-Vincent and B. Van Tiggelen, World Scientific Publishing, Singapore, 2007.
5. M. Eger, Hermeneutics and the new epic of science, in *The Literature of Science: Perspectives on Popular Science Writing*, ed. M. W. McRae, University of Georgia Press, Athens, 1993.
6. *The Public Image of Chemistry*, ed. J. Schummer, B. Bensaude-Vincent and B. Van Tiggelen, World Scientific Publishing, Singapore, 2007.
7. M. Dierkes and C. von Grote, *Between Understanding and Trust: The Public, Science and Technology*, Routledge, London, UK, 2000.
8. *Science and the Media. Delgado's Bull and the Ethics of Scientific Disclosure*, ed. P. J. Snyder, L. C. Mayes and D. D. Spencer, Academic Press, London, UK, 2009.
9. D. Knight, *Public Understanding of Science: A History of Communicating Scientific Ideas*, Routledge, Abingdon, UK, 2006.
10. *Routledge Handbook of Public Communication of Science and Technology*, ed. M. Bucchi and B. Trench, Routledge, Abingdon, 2nd edn, 2014.
11. *Communication in Chemistry*, ed. G. L. Crawford, K. D. Kloepper, J. J. Meyers and R. H. Singiser, American Chemical Society, Washington, 2019.
12. T. Velden and C. Lagoze, Communicating chemistry, *Nat. Chem.*, 2009, **1**, 673.
13. D. Brossard and B. V. Lewenstein, A Critical Appraisal of Models of Public Understanding of Science: Using Practice to Inform Theory, in *Communicating Science: New Agendas in Communication*, ed. L. A. Kahlor and P. Stout, Routledge, New York, 2010.
14. National Academies of Sciences, Engineering, and Medicine, *Effective Chemistry Communication in Informal Environments*, The National Academies Press, Washington, 2016.
15. T. Masciangioli, *Chemistry in Primetime and Online. Communicating Chemistry in Informal Environments*, The National Academies Press, Washington, 2011.

16. A. Mollett, C. Brumley, C. Gilson and S. Williams, *Communicating Your Research with Social Media: A Practical Guide to Using Blogs, Podcasts, Data Visualisations and Video*, SAGE Publications, London, 2017.
17. M. W. Bauer and P. Jensen, The mobilization of scientists for public engagement, *Public Underst. Sci.*, 2011, **20**, 3.
18. A. Dance, How museum work can combine research and public engagement, *Nature*, 2017, **552**, 279.
19. http://www.irresistible-project.eu/index.php/en/ accessed on August 27th 2020.
20. https://www.theguardian.com/science/2009/apr/21/space-raspberries-amino-acids-astrobiology accessed on August 27th 2020.
21. *Science, Worldviews and Education*, ed. M. R. Matthews, Springer Science+Business Media, 2009.
22. B. Bovenkerk, *The Biotechnology Debate – Democracy in the Face of Intractable Disagreement*, Springer, Dordrecht, 2012.
23. *Experts in Science and Society*, ed. E. Kurz-Milcke and G. Gigerenzer, Kluwer, New York, 2004.
24. J. Mehlich, Is, Ought, Should – The Role of Scientists in the Discourse on Ethical and Social Implications of Science and Technology, *Palgrave Commun.*, 2017, **3**, 17006.
25. O. Renn, Risk communication: Towards a rational dialogue with the public, *J. Hazard. Mater.*, 1992, **29**, 465.
26. O. Renn, *Risk Governance – Coping with Uncertainty in a Complex World*, Earthscan, London, 2008.
27. S. O. Funtowicz and J. R. Ravetz, Three types of risk assessment: Methodological analysis, in *Risk Analysis in the Private Sector*, ed. C. Whipple and V. T. Covello, Plenum, New York, 1985.
28. P. Laszlo, *Communicating Science: A Practical Guide*, Springer, Berlin, 2006.
29. L. Lindberg Christensen, *The Hands-On Guide for Science Communicators, A Step-By-Step Approach to Public Outreach*, Springer, New York, 2007.

16 Summary

Overview

Summary: In this last chapter, we will summarise all the aspects of *good chemistry* that we have come across throughout the book. This is a challenge because, as we have seen, the professions that chemists can occupy, the realms in which they work, and the contexts in which their chemical expertise is applied, can vary quite significantly. There is no standard way of doing good chemistry. The purpose of facing that challenge is to realise that none of the topics in this book are optional, luxury, or unimportant. In one way or the other, they all touch the daily activity of a chemical practitioner from time to time. This chapter is intended to illustrate the firm connections between methodology, professional integrity and social responsibility one more time. The examples – schedules of a busy day of selected chemical professions – may appear unrealistic owing to the density of special scenarios accumulated in one day. Yet, all the chosen events pose challenges to the chemist's decision-making and judgment ability. Some are issues that critically reflect methodological questions, others require a research or profession ethics approach, some are located at the intersection between chemical expertise (science, research, public service), society and environment, some touch all three categories at the same time. The reader is asked to transfer the quintessential conclusions from these cases to his or her real-life cases as a chemistry student, academic scientist, researcher in private sector industry, or as a chemist in public service.

Key Themes: Relevance of methodological clarity for professional integrity and social impact; professional integrity as good research practice and profession ethics; the inevitable social impact of all chemical activity; taking responsibility = taking the right action; good chemistry as a discourse skill.

Good Chemistry: Methodological, Ethical, and Social Dimensions
By Jan Mehlich
© Jan Mehlich 2021
Published by the Royal Society of Chemistry, www.rsc.org

Learning Objectives:

After this summary, more than before, the reader will:

- See all the topics introduced in this book set into perspective in respective relevant contexts.
- Understand why all three categories (methodology, research ethics, social implications) have their justification in a book on *Good Chemistry*.
- Be able to apply the acquired knowledge of this book in their own particular research field and professional niche.
- Apply all the insights from this book in order to contribute to his/her expertise to the sustainable and beneficial progress of science and technology, thus fulfilling their social responsibility as a chemist.

16.1 Good Academic Chemistry

Table 16.1 describes a full day of an academic chemist, a university professor performing research on catalysis in the context of environmental and green chemistry.

Table 16.1 Exemplary activities of an academic chemist (university professor).

Time	Event/Activity	Notes
8:00–9:00	Read the news and scan chemistry journals for latest publications	Today: check Nature, JACS, Angewandte
10:00–12:00	Group meeting, discuss thesis draft with PhD candidate, solve a dispute between two Master students (labmates).	Address student dispute in front of everyone, or better in private after the meeting?
12:00–13:00	Lunch with a potential collaborator from the biology department	Be careful! She is known for publication greed! Are animal experiments really necessary?
13:00–15:00	Finalise research paper draft, send to co-authors for comments.	Clarify co-author XY's contribution! Data sets complete? Supplementary Info?
15:00–17:00	Faculty meeting, discuss application for EU research grant ("Sustainable Chemistry for Energy Solutions")	Clarify each colleague's contribution. Consistent theme? How do we support "sustainability"?
17:00–18:00	Double-check the data analysis of postdoc's recent experimental work	Reproducible? Raw data from spectroscopy missing!
19:00–21:00	Public roundtable on urban air pollution	Prepare a 10 minute presentation on new catalysts for filter technologies. Hope they understand it!

All of the activities in this schedule, even the seemingly trivial and profane ones, can become problematic when not done appropriately. Allocating time for updating one's chemical knowledge base by scanning the chemical literature is certainly a good idea. This is a methodological aspect of good chemistry: in order to be able to design and plan meaningful research projects, to avoid repeating experiments that others have already performed, to oversee the latest developments in the ever-growing chemical knowledge pool and, perhaps, identify cross-connections between different directions of research, a good chemist should read what others report (see Chapter 3). The sheer amount of chemical publications makes it inevitable that one needs to be selective. Of course, the studied papers should be somehow relevant for one's work. Yet, when setting the margin too narrow, for example by searching articles with very specific search terms, important work may be missed. On the other hand, reading by journal, as the professor in this example does, may be a misrepresentation of current trends. Some academic scholars make the scanning of literature a group task and let students report regularly on relevant and interesting articles. In this way, the acquisition of up-to-date chemical knowledge can be more efficient. The important point is that a good chemist will prevent him- or herself from being in an expert-knowledge bubble. A broader look beyond one's own research niche can prevent bias and epistemic confinement.

The next item on the schedule, the group meeting, requires competences in mentoring (see Chapter 8, Section 8.4.1). When discussing the thesis draft, the professor needs to find the right balance between scientific rigour and empathetic assistance. Especially in a potentially exposing environment like a group meeting, the style and form of delivering critical feedback must be chosen carefully and in view of the particular student's personality. Solving a dispute between students is insofar a critical issue in the context of professional chemistry as a solution should be goal-oriented and pragmatic. It is, of course, in the professor's but also in the students' interest that they can make efficient progress in their research work. The professor could assess whether it is necessary that the two students share a lab, for example because they use the same specific lab equipment that is only available in this room. How likely is it that the dispute threatens scientific integrity, for example by sabotaging each other? The only reason for the professor to address this issue in the group meeting would be to let all group members know his style of mentoring and problem-solving with strong leadership, clarity, and trust-building. Moreover, as a good group leader, the professor will understand the dilemma that the two students face (worried about research efficiency and grade *versus* work environment and relationships) and how difficult it might be for them to enter such a conversation.

Collaborations are often the key to ground-breaking research and innovations. Inter-departmental collaborations within universities like the one in this example including a chemist and a biologist start rather informally. Yet, even an initial exploration during a lunch can decide the success of the envisioned project. If the chemist is worried that, according to rumours or experiences made my colleagues, the collaborator has interests that threaten the teamwork, the conditions of the project should be addressed clearly from the very beginning (see Section 8.4.2): what are we going to do, what is the gain for each involved party, who is in charge of what, what are the particular goals and objectives, how are the merits allocated? Make sure that the research and its goals are understood clearly by all involved researchers. Address all critical and unclear issues immediately. If the envisioned output is a co-authored research paper, discuss the conditions, authorship, publisher or journal name, and the writing process (see Chapter 7). In this example, the chemist is concerned about animal experiments (see Chapter 9). Assumingly, the biologist suggests these studies as a routine procedure in their labs. The chemist should argue for alternatives only on scientific grounds like the efficacy of animal models, the validity and significance of other study methods, or the regulatory framework of Registration, Evaluation, Authorisation and restriction of Chemicals (REACH) and the database of the European Chemicals Agency (ECHA).

Most university professors do not perform experimental work by themselves. Yet, they must know what the group members do and how they do it. This is especially evident in the process of writing research papers (see Chapter 7). The foundation of every research paper that is not violating the virtues of good scientific practice (see Chapters 5 and 6) is experimental data as the proof of all the claims made in the essay. If the principle investigator (PI), most likely the corresponding author of the paper, is concerned that a data set is incomplete or fallacious, this is a very serious issue. Although this is comparably easy to clarify if the paper is written by co-authors from the same research group, it may be time-consuming and complicated in locally more scattered collaborations. In the latter case, it may be that a collaborator assigns co-authors that another author has never heard of (for example, a masters or PhD student contributing experimental work). It is, of course, legitimate to inquire about the justification of such

authorships in view of scientific integrity and the epistemic significance of the contribution as the only legitimate factor. The professor's
remark on adding a supplementary information section should be
reflected in methodological terms: scientific work should always be
sub- and circumstantiated by data evidence. If data sets are too large
for the main article, but insightful enough to present in their entirety,
supplementary info for interested readers to study in greater detail is
well advised. It demonstrates openness and invites further discussion
and feedback.

University professors usually engage in three different activities:
teaching, research, and their contribution to academic autonomy
(participation in committees and faculty councils, *etc.*). Guidelines of
good chemical practice play a role in all three of them. Many decisions
made by faculty councils affect the academic portfolio and the research
work of the academic staff. In this example, the application of a larger
research grant requires scientific integrity that balances the monetary
and, perhaps, reputation interests of the faculty. It is no secret that
grant applications exaggerate the beneficial descriptions of objectives
and possibilities. It may be a calculated strategy to keep the theme
of the envisioned research very general (*energy solutions*) and involve
trend words (*sustainable*). Yet, the main rationale of a grant application must be a scientific one: *we deserve the funding because we have an
academic portfolio at this institution that is likely to produce viable results*!
It should be clear to all involved academic chemists that an EU research
grant implies the expectation of applicable and exploitable results that
go beyond basic research. Being entrusted with a significant amount
of tax money puts a responsibility on the scientist to use it for the
common good (see Chapter 12). Obviously, the grant in this example
is provided for developing sustainable procedures in the context of
energy production. It may have policy relevance and include a particular call for translational research with industrial partners (see Chapter
14). Here, the professional scientific rigour that chemists are used to
applying to chemical research should also be applied to define and con-
ceptualise sustainability (see Chapter 11). In order to fulfil the social
mission, it must not remain a trendy hollow jargon. In the application,
the chemists should take the chance to demonstrate their knowledge of
the broader implications of sustainability that are not limited to environmental protection.

The postdoc's data issue is another example of the importance of
the scientific virtues (Chapter 5). It may be a simple communication
issue that requires mentorship competences. The guiding principle of
the result analysis and its communication must be scientific integrity

and nothing else. It must be made sure that the data are valid and legitimately obtained. Accusations of data fabrication or manipulation can be very serious (see Chapter 6). A postdoctoral researcher is under pressure, just as students are, and may be susceptible for sloppiness and bias that may grow into scientific misconduct. The credo that scientists and their teams should be their own biggest critics describes the first and most important level of scientific rigour: discuss unclarities openly and immediately!

In contrast to the other items on this professor's schedule, the public roundtable is not a daily or regular activity. Opportunities and the demand for such events have increased (see Chapter 15). Public discourse on topics that require scientific input in the form of sophisticated knowledge (climate change, air and water pollution, energy supply, mobility, healthcare, food security, and many more) are more efficient and viable when knowledge experts participate. It should be in every scientist's interest to avoid post-factualism, demagogy, and misinformation to dominate the discourse. The professor in this example deserves praise for engaging directly in public information. Yet, the mission can fail. A presentation for an audience of chemical laymen is very different from a conference talk. The professor should not simply hope that the audience will understand his explanations but reflect on how to make them understand the crucial points even without much chemical background knowledge. This includes the avoidance of jargon, chemical details, and complex diagrams, but a focus on the relevance of chemical insights for the evaluation of different alternatives. In this example, the professor could quickly explain how chemical filter technologies work, what their pros and cons are, or how they could enrich a wider portfolio of air pollution remedies. This is the important information that public representatives need for decision-making on how to tackle the problem.

16.2 Good Private-sector Research and Development

Table 16.2 presents a typical workday of a chemist working in a medium-sized, globally operating agrochemical company with a product development unit that tests formulations, but does not synthesise new compounds.

Even though chemists are usually hired by companies in some way for *chemical* work, their daily activities are more interwoven with tasks other than experimental work or its communication and organisation.

Table 16.2 Exemplary activities of a private-sector chemist (agrochemical company).

Time	Event/Activity	Notes
9:00–10:00	Review status report from China sites	Solve labour rights and local safety regulation issues
10:00–12:00	Project team meeting. - Performance test of formulation MB2-23f - Discuss new formulations	Pay attention to patent restrictions! Product efficiency must be weighed against innovation potential!
12:00–13:00	Lunch with colleague, discuss business trip to Brazil	Coordinate company's response to environmental issues in local agriculture.
13:00–15:00	Conference call with US lawyers over herbicide dispute	Was last week's analysis report sufficient proof?
15:00–17:00	Strategy meeting with executive board. - Mid- and long-term innovation strategy	More external partners? Mention new EU environmental policies, REACH update: Action required?
17:00–18:00	Scan recent market development	Specify planned benchmarking parameters
19:00–21:00	Review literature for toxicity studies of pesticide product class XYZ.	Could be important for the approval of new formulation! Find evidence, if possible!

The chemist in this example is responsible for the collaboration with sites in Asia and South America. In the first activity of this imaginary day, the chemist must deal with seemingly non-chemical issues: labour rights and safety regulations in China. It is very likely that other colleagues (legal department, administration) work on this issue, too. What could the chemist contribute? It is very likely the case that the company operates in China in a grey zone between corporate responsibility and safety regulations (and their enforcement) that are weaker than in Europe. There might be a tension between workers expecting low exposure to harmful chemicals and the company aiming at profitability within the legal framework provided by the Chinese regulations. Finding compromises and the right balance between legal duty and social responsibility requires a sound chemical judgment of the exposure levels, toxicity of the compounds, alternative chemicals, the means to reduce exposure during transportation, storage, and usage, and so on. Here, the chemist performs well when he communicates risks clearly, fact-based rather than interest-driven, and under an awareness of the legal and social values that the operation of the chemical sites is affecting (see Chapter 13).

The experimental work conducted at this company is focused on testing formulations of pesticides, herbicides, and other agricultural chemicals, and they identify those that have a market potential thanks to their application benefits (more efficient than other products) or their economic advantages (cheaper than other products). The evaluation of formulation performances must, of course, be grounded on valid chemical judgment, that means on valid scientific methodology (Chapters 3 and 4). Yet, in the corporate context, other experts play a significant role, for example patent issues and market dynamics. Choices of compounds used in formulations are motivated not only by their chemical properties but also by their availability, intellectual property (IP) rights, the market price, environmental impact, regulations, supply security, and so forth. In an innovation team, the chemist's task is to connect the chemical information with the value considerations of the other divisions in the company.

The lunch meeting discussion illustrates another important issue that the chemist cannot escape from, even though it is not his core competence. As an epistemic expert (knowing things that other stakeholders do not understand or are not aware of), the chemist is in a position to address social and environmental externalities (the negative effects of the corporate activities). Several tensions are possible here: first, the company's executive board may be interested in the most profitable operation and aim to exploit local legal weaknesses. If the chemist has suggestions for more sustainable agriculture, the discourse must be fought with their own leadership board. Second, the local (here: Brazilian) population riots against the operation of a chemical site or the application of the company's products (see, for example, public resistance against Monsanto products). In this case, the chemist contributes to a discourse outside of the company. Third, the company's public reputation is at risk because media outlets report improper practices (for example, exporting formulations that are banned in the EU to countries where they can be sold but cause damage). In all cases the socially responsible chemist is well-advised to ground arguments in chemical facts (for example, toxicity studies) but communicate in terms of relevant values (see Chapter 15). For example, facing the executive board, the chemist may link environmental impact assessments with the economic competitiveness of the company rather than just pleading for a green corporate conscience. Facing the local ecological impact of the company's activities, the chemist may suggest alternatives and explain the chemical impact of various options. In any case, it is important to raise a voice and to not shy away from a difficult discussion. If the scientific expert doesn't participate in these kinds of debate, non-experts have the power over decisions that are better when informed by knowledge and expertise (see Chapter 14).

The next item on the schedule – communication with lawyers – shows that corporate chemical work follows objectives that are different from academic science. The chemist will feel the exposure to market pressures and competition. Other companies try to limit this company's market power by suing it for copyright infringements or violation of regulations. The purpose of a conference call with lawyers in the USA is to deliver proof that the legal accusations are untenable, for example by detailed analysis of the chemical formulations or life-cycle assessments. Despite the interest-oriented character of the dispute, the chemist should stick to sound scientific reasoning that derives its conclusions directly from scientifically acquired data.

Conflicts of interest may also be abound in the communication between chemical experts and the executive board. Entrepreneurial, managerial, marketing, and economic thinking often stand in sharp contrast to scientific, analytical, and empirical thinking. Discussing innovation strategies can reveal these differences quite drastically. A good chemist, in this context, is one who is not corrupted in his or her scientific integrity but accepts and embraces that other players have legitimate interests that are beyond scientific integrity. The REACH example illustrates this aptly: the chemist notes in the schedule *'Action required?'*. The executive board is likely to avoid anything that confines the profit margins of the company. Yet, the mindful chemist knows that the policy considerations of REACH follow a rationale that bridges chemical regulation, sustainability, economic competitiveness, and social interest in health and safety. Arguments on these grounds must be brought forward when discussing future innovation activities.

Benchmarking is a type of research that university chemists seldom get in touch with. In globally operating corporations it is a standard procedure to increase market power and improve operational and strategic knowledge management. The goal of a benchmarking process is to understand competitors and their performance in order to align their own innovation strategy, identify promising market niches, improve products, or coordinate collaborations and ventures. A chemist's task in such benchmarking studies is to help identifying what parameter is benchmarked (for example, environmental impact, chemical prices, supply chain issues, market segments) and how the company-internal competences can respond to it. Here, the chemist works closely with the management division. As a common theme described throughout this book, the knowledge-intensive benchmarking procedure works more efficiently when chemists contribute their viable knowledge and don't shift responsibility for this non-experimental work to other stakeholders.

The imaginary day ends with another activity within the core competences of the chemist: studying the chemical literature and extracting information that is judged as relevant and impactful. The production and dissemination of agrochemical products is a master example of the tension between chemical knowledge and profit interests. On the methodological level, a chemist may be biased and lack scientific standards. The example chemist's schedule remark (*'Find evidence, if possible!'*) indicates such a bias. Despite all other interests and even if dedicated disinterestedness cannot be maintained in private-sector industry contexts, the scientific virtues of objectivity and truthfulness are of utmost importance, owing to its social and environmental impact even more than in academic science, perhaps.

16.3 Good Chemistry in Public Service

Table 16.3 describes the activities of a chemist working in the environmental protection agency (EPA) of a larger city with a surrounding county attached.

Table 16.3 Exemplary activities of a public service sector chemist (environmental protection agency).

Time	Event/Activity	Notes
9:00–10:00	Mail check	Important: Policy updates from ministry of environment; citizen panel invitations
10:00–12:00	Lacquer factory case: Compare expert testimonies, plan next steps	Why is our report not convincing to the Greenpeace guys?! Why do they still criticise our risk assessment?
12:00–13:00	Lunch with unit head	Mention office issues: Competence distributions, collaboration.
13:00–15:00	Lobby group meeting	Beware of misunderstandings between company XZY's CEO and Chamber of Commerce president! Facts!
15:00–17:00	Highway re-construction case: Check underground condition test → special protection layer necessary?	Local chemical industry → many chemical transports! Prepare argument for the planning and building board
20:00–22:00	Read up on REACH!	Latest amendment could be important for future decisions!

In principle, the public service chemist follows the same considerations on good professional practice as the academic and the private-sector chemist: Based on a solid methodological foundation, chemical issues are handled and communicated under compliance with virtues of professional integrity and an awareness of the societal and environmental implications of decisions that are in any way informed by or related to chemical knowledge. Yet, chemical expertise in public policy, in agencies and offices, in scientific consultancy, and in regulatory bodies is attributed responsibility that differs significantly from that of academic chemists and industrial research and design (R&D). Societal impact is more immediately relevant than for the other professional fields. Multi-stakeholder discourses occur more frequently and are part of the job description. Research results may have direct policy impacts (see Chapter 14), testimonial relevance (for example, in the court), or an economic dimension (for example, when deciding over patents). The chemist working for the EPA, for example, will communicate with stakeholders that are very different from the academic and industrial chemist. As an EPA is firmly embedded in the regulatory framework of the national or international (EU) legislation, being up to date on policies and regulations and communication with public stakeholders are essential elements of the work. What could go wrong when inviting public representatives for a citizen panel on, say, urban air pollution? The chemist understands the importance of scientific knowledge and its role in countering post-factualist trends. At the same time, public communication can only be successful when the pure scientific stance is given up in favour of evaluative and normative discourses on risks and benefits. In terms of Chapter 14, policy and otherwise relevant chemical knowledge requires the communication approach of the *honest broker* rather than that of the *pure knowledge exponent*. Thus, the public service chemist invites the good chemistry professor from Example 16.1, who is known for being able to explain complex issues in understandable ways and relating them to the discussed options – an indication that networking is extraordinarily important for public service chemists.

An EPA chemist will most likely be hired for dealing with cases in which chemical expertise is required for solving problems arising from clashing interests. In this example, a local lacquer manufacturer is accused by Greenpeace of releasing toxic substances into the environment. The EPA takes the role of a mediator, collects evidence, mandates experts to provide testimony, and organises risk communication channels. Apparently, Greenpeace is not convinced by the provided assessment. This may be the case because two different testimonies come to different conclusions. To clarify this issue, the scientific soundness of the testimonies need to be compared in terms of their methodologies, logic plausibility, and scientific rigour (see Chapter 4). It is also possible that Greenpeace is not questioning the scientific validity of the reports but the relevance concerning the issues at stake. In terms of Chapter 13, the complexity-induced risk approach, with its instrumental discourse on cognitive conflicts, fails to solve the real issue, for example an ambiguity-induced risk arising from clashes between environmental integrity and profit interests. If the goal of the EPA is to facilitate a *sustainable* solution for the case, the chemist should be able to formulate sustainability goals and incentives that are practical, holistic (covering more than the environment, economy, and society as individual spheres of interest), and feasible (see Chapter 11).

The public service chemist, too, works in teams. Frictions and conflicts need to be handled with professionalism and integrity. Again, even if different from academic and corporate settings, the same principles for successful communication apply. When co-workers argue over competence distributions and work process issues, the only relevant factor to inform arguments is the viability of decisions concerning a goal that must be reached. For scientists, this pragmatism implies that personal interests, character dispositions and clashes, preferences and interests must stand behind a scientific rationality informing arguments. What is research ethics for academic scientists and business ethics for private-sector employees – guidelines concerning professional conduct and the organisation of work – is legal and social ethics for public service personnel. 'Office issues' are solved with appropriate communicative strategies (possibly, a lunch with a unit head as in this example) in view of the service goals of the organisation.

As indicated above, the public service sector involves more diverse stakeholders than the previous two professional realms. The listed lobby group meeting may be understood as representing this diversity in which interests clash. The remark gives a hint that our chemist observes a disagreement between the chief executive officer (CEO) and the Chamber of Commerce (CoC) president that is rooted in a lack or ignorance of facts. Generally, chemists as epistemic agents (a term used here for professional actors whose main competence is the acquisition and provision of knowledge) should inform discourses with relevant knowledge and insist on the credibility and legitimacy of this role (see Chapter 14).

The highway reconstruction case that the chemical expert at the EPA is entrusted with indicates how important it is to see the larger picture in even the simplest risk assessment. The question of whether investment in special protection underneath a road is needed depends on many chemical and non-chemical factors. The insight that a local chemical industry means a higher-than-usual frequency of chemical transports is an important argument in the discourse with the planning and building board. Chemical lifecycle assessments combined with larger picture elements like economic circumstances, societal trends, stakeholder interests, and related values, enables the chemist to respond to the decision-making panels' normative decisions. The chemist in this example will, thus, prepare environmental impact assessment reports alongside benefit-cost analyses of infrastructure construction projects, the economic structure of the local businesses, the dependence of local settlements on ground water quality, the local flora and fauna, and so forth.

The example chemist is so motivated that even the evening reading session is dedicated to job issues. Being up to date on job-related knowledge requires such a commitment, perhaps. In public service, a chemist won't need to read highly specific research papers because this knowledge is provided by chemical experts in the wider network (see Chapter 14). Yet, the EPA employee's competence is to judge what kind of specific knowledge is needed for what purpose. Understanding chemical policy and regulation facilitates this judgment. Chemical risk assessments, precautionary measures, multi-stakeholder discourses and science and technology (S&T) governance for increased sustainability of progress benefit greatly from chemists applying their expertise in public service contexts, even if that requires the acquisition of knowledge that is beyond the chemistry textbooks studied at university.

16.4 Book Summary

The most important mission of this book starts now. After working through 15 chapters in three thematic parts, hopefully, the reader is equipped with knowledge and skills that enable him or her to reflect on their own cases and own decision-making. It is impossible to cover all possible issues and situations of good chemistry in this book. Yet, with the provided examples, the discussed principles and strategies, and the opportunities to practice discourse and evaluation of arguments, any occurring case can be mastered with confidence giving a clearer vision of what is at stake and what should be done to achieve a good outcome. As claimed in the introduction, this book is not about ethics. Yes, ethical judgment is part of it, but by far not all. We started with three chapters on methodology. Mastering the very foundation of one's profession is undeniably at the core of *doing a good job*. Building on that, we distinguished two spheres of professional responsibility: the internal level including professionalism, research ethics, good scientific practice, research integrity, and related guidelines for academic, industrial, and public service chemists; and the external level that calls for an assessment of the societal and environmental impact of scientific and R&D activities within the normative framework provided by sustainability concepts and with chemists as participants of a multi-stakeholder discourse on possible futures.

We started this book with a clarification on the role of knowledge and expertise in normative discourses. Chemists don't have to be ethical experts to act ethical and professional. Chapters 2 to 15 honoured this promise by illustrating that good chemistry is, above all, a matter of attitude and expertise. Although the latter is acquired *via* tertiary education and during the vocational adjustment period, the former is sharpened by continuous reflection and exercise. The means for that have been provided in this book.

The most general way of explaining what good chemistry is would be to define it as the responsible handling and application of chemical expert knowledge. Thus, the first part was dedicated to scientific methodology. Here, we found the justifications of all further claims, from the formulation of scientific virtues to the refutation of the neutrality thesis to the role of chemists in public discourse on scientific progress. We have questioned some of the basic dogmas that are dear to many chemists – scientific realism, naturalism, reductionism – and have built scientific methodology on the epistemological premise of constructivism. Knowledge acquisition with scientific means is the core of chemical activity and, thus, informs the codes and guidelines of what counts as good chemical practice.

These guidelines have been conceptualised as virtues of professionalism. The impact of ethics in this context is rather minimal. In many conflict cases, the judgment is based on very simple common-sense morality, or on the ends of scientific inquiry. The rules are rather a matter of compliance than of argument. With this rationale in mind,

it is evident that scientific integrity is not a random matter of preference or taste, but the result of rational reflection and agreement (with
oneself, or with others). The longest chapter in this book is Chapter 8,
which discussed all those issues that arise from the fact that chemistry is in all its forms a network activity. Conflicts of interest,
intellectual property rights allocation, asymmetric relationships
(student-supervisor, employee-boss), interdisciplinarity – these are
very common issues that cause clashes, arguments, disagreements (of
the non-scientific kind), and an inhibition of effective chemical work.
Conduct and discourse in the context of animal experimentation are
discussed in Chapter 9. Here, too, the main objective was the acquisition of a very practical skill: how to make a good (= scientifically
informed) decision and vouch for its plausibility and reason.

Part 3 went one step further and left the realm of the laboratory and the desk at which the lab output is processed. Chemistry is good when its impact on society and the environment is, in
one way or another, beneficial. As all scientific and technological achievements and innovations have a dual-use potential, the
overall approach of this part was to empower chemists to participate effectively in exactly that discourse that aims at reducing
risks and promoting benefits. As, assumingly, chemists are not as
familiar with this topic as with the previous two, a strategy has
been followed that elaborates the argument step by step: first, it
is necessary to understand that scientific progress is firmly intertwined with technological progress and follows similar mechanisms. Chapter 10 introduced the social constructivist view and
promoted the parlance of socio-techno-scientific systems rather
than science as the sphere in which chemistry operates. Second,
with value judgements being enabled, a normative framework for
evaluations is necessary. Chapter 11 provided this framework in
the form of the sustainability concept. For an effective application in academic and economic chemistry contexts, the common
triple-bottom-line approach has been sided with an alternative
approach that understands sustainability as value co-creation.
Third, it is necessary to define the role of chemists in this framework. Chapter 12 introduced a responsibility approach that allows
tenable responsibility ascriptions to chemists and, at the same
time, underlines the call for action of chemists as important epistemic agents in the S&T discourse. Fourth, while the previous three
premises are rather abstract, a practical guideline must be provided. Chapter 13 extended the familiar risk assessment concept
to a larger picture strategy that sets chemical knowledge into the

context of normative discourses. Fifth, it must be shown that these discourses have a real-world impact as, otherwise, they would be waste of time. Thus, Chapter 14 introduced current forms of S&T governance that enable the engagement of chemists to contribute their expertise to sustainable development and to tackle dual-use threats much easier. Additionally, Chapter 15 provided further considerations concerning public communication of chemistry.

The most important question is: *now, at the end of this book, does it make anything different?* If the reader is a chemistry student, the issues discussed in the first nine chapters may have a more direct and visible effect. These are considerations that can be taken directly to the lab and constitute important factors for decision-making and actual professional conduct. The chapters on societal impact may seem rather abstract at this stage of the chemical career. Yet, hopefully, the student reader will recall and benefit from this book in five or ten years from now, when having a job and influential position in industry, in regulation or policy-making, in agencies, or at academic or private research institutions, reflecting on the social impact of the particular professional activities and their own personal responsibility to act in certain ways. Even though the book discussed several real and hypothetical example cases, it cannot provide answers and solutions to all possible kinds of ethically relevant cases. Thus, the chosen strategy – a mix of cases, concepts, principles, and strategies – should have sharpened the reader's critical thinking, judgment ability, awareness of normative dimensions of chemical activity, and the attitude to approach these not purely chemical issues with an open mind. If anything in this book helps a professional reader to gain clarity and confidence concerning a future course of action, then its mission was successful!

Solutions for the Exercise Questions

Chapter 1:

1C, 2A, 3C, 4C, 5D, 6D, 7A, 8D, 9A, 10C, 11B, 12D, 13B, 14C, 15B, 16C, 17D, 18B, 19B, 20B

Chapter 2:

1D, 2A, 3B, 4D, 5C, 6C, 7A, 8C, 9B, 10C, 11B, 12A, 13C, 14A, 15B, 16D, 17D, 18B, 19A, 20C

Chapter 3:

1B, 2C, 3A, 4B, 5C, 6D, 7C, 8C, 9D, 10C, 11A, 12B, 13A, 14C, 15A, 16C, 17B, 18D, 19C, 20D

Chapter 4:

1C, 2A, 3B, 4D, 5A, 6C, 7D, 8C, 9C, 10B, 11C, 12A, 13D, 14C, 15A, 16B, 17B, 18A, 19C, 20D

Chapter 5:

1C, 2D, 3B, 4B, 5A, 6C, 7B, 8D, 9C, 10D

Chapter 6:

1A, 2D, 3C, 4D, 5C, 6B, 7C, 8B, 9D, 10C

Good Chemistry: Methodological, Ethical, and Social Dimensions
By Jan Mehlich
© Jan Mehlich 2021
Published by the Royal Society of Chemistry, www.rsc.org

Chapter 7:

1C, 2B, 3C, 4A, 5B, 6A, 7C, 8A, 9A, 10C, 11D, 12C, 13C, 14D, 15C, 16A, 17D, 18B, 19B, 20A

Chapter 8:

1A, 2C, 3D, 4C, 5A, 6C, 7B, 8D, 9B, 10A, 11C, 12D, 13A, 14B, 15C, 16B, 17C, 18D, 19A, 20B, 21D, 22A, 23C, 24B, 25C

Chapter 9:

1B, 2C, 3A, 4D, 5B, 6A, 7C, 8B, 9A, 10C, 11D, 12A, 13B, 14C, 15B, 16D, 17B, 18C, 19D, 20A

Chapter 10:

1C, 2A, 3B, 4B, 5D, 6A, 7D, 8C, 9C, 10D, 11B, 12D, 13A, 14B, 15D, 16C, 17A, 18D, 19C, 20D

Chapter 11:

1B, 2A, 3C, 4C, 5A, 6D, 7A, 8B, 9D, 10A, 11C, 12B, 13C, 14A, 15D, 16A, 17D, 18C, 19B, 20C

Chapter 12:

1A, 2D, 3D, 4B, 5C, 6B, 7A, 8D, 9C, 10D, 11A, 12D, 13B, 14A, 15C, 16D, 17A, 18C, 19B, 20D

Chapter 13:

1B, 2B, 3C, 4D, 5A, 6C, 7C, 8A + D, 9B, 10B, 11B, 12A, 13D, 14D, 15C, 16B, 17B, 18A, 19C, 20A

Chapter 14:

1D, 2C, 3B, 4A, 5B, 6D, 7B, 8D, 9A, 10C, 11D, 12C, 13A, 14B, 15A, 16C, 17C, 18D, 19A, 20D

Chapter 15:

1C, 2A, 3A, 4C, 5D, 6B, 7D, 8C, 9D, 10A, 11C, 12B, 13C, 14C, 15B, 16A, 17D, 18B, 19D, 20B

Subject Index

Figures appear in *italic;* tables in **bold**

400

fabrication 139, 140–141, *141*
fabrication, falsification and
    plagiarism (FFP) 139–142
fairness 124, 278
fake journals 168–170
fallacies 16, 100–102, 103–104, 143
    ambiguous 99
    formal 97–98
    informal 98–99
    of relevance 100
falsification
    definition 139
    examples 140
Feynman, Richard 50
foundationalism *33*, 34
fraud 138
frequentist statistics 107–108, 324
functionality 279
funding 198–200

Galileo 124
ghost authorship 162
gift authorship 162
good chemistry
    definitions 3–5
    ethics 14–20
    examples 2–3
    exercise questions 23–26
    interested parties 11–14
    knowledge 247
    learning objectives 20–22
    methodology 5–8
    professional integrity 8–9
    societal impacts 9–11
good will 143
governance *see* policy and
    governance
grant applications 386
Greece, ancient 120
green chemistry 10, 84, 274–277,
    306, 383
grey zone, scientific misconduct
    138, 142–145
group-internal behaviour 182, *183*
group-internal communication 149
guidelines, professional 8

h index 168
Hague Ethics Guidelines 127
harassment 184, 189
hazards 318
health 279
health risks 318, 319
heuristic analyses 104–106

hierarchic structures 183, 296
Hippocratic Oath 127
historical-comparativist method 105
holism 223
holistic systems thinking 54
honesty, intellectual **121**, 121, 128,
    159, 370
honorary authorship 162, 167–168
Human Genome Project 179
human rights 282–283
human values 8
hypothesis formulation 65, 71–72
hypothesis revision 80–81
hypothesis testing, type I and type II
    errors 108–109
hypothetico-deductive method 105

ideal scientist 121
Imanishi-Kari, Thereza 134–135, 144
impact factors, journals 167–168
in vitro vs. *in vivo* experiments 216
individual attitudes, teamwork
    182, *183*
indoctrination 44
inductive logic **94**, 95, 103, 104–105,
    179
    fallacies 98–99
    predictions *103*
    statistics 106
industrial funding, academia 197
industry 13, **15**
information collection and
    analysis 73–75, 75–77
innovation 248–253
institutional pressure 146–147
institutions *179*
instrumentalism 245–246
insurances, risk 318
integrity, professional 5, **6**, 8–9, 11,
    296, 304, 392
intellectual contributions 160–161
intellectual honesty **121**, 128,
    159, 370
intellectual property (IP) rights
    186–188, 389
inter-disciplinary collaborations 258
    networking 179, 191–194, *192*
    risk 328
    sustainability 277
    technology assessment 340,
        341–343, 345, 346,
        353, 356
internal responsibility 296
internalism 33